S0-CAH-359

ASPECTS OF HOMOGENEOUS CATALYSIS

A Series of Advances

EDITED BY
RENATO UGO

ISTITUTO DI CHIMICA GENERALE ED INORGANICA
UNIVERSITÀ DI MILANO

VOLUME 5

D. Reidel Publishing Company

A MEMBER OF THE KLUWER ACADEMIC PUBLISHERS GROUP

Dordrecht / Boston / Lancaster

The Library of Congress Cataloged the First Issue of this Work as Follows:

CIP

Ugo, Renato (ed.)
Aspects of homogeneous catalysis. vol. 1.
1970-
Milano, C. Manfredi.
v. illus. 25 cm. annual.
'A Series of Advances'.
Editor: 1970- R. Ugo.
1. Catalysis–Periodicals. I. Ugo, Renato (ed.).
QD501.A83 541′395 72-623953
ISBN 90-277-1738-9

Published by D. Reidel Publishing Company,
P.O. Box 17, 3300 AA Dordrecht, Holland

Sold and distributed in the U.S.A. and Canada
by Kluwer Academic Publishers,
190 Old Derby Street, Hingham, MA 02043, U.S.A.

In all other countries, sold and distributed
by Kluwer Academic Publishers Group,
P.O. Box 322, 3300 AH Dordrecht, Holland

Printed in The Netherlands

Editorial Board

Contents of Volume 5

ASPECTS OF HOMOGENEOUS CATALYSIS

Volume 5

Telomerization of Dienes by Homogeneous Transition Metal Catalysts

ARNO BEHR

Institut für Technische Chemie und Petrolchemie der Rheinisch-Westfälischen Technischen Hochschule Aachen, Worringer Weg 1, D–5100 Aachen, F.R.G.

R. Ugo (ed.), Aspects of Homogeneous Catalysis, Vol. 5, 3–73.

1. INTRODUCTION

In homogeneous catalysis telomerization is defined as the oligomerization of dienes under the concomitant addition of a nucleophilic reagent. For example, two molecules of butadiene react with one nucleophile HY to form the dimeric telomers **1** and **2**. This reaction (Equation 1) is catalyzed by various organometallic compounds

2 [butadiene] + HY —catalyst→ **1** + **2** (1)

of the transition metals, especially by palladium complexes. The nucleophiles – for example, water, alcohols, amines or acids – are introduced mainly at the terminal position of the dimeric molecule giving excellent yields of the product **1**.

Telomerization was discovered independently by Smutny [1–4] of Shell and by Hagihara and Takahashi [5] of Osaka University. Various new syntheses produced via telomerization have been made known and since 1967 this reaction has been studied extensively by a large number of authors. Some reviews have already been published by Smutny [1], Keim [3], Baker [6], Tsuji [7, 8] and Rylander [9].

In this review the telomerization reactions and their applications are surveyed based on the classification by dienes applied and the nucleophiles used. First a short survey will be given on the active catalysts and the probable mechanisms involved.

2. TELOMERIZATION OF BUTADIENE

2.1. Mechanistic aspects

The telomerization of 1,3-butadiene yields 8-substituted 1,6-octadienes **1**, and 3-substituted 1,7-octadienes **2** as a minor product. A great number of homogeneous complexes catalyze this reaction. The most important ones are palladium and nickel catalysts, but rhodium, platinum and other noble metal compounds are also active catalysts.

2.1.1. PALLADIUM COMPOUNDS

The best known telomerization-catalysts are palladium complexes. Smutny and his co-workers [1] discovered the telomerization of butadiene with phenol by using a $PdCl_2$/pyridine system. At first the 8-phenoxy 1,6-octadiene could be isolated only in a 3% yield. By substituting for the pyridine with the phenoxide anion, the conversion of phenol was increased to 96%. This example shows that the telomerization is a very sensitive reaction regarding the catalyst and its modifiers.

Hagihara [5] discovered telomerization by using palladium bis(triphenylphosphine)(maleic anhydride). In the meantime many other phosphine- and phosphite-complexes of palladium have been well investigated. Not only zero-valent palladium-complexes, but also bivalent palladium compounds used in combination with a phosphorous ligand have been found to be active telomerization-catalysts. For example the compounds palladium acetate or palladium acetylacetonate, which are air-stable and easily available are very suitable *in situ* catalysts.

Some important palladium catalysts already used by Smutny [4] during his work on the butadiene/phenol-telomerization are summarized in Table 1.

Table 1
PALLADIUM CATALYSTS

$[PdCl_2(pyr)_2]$	$[PdCl_2]/NaOPh$
$[Pd(PPh_3)_4]$	$[PdCl_2(PhCN)_2]/NaOPh$
$[Pd(C_3H_5)_2]$	$[\{Pd(C_3H_5)Cl\}_2]/NaOPh$
$[Pd(C_3H_5)OAc]$	$[\{Pd(C_3H_5)Cl\}_2]/NaBH_4$
$[Pd(OAc)_2]$	

The ligand has an important influence on the selectivity and the activity of the palladium catalyst. Hagihara [5] telomerized butadiene and acetic acid with a palladium triphenylphosphine catalyst which produced a yield of 30%; Rose [10] carried out the same reaction by using a palladium acetylacetonate/tris(*o*-methylphenyl)phosphite system and found an almost quantitative conversion. The chloride ligand seems to disturb the telomerization. However, if the chloride ion is moved away from the coordination sphere of the palladium, for example by an excess of a base such as sodium phenoxide, the chloride-containing complex will also become active.

Polymer-bound palladium(0) complexes are also selective telomerization catalysts. Pittman discussed the effects of the phosphine loading in the polymeric palladium complexes on the telomerization reactivity [11–13]. Teranishi used a polymeric complex which was prepared by reducing a phosphinated polystyrene-bound palladium(II) chloride complex with hydrazine hydrate and studied the telomerization and the catalyst recycling [14].

The mechanism of the palladium-catalyzed telomerization has not yet been completely understood. In principle three mechanisms can be envisioned: a mono- and a dipalladium bisallyl mechanism and a Pd(II)-hydride mechanism. The bisallyl monometallic reaction route is presented in Scheme 1. Two molecules of butadiene, the nucleophile HY and the palladium species form the complex **3**. The bis-allylic complex **4** occurs by linear connection of the two butadiene units. If the nucleophile attacks this complex at a terminal carbon atom, the telomer **1** is formed while an attack at the carbon atom 3 yields the secondary product **2**. If no nucleophilic attack takes place, the dimer 1,3,7-octatriene **5** leaves the complex as a by-product of the telomerization.

This mechanism has been confirmed by experimental results by Jolly [15]: by reacting bis(η^3-allyl)palladium with trimethylphosphine and butadiene the η^1,η^3-octadienediylpalladium complex **6** was able to be prepared as stable, air-sensitive crystals. Complex **6** reacts with methanol at $-80°C$ to produce the η^2,η^3-octadienediylpalladium methoxide complex **7** which forms at $-35°C$ complex **8** in which the telomer of butadiene and methanol is bonded as an η^2, η^2-organic

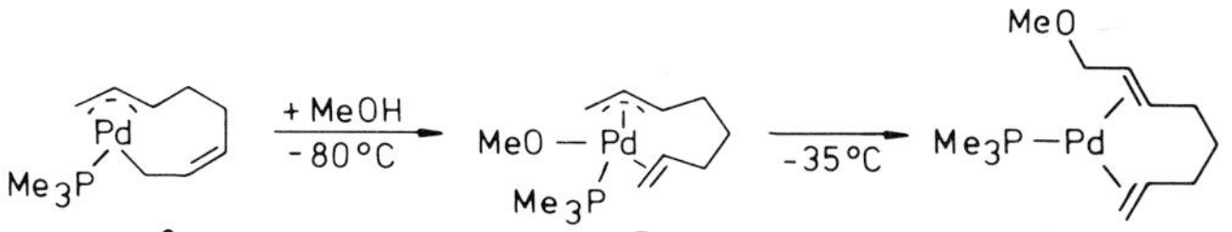

$$\mathbf{6} \xrightarrow[-80\,°C]{+\,MeOH} \mathbf{7} \xrightarrow{-35\,°C} \mathbf{8} \qquad (2)$$

ligand to palladium. The second monometallic mechanism was proposed by Maitlis [16]. He explains the telomerization by a step by step reaction which involves the primary formation of the Pd(II)-hydride complex **9** from a Pd(0) complex by an oxidative addition of the nucleophile HY (scheme 2).

With respect to this mechanism it is remarkable that cationic square-planar palladium hydride complexes like compound **10** act as an efficient catalyst for the telomerization of butadiene and methanol [17].

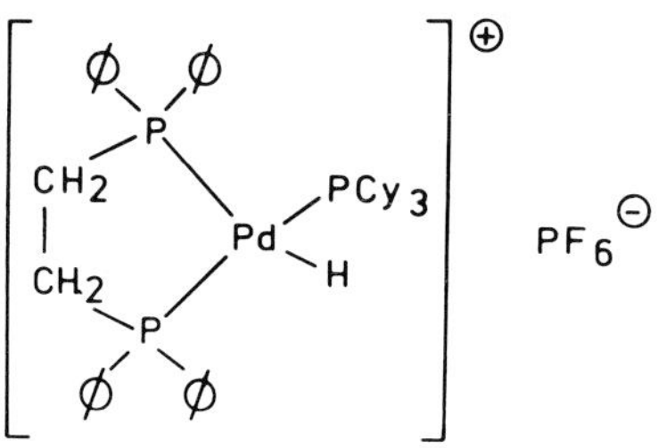

$[(\phi_2PCH_2CH_2P\phi_2)Pd(PCy_3)H]^{\oplus}\ PF_6^{\ominus}$

10

The third possible mechanism contains a bridged bimetallic palladium complex as the intermediate. This mechanism is a very plausible one because bridged

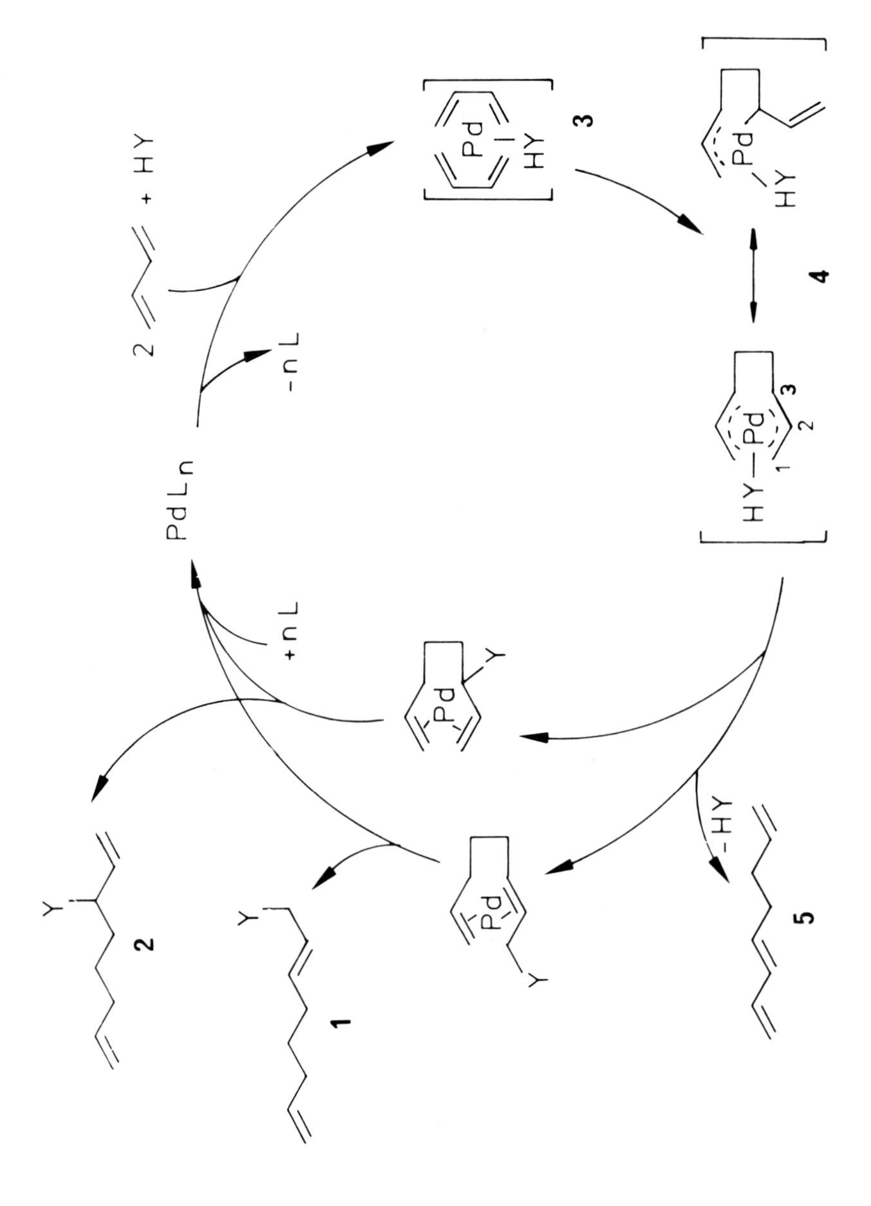

Scheme 1.

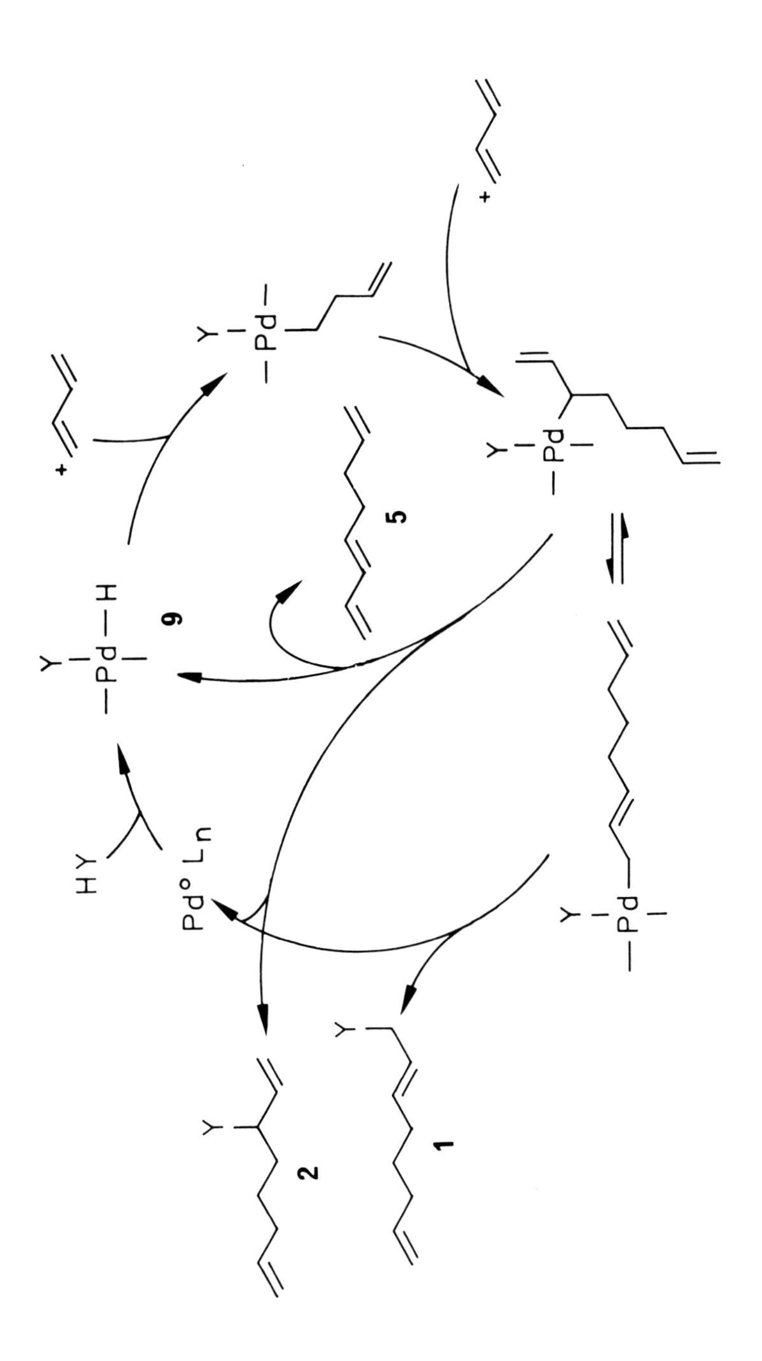

Scheme 2.

complexes with the structure **11** have been isolated from reaction mixtures of butadiene, bis-allyl palladium and the nucleophiles phenol and acetic acid respectively [3]. The same bridged complexes were able to be prepared by the reaction of the dimeric palladium allylacetate with butadiene [18].

11

In complex **11** the palladium atoms are formal in the oxidation state I. Within the last few years various similar bimetallic and bridged palladium(I)-complexes have been able to be prepared [19–22]. The bimetallic mechanism is shown in Scheme 3. The reaction steps are similar to those described in Scheme 1.

Pittman [11, 12] tried to distinguish between the mono- and the bimetallic reaction path by telomerizing butadiene and acetic acid by using polymer-anchored palladium complexes. As is shown in Figure 1, it is very likely that within the cross-linked resin matrix the palladium atoms are isolated from one another and that no bimetallic intermediates exist.

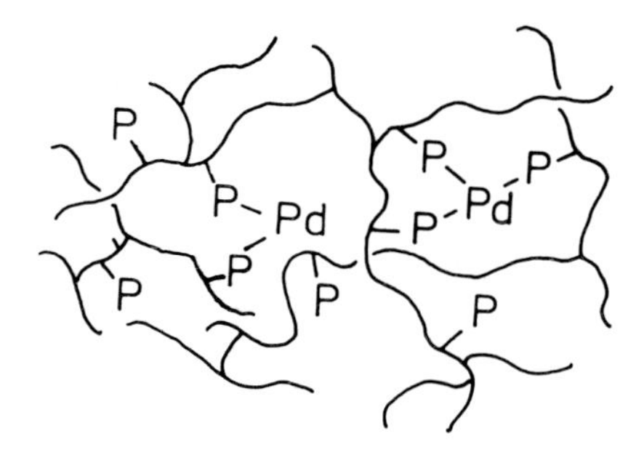

Fig. 1.

Pittman showed that the product distribution for telomerizations catalyzed by such anchored palladium catalysts was the same as those in the homogeneous phase when compared at equal P/Pd ratios. These observations suggested both reactions proceed by the same pathway. Nevertheless, this information does not rule out a bimetallic intermediate in homogeneous telomerization. In view of the extreme reluctance of palladium to go to a coordination number greater than four, a mono-palladium bisallyl-complex of the type **4** might be difficult to form within a resin matrix. Therefore it is possible that different mechanisms exist in the homogeneous and in the polymer anchored telomerization catalysis.

Another unsolved problem of the bisallylic mechanism which is independent of the detailed structure of the intermediates is the question of how the nucleophile

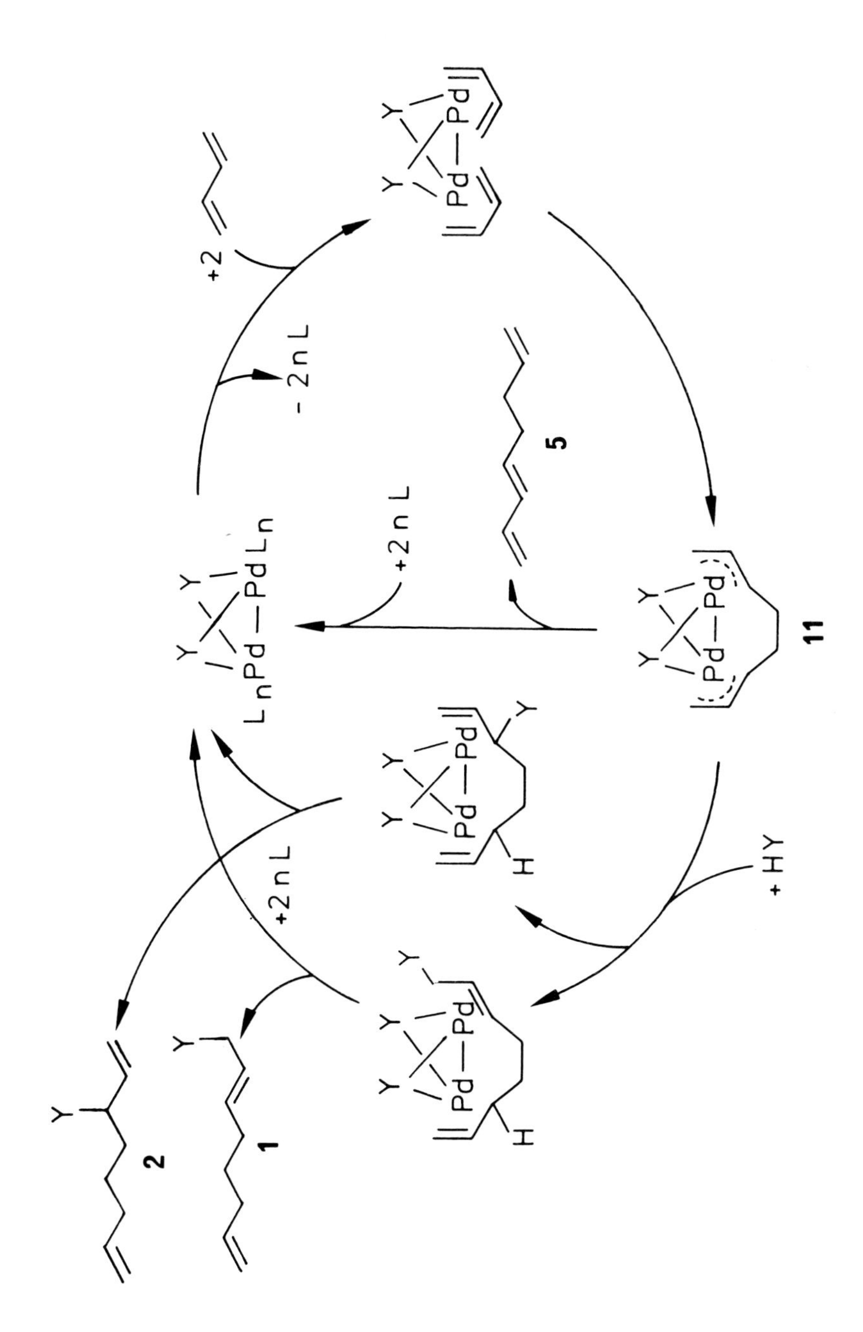

Scheme 3.

attacks the palladium complex. On the one hand it is possible that the nucleophile is added in a first step on the palladium by oxidative addition and that it must shift to the octadienyl-chain in a second step. But it is also feasible that the nucleophile can directly attack the coordinated butadiene dimer. These two possibilities are shown – in the case of the bimetallic mechanism – in Equation 3 and Formula **12**.

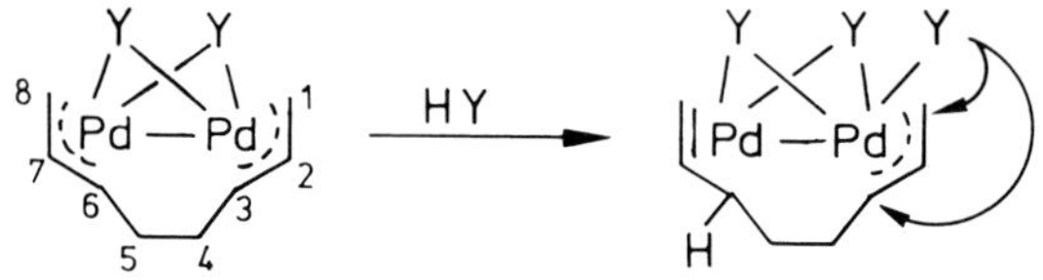

(3)

11

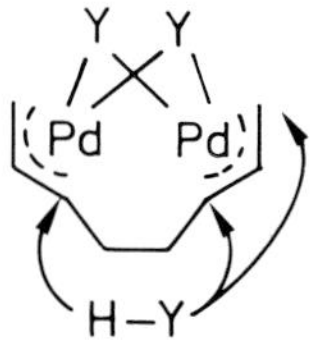

12

It is, however, not equivocal that the proton of the nucleophile is added on the carbon atom 6. Experiments with CH_3COOD and CH_3OD have proved, that only products with the deuterium in position 6 are formed [5, 23].

2.1.2. NICKEL COMPOUNDS

Nickel compounds are also able to catalyze the telomerization, but contrary to the palladium catalysis, nickel catalysts give rise to more by-products.

Heimbach [24] was one of the first who discovered the nickel-catalyzed telomerization. In 1968 he investigated the dimerization of butadiene by using a nickel-triethylphosphite/morpholine catalyst. In addition to the dimer 1,3,6-octatriene he found the telomeric octadienylmorpholine. At the same time the nickel-catalyzed telomerization of butadiene and phenol has been studied [25]. In 1971 Shields and Walker [26] prepared the methoxyoctadienes from butadiene and methanol in a 40% yield by using a $Ni(acac)_2/PhP(OPr^i)_2/NaBH_4$-system, but they received many by-products, especially 1,3,7-octatriene and the 1 : 1-adducts, the methoxybutenes.

Some important nickel catalysts which are active in the telomerization of butadiene are summarized in Table 2. In Table 2 it is remarkable that there are complex ligands which occur very frequently also in the nickel-catalyzed telomerization: phosphines and phosphites. What is very important is the group of the aryl-dialkoxyphosphines. The catalyst system $Ni(acac)_2/NaBH_4$ can only be activated by this type of ligand and not by simple phosphines or phosphites [26, 28]. At first

Table 2
NICKEL CATALYSTS

Catalysts	Ref.
[$NiCl_2$], diphos, $NaBH_4$	27
[$Ni(acac)_2$] $PPh(OPr^i)_2$, $NaBH_4$	26, 28, 29
[$NiCl_2$], PR_3, BuLi	30
[$NiBr_2$], PR_3, NaOPh	31
[NiL_4]	32
[$NiCp_2$], PR_3	32
[$Ni(cod)_2$], PR_3	2
[$Ni\{P(OPh)_3\}_4$]	2
[$Ni\{P(OPh)_3\}_2(CO)_2$]	2
[$Ni(acac)_2$], $P(OR)_3$, $Al(OEt)Et_2$	24
[$Ni(acac)_2$], $PhP(OPr^i)_2$	33

it was supposed that this catalytic activity could be explained by the fact that such alkoxyphosphanes can reduce Ni(II)- to Ni(0)-compounds, but Rose [29] was able to prove that the reducing effect itself cannot be the only reason: the also reducing alkoxyphosphanes $MeP(OPr^i)_2$ and $PhP(OMe)_2$ are inactive ligands in the nickel-catalyzed telomerization of butadiene with diethylamine.

The mechanism of the nickel-catalyzed telomerization [32] (scheme 4) was postulated following the well-known mechanism of the nickel-catalyzed oligomerization [34]. The mechanism proposed in Scheme 4 can explain both the telomeric

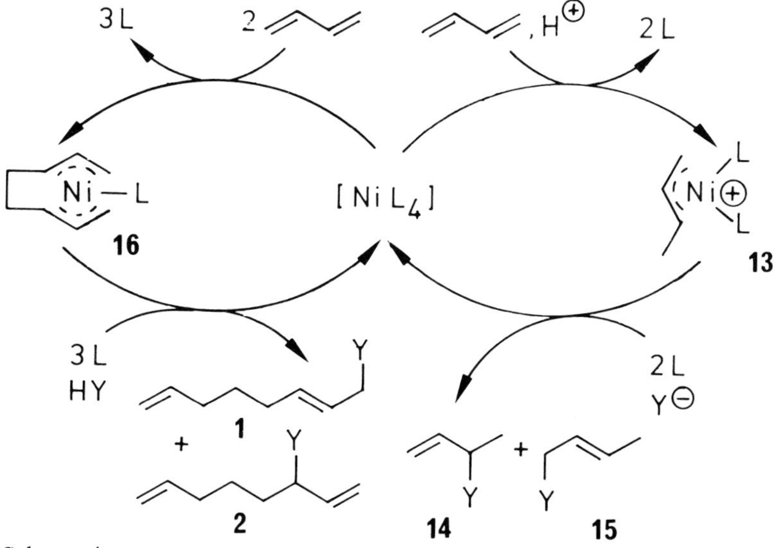

Scheme 4.

octadienyl derivatives as well as the butenyl derivatives which are often the main products of the nickel-catalyzed telomerization. It is suggested that the two types of products are formed in separate but interrelated pathways.

On the one hand the catalyst NiL_4 can react with one molecule of butadiene and a proton to form the intermediate **13**. This monoallyl complex may react with a nucleophilic anion and lose the butenyl derivatives **14** and **15** to reform the nickel(0) species. On the other hand the complex NiL_4 can also start a catalytic cycle leading to the telomers. With the addition of two molecules of butadiene, the bisallyl complex **16** is formed. This complex can be attacked by the nucleophile upon leaving the octadienyl derivatives.

The two-cycle-mechanism is confirmed by the facts that the 1 : 1-adducts
- do not reach a maximum at some point, as is typical for a consecutive reaction;
- do not react in separated experiments with butadiene forming the 2 : 1-adducts;
- are formed by an excess of ligands in a great yield.

2.1.3. RHODIUM COMPOUNDS

The first telomerization catalyzed by rhodium is described by Dewhirst [35]. He used $RhCl_3 \cdot 3\,H_2O$ as a catalyst in the reaction of butadiene with ethanol and synthesized the two 1 : 1-adducts already mentioned in Scheme 4, and in small amounts synthesized the unusual 2 : 1-adducts 2-ethoxy-3-methylidene-5-heptene. Other approved rhodium-catalysts are shown in Table 3.

Table 3
RHODIUM CATALYSTS

Catalysts	Ref.
$[RhH(PPh_3)_4]$	2
$[Rh(OPh)\,(PPh_3)_3]$	36
$[RhCl(PPH_3)_3]$	37
$[RhCl(CO)\,(PPh_3)_2]$	37
$RhCl_3 \cdot 3H_2O$, PR_3	37
$[\{RhCl(cod)\}_2]$, PR_3	38

The telomerization of butadiene with phenol can be achieved with a $[Rh(OPh)(PPh_3)_3]$ catalyst [36] and the reaction with diethylamine with a $[\{RhCl(cod)\}_2]/PR_3$ system [38]. In the telomerization of butadiene with morpholine with using rhodium trichloride as catalyst, a guiding effect of added ligands was observed [37]: without PPh_3 and with a great excess of PPh_3 (1 : 9) only the products substituted by butenyl are prepared, whereas a 1 : 1-ratio of $RhCl_3 \cdot 3H_2O$ and PPh_3 yielded a mixture which contained a large amount of octadienylamines (compare Scheme 4).

This suggested that the intermediates **17** and **18** were involved in the production of these two types of products.

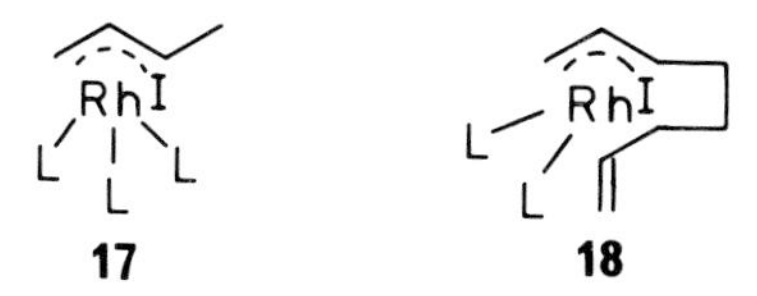

Formation of telomers by way of the intermediate **18** requires the presence of a stabilizing ligand, but a large increase in the amount of PPh_3 makes the formation of **18** less likely due to the competition between phosphine and butadiene for the available coordination sites. This argumentation is very similar to that used in nickel catalysis: the rhodium intermediates **17** and **18** are generally comparable to the nickel complexes **13** and **16**.

2.1.4. PLATINUM AND FURTHER GROUP VIII METALS

Platinum compounds are also described as active telomerization catalysts (Table 4).

Table 4
PLATINUM CATALYSTS

Catalysts	Ref.
$PtCl_2$/NaOPh	2
$PtCl_4$/NaOPh	39
$[Pt(C_3H_5)_2]$	40
$[Pt(PPh_3)_4]$	5, 41
$Pt(PPh_3)_2CH_2{=}CH_2$	41
$[Pt(cod)_2]$	42

The telomerization of butadiene with diethylamine is catalyzed by allyl-platinum systems [40, 43–45]. Platinum chloride-sodium phenoxide catalysts were found to be active in the reactions with phenol [2] and acetic acid [39], whereas $[Pt(PPh_3)_4]$ catalyzed the reaction with alcohols. The platinum compounds are generally less active than the comparable palladium compounds.

The "ligand-free" platinum complex $[Pt(cod)_2]$ was studied by Stone [42]. He discovered that this catalyst is a very selective one for the telomerization of butadiene with acetic acid to produce a 100% conversion into the telomer with the structure **1**. However, this complex is much less active in the reaction with secondary amines and is even inactive with alcohols and aldehydes. Stone considered the hydride mechanism in order to explain his results (compare Scheme 2). The oxidative addition of the nucleophile, the first step in this mechanism, is very favourable for platinum complexes because the formation of metal-hydride bonds in this manner would increase the subgroup Ni < Pd < Pt.

The activity of platinum compounds, which is generally less, combined with a simultaneously higher formation of the butadiene dimer can be explained by the preferred elimination of the $[PtHYL_n]$-type species in the final step. This is due to the higher stability of PtH compared with PdH (Equation 4).

Y
—Pt— → —Pt—H + (octatriene) (4)

The complexes of further noble metal compounds are of no importance in telomerization. Smutny and his co-workers [1, 2, 39] studied such catalysts as $[IrH_3(PPh_3)_3]$, $OsCl_3/NaOPh$ or $RuCl_3/NaOR$ and observed only small conversions.

2.2. Telomerization with O-H-nucleophiles

A short survey of the telomerization of butadiene with hydroxy-nucleophiles is given in Table 5.

Table 5
TELOMERIZATION OF BUTADIENE WITH O-H-NUCLEOPHILES

Nucleophile		Telomer	
alcohols phenols	R—OH	ethers	R—O—C_8
silanols	R_3SiOH	ethers	R_3Si—O—C_8
carboxylic acids	R—C(=O)—OH	esters	R—C(=O)—O—C_8
hydroxy-carboxylic-esters	R(COOR′)—OH	ethers	R(COOR′)—O—C_8
boric acids	$B(OH)_3$	esters	$B(O-C_8)_3$
	R_2B—OH		R_2B—O—C_8
water	H—OH	alcohols	HO—C_8
		ethers	C_8—O—C_8

2.2.1. ALCOHOLS

Primary and short-chained alcohols react very easily with butadiene to form ethers. Hagihara [5] showed that the telomerization of methanol proceeds smoothly even at low temperatures (40–100°C) and with a short reaction time (1–3 h). The primary telomer **1** is always the main product which is accompanied by the secondary telomer **2** and small amounts of the by-product 1,3,7-octatriene.

By using the catalysts [$Pd(PPh_3)_2$ (maleic anhydride)], [$Pd(PPh_3)_2$ (dimethylfumarate)], [$Pd(PPh_3)_2$ (*p*-benzoquinone)] and [$Pd(PPh_3)_4$] the telomer **1** was formed with a yield of 85–90% and telomer **2** with a maximum yield of 8%. Other palladium catalysts which have been proven are [$PdCl_2(PPh_3)_2$]/$NaBH_4$ [46], [$Pd(OAc)_2$]/PPh_3 [47] and [$Pd(OAc)_2(PPh_3)_2$]/KOH [48].

The same reaction can also be catalyzed by palladium catalysts anchored in resin which are more active than their homogeneous counterparts at comparable P/Pd ratios [13]. Another interesting technique used to carry out the methanol telomerization is the use of a $PdCl_2$/alkylsulfinate catalyst (i.e., without any phosphorous ligand) [40]. For example the sodium neophylsulfinate **19** is a very effective and selective cocatalyst: after a reaction time of 20h at an ambient temperature the telomer **1** was formed with a yield of 94%. The addition of a base like methoxybenzyltrimethylammonium also increases the yield of the telomeric ethers [53].

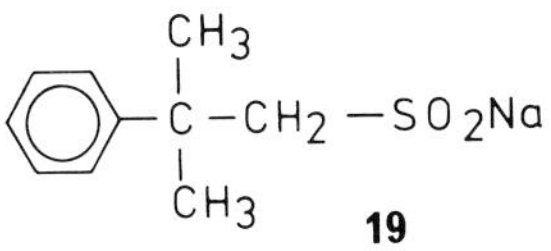

$$C_6H_5-C(CH_3)_2-CH_2-SO_2Na$$

19

Commereuc and Chauvin [31] studied the effect of ligands and alcoholates on the methanol telomerization. They investigated both monodentate as well as bidentate ligands, namely PPh_3, trialkylphosphines and diphos. With a $PdCl_2/PPh_3/NaOPh$ system they found only the methoxyoctadienes. The amount of unknown oligomers increases by changing the halogen in the palladium compound from chlorine to bromine and iodine. The change of the phosphorous ligand is also remarkable: the 1 : 1-adducts, the methoxybutenes, are also formed both with the monodentate triethylphosphine and with the bidentate diphos.

Hence in the telomerization catalyzed by palladium there also occurs an intermediate in which so many coordination sites of the metal are occupied by strong ligands, that a second molecule of butadiene cannot move into the coordination sphere (compare the nickel catalysis in Scheme 4).

Another group of phosphorous ligands was recently studied by Kuntz [38]. The telomerization of butadiene and methanol was catalyzed by the water soluble ammonium or sodium salts of sulfonated triphenylphosphines.

In the reaction of methanol and butadiene with cationic allylpalladium complexes – which may be prepared *in situ* through the reaction of tris(dibenzylideneacetone)dipalladium with an allyloxyphosphonium salt such as $CH_2=CMeCH_2OP^+(NMe_2)_3PF_6^-$ – the telomers C_8—OMe and C_{12}—OMe are formed as well as C_{16}—OMe products [50, 51].

The telomerization of butadiene with methanol catalyzed by a nickel complex was first studied by Shields and Walker [26]. A $Ni(acac)_2/NaBH_4$ catalyst which was stabilized by arylalkoxiphosphines as ligands yielded 34% 1 : 1-adducts, 18% octatriene and 41% telomers. Very similar results were obtained with a $NiBr_2/PEt_3$-system [31]. The zero valent nickel complex $Ni(cod)_2$ activated by phosphines is also an active catalyst system [52].

An intensive study of the butadiene/methanol telomerization catalyzed by nickel was carried out by Beger [30]. He used Ziegler-Natta-type catalysts which were formed by a nickel(II) complex such as $[Ni(acac)_2]$, an organolithium compound and phosphorous compounds such as $P(NMe_2)_3$ or PBu^n_3. He showed that an increasing amount of alcohol raises the yield of the telomers **1** and **2** and surprisingly enough not the yield of the butenyl derivatives **14** and **15** (see Scheme 4 and Table 6).

Table 6
INFLUENCE OF THE BUTADIENE/ALCOHOL RATIO [a]

CH_3OH mmol	PR_3 R =	Yield (%) 1 + 2	14 + 15
125		12	46
150	NMe_2	13	50
250		66	28
175		56	38
200	Bu^n	85	10
500		90	8

[a] 250 mmol butadiene; cat.: $[Ni(acac)_2]/Bu^nLi/PR_3$; t = 1 h; T: 50°C.

The telomerization of various higher alcohols has also been carried out. The primary alcohols react most easily with butadiene, whereas secondary and tertiary alcohols yield only small amounts of telomers. Steric hindrance alone cannot be the only factor responsible for the reactivity because the voluminous 2,2-bis(trifluoromethyl)benzyl alcohol reacts smoothly to produce the 2,7-octadienyl ether [4]. Some alcohols which have been investigated are summarized in Table 7.

The telomerization of butadiene with ethanol can be carried out by a $[Pd(acac)_2]$/ligand-system [56]. The change of the ligand has a great influence on the conversion of butadiene and on the relative composition of the product mixture: the yields of telomers were satisfactory with alkyl- and arylphosphines;

Table 7
TELOMERIZATION OF BUTADIENE WITH HIGHER ALCOHOLS

Pd-Catalysis	Ref.	Ni-Catalysis	Ref.
CH_3CH_2OH	5, 49, 53–56	CH_3CH_2OH	26, 30
CF_3CH_2OH	4, 57, 58	$CH_2{=}CHCH_2OH$	30, 60
$CH_3CH_2CH_2OH$	49, 56	$PhCH_2OH$	30
$(CH_3)_2CHOH$	5, 48, 49, 53, 54, 56		
$CH_2{=}CHCH_2OH$	4, 56		
Bu^nOH	53, 56		
Bu^tOH	53, 56		
$PhCH_2OH$	4		
$PhC(CF_3)_2OH$	4, 59		

Table 8
INFLUENCE OF THE LIGAND[a]

Ligand	Conversion (%)	Ratio (%) Dimer 5	Telomers 1	2
PBu^n_3	100	8	89	3
PPh_3	100	24	69	7
$AsPh_3$	75	14	84	2
diphos	75	38	60	2
$P(OBu^n)_3$	12	71	29	0

[a] C_4H_6 + EtOH; cat.: $[Pd(acac)_2]/L$; t = 4 h; T: 60°C.

modest with arsines and bidentate phosphines; and bad with alkyl and arylphosphites (Table 8).

The length of the alcohol chain influences the reactivity of the nucleophile. Whereas the telomers of methanol are formed in a 88% yield, propanol and nonanol produce yields of only 65% and 21% respectively. The telomerization of the branched alcohols is rather difficult. As is shown in Table 9, the *n*-butanol produces a yield of 64%; the secondary butanol of 7%; whereas the tertiary butanol produces no reaction.

Table 9

INFLUENCE OF THE ALCOHOL [a]

ROH	Conversion (%)	Ratio (%) 5	1	2
CH_3OH	100	12	78	10
n-C_3H_7OH	100	35	58	7
n-$C_9H_{19}OH$	74	72	28	0
n-$C_{12}H_{25}OH$	25	96	4	0
n-C_4H_9OH	100	36	58	6
i-C_4H_9OH	100	46	52	2
s-C_4H_9OH	71	90	10	0
t-C_4H_9OH	69	67	0	0

[a] Cat.: $[Pd(acac)_2]$, PPh_3, t = 10 h, T: 60°C.

It is remarkable that unsaturated alcohols are bad nucleophiles. Allyl alcohol produces 46% telomers with a nickel- [30] and only 10% with a palladium catalyst [56].

Propargyl alcohol gives no reaction. It is possible that these alcohols hardly bind to the metal like other unsaturated ligands, and that for this reason the telomerization is blocked. Also cyclic alcohols like cyclohexanol are inactive in telomerization which is probably a steric effect of the chair form of the molecule. In addition to the steric influence of the nucleophile discussed above an electronic effect can be observed: fluorinated alcohols are more reactive than their counterparts which contain hydrogen. CF_3CH_2OH, $CHF_2(CF_2)_3CH_2OH$ and $CHF_2(CF_2)_5CH_2OH$ have been telomerized with butadiene in good yield by using a $PdCl_2/PPh_3/PhONa$ catalyst [57].

2.2.2. PHENOLS

Phenol reacts with butadiene to produce the octadienyl phenyl ethers in high yields [1, 2, 9, 25]. The reaction catalyzed by palladium has been carried out in chloroform or acetonitrile, but this has also been done successfully without a solvent. Phenol and butadiene which were heated together at 100°C in the presence of $PdCl_2$ and PhONa produced *trans*-1-phenoxyoctadiene in 91% yield which was accompanied by 4% of the cis isomer and 5% of 3-phenoxy-1,7-octadiene (way A in Scheme 5). The condensation of butadiene and phenol which is carried out in the presence of a tertiary aromatic phosphine or of excess phenol takes a different course and produces the *ortho*- and *para*-octadienylphenols (way B in Scheme 5).

Substituted phenols such as p-chlorophenol, p-methylphenol and α-naphthol,

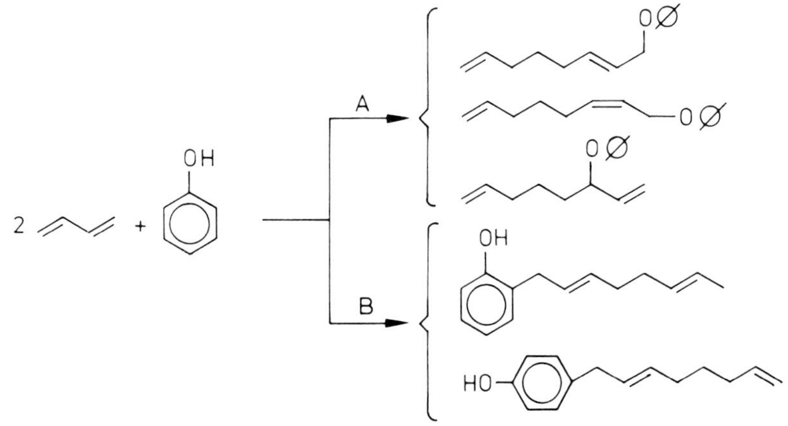

Scheme 5.

are nucleophiles which are also very reactive. Besides the palladium catalysts, nickel complexes could be also applied. Beger [30] optimized the synthesis of the phenoxioctadienyls (62%) by using a $[Ni(acac)_2]/P(NMe_2)_3/Bu^nLi$ catalyst. Weigert [32] optimized the formation of the phenoxybutenes which are favoured by electron-donor ligands by using of the phosphorous ligand in excess and by using high phenol concentrations.

2.2.3. SILANOLS

Trimethylsilanol reacts with butadiene [61–63] with a $[Pd(PPh_3)_2$ (maleic anhydride)] catalyst to yield 8-trimethylsiloxy-1,6-octadiene. This product could be converted to *n*-octanol by hydrogenation and hydrolysis. This saturated octanol can react with phthalic anhydride to yield excellent plasticizers (Scheme 6).

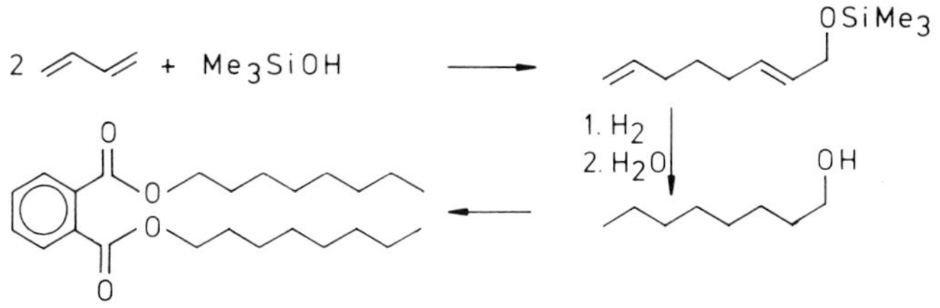

Scheme 6.

2.2.4. CARBOXYLIC ACIDS

Carboxylic acids proved to be excellent nucleophiles for the telomerization with butadiene. However, no telomerization takes place with formic acid. A reaction of formic acid with butadiene which is catalyzed by a $[Pd(OAc)_2]/NEt_3$ system leads only to the formation of 1,6-octadiene; reactions using platinum salts produced 1,7-octadiene in addition to the 1,6-diene [64–66]. The formic acid seems to behave as a reductant rather than as a nucleophile (Scheme 7).

2 + HCOOH $\xrightarrow{-CO_2}$

Scheme 7.

A highly selective catalyst system for hydrodimerization of butadiene to 1,7-octadiene in which formic acid serves as the hydrogen source was found recently by Pittman [67]. Palladium bisacetate, triethylamine and triethylphosphine catalyze the formation of 1,7-octadiene with a 93% selectivity. The produced diterminal olefin may function as a precursor for sebacic acid, 1,10-decanediol or 1,8-dicyano-octane.

Acetic acid and butadiene yield acetoxyoctadienes in excellent conversions and selectivities. In the presence of palladium acetylacetonate and *o*-alkyl or *o*-aryl substituted triarylphosphites in a 1 : 1 molar ratio an almost quantitative telomerization occurs [10]. When other palladium catalysts such as $[Pd(PPh_3)_2(MSA)]$ or $[Pd(PPh_3)_4]$ are used [5], the conversion to acetoxyoctadienes decreases.

A remarkable effect on the telomerization with acetic acid was observed when amines were added to the catalyst [68]. Without amines, a $[Pd(acac)_2]/PPh_3$ catalyst produces a 45% conversion of butadiene. About 65% of the product mixture were octadienyl acetates; 15% butenyl acetates; and 20% octatriene. When the same reaction was carried out in the presence of molar quantities of tertiary amines, such as triethylamine, *N*-methylmorpholine, or triethylendiamine, an increase in the rate of the reaction and in the selectivity to the octadienylacetates was observed. For example, the reaction in the presence of 2-(*N,N*-dimethylamino)ethanol essentially produced a complete conversion of butadiene with a 92% yield to telomers. Other cocatalysts can also have an effect on the reaction: the presence of additional acetate ions, for example, added as NaOAc, favours a high isomer ratio of the primary to the secondary telomer. This ratio is also influenced by an isomerization catalyzed by palladium, which occurs parallel to the telomerization (Equation 5). Rose and

OAc $\rightleftharpoons$ OAc (5)

Lepper [10] found that when PPh_3 is used as phosphorous ligand, the pure primary telomer is isomerized within 12 hours to an isomer ratio of about 4 : 1. A comparable experiment using tris(*o*-methylphenyl)phosphite produced a ratio of 17 : 1. This indicates that the high yields of primary acetoxyoctadiene which are found by using tris-*ortho*-substituted triphenyl phosphite ligands are due to the slowness of the isomerization reaction.

Other carboxylic acids can also participate in the telomerization [4]. For example, pivalic, benzoic and methacrylic acid react with butadiene to form the octadienyl products in more than a 40% yield.

Dicarboxylic acids also telomerize with butadiene [69, 70]: phthalic acid reacts with butadiene to form octadienylphthalates which are potential precursors for plasticizers. This telomerization can be carried out in a continuous way by using a two-phase reaction system: butadiene and phthalic acid react under palladium catalysis in a dimethylsulfoxide phase. The products, the octadienylphthalates, can be extracted by *iso*-octane, whereas the catalyst, which is solved in DMSO, can be recycled to the reactor. The flow scheme of this reaction is shown in Figure 2.

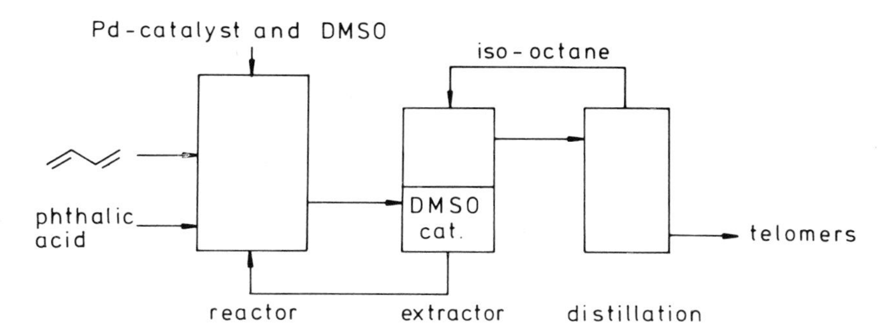

Fig. 2.

2.2.5. HYDROXYCARBOXYLIC ESTERS

Glycolic acid butyl ester reacts with butadiene in the presence of a palladium catalyst to form the 2,7-octadienyl ether [71]. The butenoxy-acetic ester occurs only in small amounts. Lactic acid ethyl ester, a nucleophile with a secondary alcohol group, reacts – surprisingly – very smoothly with butadiene: the telomer was formed in a yield of 76% during the reaction time of 30 hours and at room temperature (Equation 6).

$$2\ \text{CH}_2{=}\text{CH}{-}\text{CH}{=}\text{CH}_2 + \text{HO}{-}\text{CH}(\text{CH}_3){-}\text{COOEt} \longrightarrow \text{CH}_2{=}\text{CH}{-}(\text{CH}_2)_3{-}\text{CH}{=}\text{CH}{-}\text{CH}_2{-}\text{O}{-}\text{CH}(\text{CH}_3){-}\text{COOEt} \qquad (6)$$

2.2.6. BORIC ACIDS

Treating butadiene with a 20% aqueous solution of boric acid and a catalyst system of Pd(II), PPh_3 and $AlEt_3$, yields unsaturated boric esters as intermediates [72]. After hydrolysis of these esters, two octadienols and one dodecatrienol could be obtained in an 80% combined yield. Using Ph_2BOH for H_3BO_3 produced only octadienols via the intermediate diphenylborinate esters.

2.2.7. WATER

The direct formation of octadienols by telomerization of butadiene and water has attracted much attention as a novel synthesis for *n*-octanol which is a chemical of certain industrial importance. However, the nucleophile water has only low reactivity compared to the other molecules containing the hydroxy group.

In the presence of palladium/phosphine catalysts the octadienols are formed only in traces if no cocatalyst is used [73]. For example, the reaction of butadiene and water in the solvent isopropanol with a $[Pd(PPh_3)_4]$/KOH catalyst produced the octadienols in only small amounts [48]. This telomerization is also described in the patent letters of Romanelli [53]: starting from more than 300 mmols of butadiene, only two mmols of octadienols were formed; in addition, 1,3,7-octatriene and bis(2,7-octadienyl)ether occur.

No telomers have been obtained by using a palladium catalyst which is free from phosphine ligands, for example, by using $[Pd(cod)_2]$ [42]. The same behaviour has been observed in nickel catalysis: if the $[Ni(acac)_2]/P(NMe_2)_3/n$-BuLi-system is used, which is a very reactive one in the telomerization of alcohols, no reaction of butadiene with water takes place [30]. Some efforts were made on account of these difficulties to synthesize the octadienols indirectly. As already mentioned, it is possible to form the trimethylsilyloctadienyl ether by starting from butadiene and trimethylsilanol, or the esters of carboxylic acids or boric acid. The octadienols can be set free after hydrolysis of these compounds. The acidity of the reaction medium seems to be of great importance in the direct syntheses of the octadienols. The primary octadienol is formed in a dioxane solution buffered by a mixture of mono- and disodium phosphate [74]. In the presence of formic acid the palladium-catalyzed conversion of butadiene and water is also increased in a very evident way [75].

A breakthrough in the water telomerization was found by Atkins [73, 76]: the octadienols are formed in the presence of carbon dioxide as a cocatalyst in a yield of more than 80%. The dioctadienyl ethers were obtained only in small amounts. Only 1,3,7-octatriene was formed in the absence of carbon dioxide.

The role of the carbon dioxide is not well understood. One explanation which was already discussed by Atkins rests on proposing the formation of the carbonic acid half esters **20** and **21** by the reaction of butadiene with carbonic acid. These

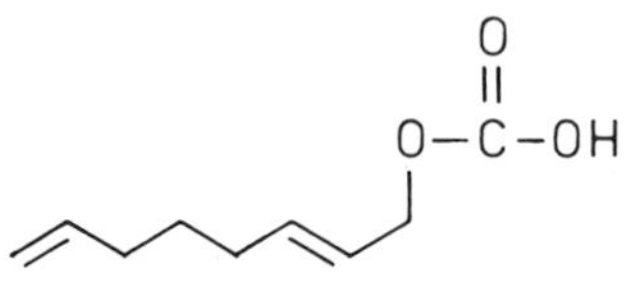

O–C–OH (with C=O) **20** O–C–OH (with C=O) **21**

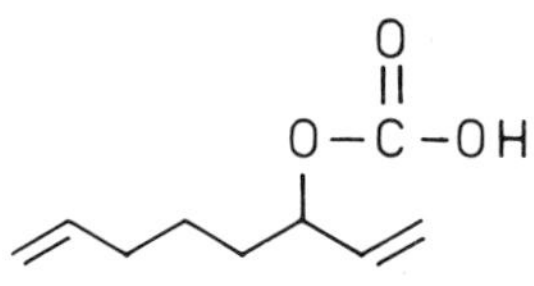

possible intermediates are unstable in polar solvents like *t*-butanol, acetone or acetonitrile, and decompose immediately into the octadienols and CO_2.

The solvent is of great importance for the water telomerization: the reaction which is catalyzed by a $Pd/PPh_3/CO_2$ system and carried out without a further solvent in an aqueous emulsion yields only the bis(octadienyl)ethers **22** and **23** and

22 **23**

no octadienols. A conversion of more than 40% can be obtained by the addition of detergents such as sodium oleate [80].

Another feasible interpretation of the "carbon dioxide effect" is the interaction of CO_2 with the catalyst: carbon dioxide acts as a ligand which has an activating influence on the coordination sphere of the palladium. The increasing and guiding effect of carbon dioxide is also well known in the dimerization reaction of butadiene: without CO_2, only 4-vinylcyclohexene is formed; the linear dimers 1,3,7-octatriene or 2,4,6-octatriene can be obtained with CO_2 [77–79]. The fact that low amounts of carbon dioxide already enhance the reaction rates – both of dimerization as well as water telomerization – supports the speculation that transition metal–carbon dioxide complexes are formed *in situ*. These complexes must be rather unstable since all attempts at isolation have been unsuccessful until now.

The telomerization of butadiene with water was recently studied by Yoshimura [81, 82]. A mixture of butadiene and carbon dioxide was contacted with palladium bis-acetate, 3-$(Ph_2P)C_6H_4SO_3Na$ and triethylamine in a water-sulfolane solution to yield 2,7-octadien-1-ol as the main product. Another research group studied the recovery and re-use of the palladium catalyst in water/butadiene telomerization [83].

2.3. Telomerization with N-H-nucleophiles

2.3.1. AMINES

A variety of telomers and 1 : 1-adducts can be envisaged when one considers the various potential primary and secondary amines (Equation 7). Both aromatic and aliphatic primary and secondary amines react with butadiene.

$$\begin{aligned} R_2NH + n\ \text{butadiene} &\longrightarrow R_2N{-}C_8 + R_2N{-}C_4 \\ RNH_2 + n\ \text{butadiene} &\longrightarrow RNH{-}C_8 + RN{-}(C_8)_2 \\ &\qquad RNH{-}C_4 + RN{-}(C_4)_2 + RN(C_4)(C_8) \end{aligned} \tag{7}$$

By using *palladium* compounds which contain monodentate phosphine ligands octadienylamines are formed in high yields [5, 55]. Only the 1 : 1-adducts occur in the presence of palladium catalysts containing bidentate ligands such as diphos [84]. This effect probably arises through changes in the number of coordination sites which are available to the reactants. The catalytic activity is remarkably enhanced with the addition of a small amount of phenol. A larger amount of phenol affects the yields of the products and the ratio of the two isomers **24** and **25**: with a great excess of phenol, the *N*-(2-butenyl)isomer **25** is the predominant product.

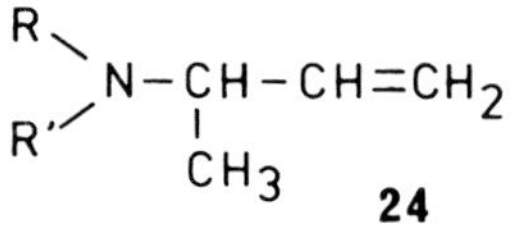

$$RR'N-CH(CH_3)-CH=CH_2 \quad \mathbf{24} \qquad RR'N-CH_2-CH=CH-CH_3 \quad \mathbf{25}$$

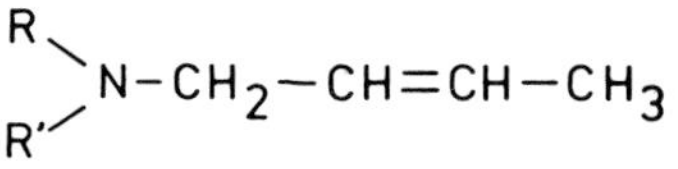

With the "ligand-free" palladium catalyst $[Pd(cod)_2]$, only the octa-2,7-dienylamines were able to be obtained and no butenylamines were formed [42]. The addition of one mole of triphenylphosphine per mole of $[Pd(cod)_2]$ produced only a small increase in the conversion of the amines.

The importance of the ligands in the butadiene-amine telomerization catalyzed by palladium was recently underlined by Beger [85, 86]: in the presence of $[Pd(acac)_2]$, triarylphosphates are the most efficient phosphorous ligands, followed by triarylphosphites and -phosphines.

The *platinum* catalysts have been rarely studied in amine telomerization. The treatment of butadiene with $[Pt(cod)_2]$ in the presence of morpholine yields octa-1,3,5-triene as the only product. However, it is interesting that the addition of one mol equivalent of the Lewis acids $Al(OPr^i)_3$ or $Al(OBu^t)_3$ to the reaction system yielded a mixture of the primary and the secondary telomer in a 75% yield.

The addition of triphenylphosphine encourages the formation of the octadienyl adducts by using *rhodium* catalysts [37]. For example, the telomerization of butadiene and morpholine in the presence of the rhodiumtrichloride yielded 1 : 1-adducts, whereas with the addition of PPh_3, the octadienylamines predominated. Rhodium(I)-phosphine complexes such as $[RhCl(PPh_3)_3]$ also favour the octadienyl adduct formation.

Butenyl derivatives are very often formed with *nickel* catalysts [28, 29, 33, 87]. In some cases the dodecatrienylamines were also able to be obtained [24, 44]: Heimbach found that nickel allyl complexes such as **26** react with three molecules of butadiene and one molecule of amine to form the trimeric product **27**.

$$\mathbf{26} \xrightarrow[+3\ \text{butadiene}]{+RR'NH} \mathbf{26} + RR'N{+}CH_2-CH=CH-CH_2{+}_3H \quad \mathbf{27} \tag{8}$$

Many secondary and primary aromatic and aliphatic amines react with butadiene. Diethylamine has been very intensively investigated in the group of the aliphatic amines. Some important telomerizations with diethylamine are summarized in Table 10.

Table 10

TELOMERIZATION OF BUTADIENE WITH DIETHYLAMINE

Catalysts	Products	Ref.
P-containing Pd-complexes fixed to an SiO_2 surface	C_8-NEt_2	88
Ni, Pd, Pt-allyl halide complexes + PPh_3: Ni < Pt < Pd	C_8-NEt_2	43
Ni (II) Laurate + $NaBH_4$ + $PhP(OR)_2$ or $Ph_2P\ OR$	C_8-NEt_2 C_4-NEt_2	29
Pd-Sn-complexes obtained through interaction between Pd-allyl compounds and Sn-acetate, fixed to an SiO_2 surface	C_8-NEt_2	89
Pd-, Pt- and Rh-compounds + sulfonated phosphines	C_8-NEt_2 C_4-NEt_2	38
Ni, Pd, Pt-allyl complexes + PPh_3	C_8NEt_2 $C_{12}-NEt_2$	44
Pd + Et_2N-C_8-NEt_2	C_8-NEt_2	90
[Pd(acac)$_2$], PPh_3, HOAc	C_8-NEt_2	68
[Pd(acac)$_2$]	C_8-NEt_2	91

Secondary amines with longer chains show the effect of steric hindrance [86]: whereas diethylamine can be converted quantitatively, the di(*n*-propyl)- and the di(*n*-butyl)-amine are less active (95% and 84% conversion of butadiene).

This steric effect is of little importance for primary amines: mono-*n*-butyl- and mono-*n*-octylamine produce the same yield of about 85%. However, branched amines such as secondary butyl and tertiary butyl amine react with only one N-H-group.

A systematic examination of the secondary cycloaliphatic amines was done by Dzhemilev [92]: no reaction took place with ethylenimine **28**, whereas acetidine **29**, pyrrolidine **30**, piperidine **31** and hexamethylenimine **32** yielded the octadienyl derivatives.

NH NH NH NH NH

28 **29** **30** **31** **32**

Other well-studied amines are shown in Table 11. Morpholine and aniline, in particular, have often been investigated. In aromatic amines the group substituted

Table 11

TELOMERIZATION OF BUTADIENE WITH HIGHER AMINES

Amines	Catalysts	Products	Ref.
morpholine piperidine dialkylamines	$[Pd(cod)_2]$ $(+PPh_3)$; $[Pt(cod)_2]$ + $Al(OR)_3$	C_8-NRR', no C_4-NNR'	42
morpholine dialkylamines	Ni/P$(OEt)_3$: Ni/without ligand:	C_8-NRR' $C_{12}-NRR'$	24
morpholine piperidine carbazole aniline and others	$[Pd(PPh_3)_2$ (maleic anhydride)]	C_8-NRR' $(C_8)_2NR$	5
aniline	$[Pd(PPh_3)_2$ (maleic anhydride)]	$C_8-NH-Ph$ $(C_8)_2N-Ph$	55
morpholine aniline *n*-butylamine piperidine	Pd-diphosphine- complex + addition of phenol	C_4-NRR'	84
piperidine aniline *N*-methyl- aniline	Rh(I)-complexes, containing phosphine- ligands and H, OPh, OAc, alkyl, aryl or allyl ligands	C_8-NRR' $(C_8)_2-NR$	36
morpholine	(a) $[PdCl_2(allyl)_2]$, PPh_3 $NaBH_4$ (b) $[Pd(PPh_3)_4]$	C_8-NRR'	58
aniline *N*-methyl-aniline and others	Pd, Pt or Ru-com- pounds with a phenoxide promotor	C_8-NRR'	93
morpholine di-*n*-propylamine pyrrolidine aniline	(a) $NiCl_2$-PPh_3-$NaBH_4$ (b) $CoCl_2$-PPh_3-$NaBH_4$- $Al(OPr^i)_3$ (c) $IrCl_3$ (d) $RhCl_3$	C_4-NRR' C_8-NRR'	87
morpholine piperidine pyrrolidine *p*-anisidine and others	$[Ni(acac)_2]$ $PhP(OPr^i)_2$	C_4-NRR' C_8-NRR'	33

Table 11 (continued)

Amines	Catalysts	Products	Ref.
morpholine di-*n*-propylamine aniline	(a) $RhCl_3\ 3H_2O\ (+PR_3)$ (b) $[RhCl(PPh_3)_3]$ (c) $[RhCl(CO)\ (PPh_3)_2]$	C_4-NRR' C_8-NRR'	37
morpholine di-*n*-propylamine aniline *n*-butylamine	$[Ni(acac)_2]$- $PhP(OPr^i)_2$-$NaBH_4$	C_4-NRR' C_8-NRR'	28
piperidine morpholine dialkylamines monoalkylamines	$[Pd(acac)_2]$ + (a) triarylphosphates (b) triarylphosphites (c) triarylphosphines + ROH/H_2O	C_4-NRR' C_8-NRR' $(C_8)_2NR$	85, 86

influences the telomerization [5, 94]. The reactivities of aniline derivatives were found to increase in the order shown in Equation 9.

$$CH_3OC(=O)\text{-}C_6H_4\text{-}NH_2 < Cl\text{-}C_6H_4\text{-}NH_2 < C_6H_5\text{-}NH_2 < CH_3\text{-}C_6H_4\text{-}NH_2 < CH_3O\text{-}C_6H_4\text{-}NH_2 \qquad (9)$$

These facts suggest that there is a strong relationship between the reactivity and the basicity of the nucleophiles. This presumption was able to be confirmed [84]: the more basic amines show higher reactivity. For example, the reaction of morpholine (pK = 9.61) and aniline (pK = 9.42) produced the butenyl adducts in 79% and 67% yields respectively, whereas the reaction of the less basic amines *n*-butylamine (pK = 3.39) and piperidine (pK = 2.80) produced these products only in 19% and 29% yields respectively.

Moberg and Åkermark studied the reaction of butadiene with methylamine in the presence of catalytic amounts of $Ni(cod)_2$, tributylphosphine and boron trifluoride etherate [95]. They found that under these defined conditions the dioctadienylamines were obtained as main products.

Tertiary allylic amines such as *N,N*-dimethylallylamine react with butadiene in the presence of palladium(0) catalysts with cleavage of the carbon–nitrogen bond of the amine and formation of the amine **33** (Equation 10). This unexpected product was formed with an optimum yield of 81% [96].

$$CH_2{=}CHCH_2NMe_2 + 2\ CH_2{=}CH{-}CH{=}CH_2 \longrightarrow \mathbf{33}\ (\text{-NMe}_2) \qquad (10)$$

33

Dzhemilev and his coworkers have intensively studied the telomerization with the group of N-H-nucleophiles [100, 101]. In the telomerization of butadiene with furfuraldimines, 8- and 14-membered nitrogenous heterocycles were formed [97]. The reaction of 3,3-dimethyldiaziridine with butadiene produced *N*-octadienyl-substituted diaziridines in a total yield of about 40% [98]. Analogous compounds have been obtained by using substituted pyrazolines [99]. The telomerization of butadiene with secondary amines was also carried out with nickel compounds activated by optically active phosphites, which were obtained from halophosphines and naturally occurring alcohols such as menthol or cholesterol [102, 103].

2.3.2. AMMONIA AND AMMONIUM SALTS

Whereas the addition of ammonia to monoolefins has not been reported until now, the telomerization of ammonia with butadiene is known. If a mixture of aqueous ammonia and butadiene in acetonitrile is allowed to react in the presence of palladium acetate and triphenylphosphine in addition to a small amount of diocta-2,7-dienylamine **35**, the triocta-2,7-dienylamine **36** and the di(2,7-octadienyl) (1-vinyl-5-hexenyl)amine **37** are formed as the main products [104–106].

+ NH_3 ⟶ [H_2N … **34**] ⟶(+ butadiene) HN … **35** ⟶(+ butadiene) N … **36** + N … **37** (11)

The reaction seems to proceed step by step, but the primary amine **34**, which is formed in the first step, was able to be detected only in negligible amounts because it is more reactive than ammonia itself. In telomerization the branched products are usually less than 20%. The formation of the branched telomer **37** in nearly the same proportion as **36** is therefore somewhat surprising. An essential factor in this reaction is the presence of water. The telomerization proceeded smoothly in a mixture of aqueous ammonia and organic solvents, whereas almost no reaction took place when liquid ammonia and an organic solvent were used in the absence of water. However, there is no information which can explain this important role of water.

The reaction of butadiene with ammonium bicarbonate [107] in the presence of a catalytic system of $[Pd(acac)_2]$, $AlEt_3$ and trifluoroacetic acid also yields the trioctadienylamine **36**.

2.4. Telomerization with C-H-nucleophiles

Compounds with methylene and methyne groups in the neighbourhood of two electronegative groups X and Y undergo telomerizations with butadiene very smoothly: one or two acidic hydrogens can be replaced with the octa-2,7-dienyl group to produce the mono- and disubstituted compounds **38** and **39** respectively.

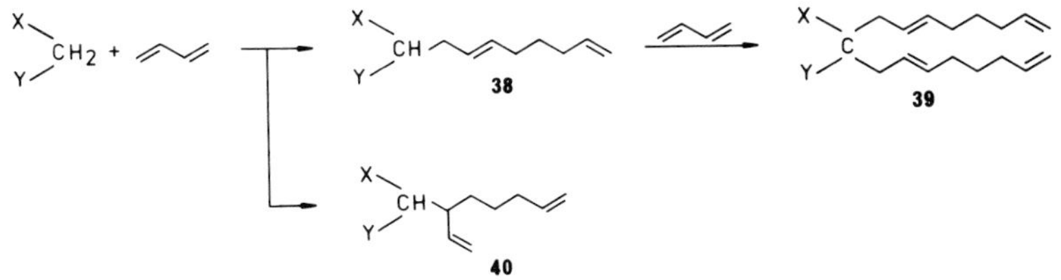

(12)

In addition to this branched octadienyl and linear as well as branched butenyl compounds are also formed as by-products, depending importantly on the catalytic system. The electronegative groups X and Y can be carbonyl, nitro, cyano and sulfonyl.

2.4.1. CARBONYL COMPOUNDS

In general, the methylene and methyne groups must be activated by two carbonyl groups. The three molecule fragments, the carbonyl, formyl and carboxyl group, can be combined to the following highly active nucleophiles: β-diketones, malonates, β-ketoesters or α-formyl ketones. All these compounds are in equilibrium with keto and enol forms. Therefore it might be possible that the hydroxy groups of the enol forms react with butadiene to yield products in which the octadienyl group is bonded to the oxygen. However no such products were observed.

The methylene compounds already enumerated change into methyne compounds by substituting one hydrogen by an aliphatic or aromatic group.

This means that a great number of C-H-nucleophiles can be studied in telomerization. This reaction has been thoroughly investigated in two laboratories: Hata and his coworkers [84, 108] discovered this telomerization in 1969 by using palladium complexes with unidentate tertiary phosphines. For example, the ethyl acetoacetate was telomerized with butadiene to form the 1 : 2 adduct **38** in a 78% yield and the 1 : 4 adduct **39** in a 12% yield when a $[PdCl_2(PPh_3)_2]$/NaOPh catalyst was used. Platinum catalysts showed a little different behaviour. $[PtCl_2(PPh_3)_2]$/NaOPh catalyzed the reaction of acetylacetone with butadiene to yield the same products as formed in the palladium-catalyzed reaction; but the amount of the branched 1 : 2-adduct **40** is much larger. The 1 : 1- and the 1 : 3-adducts could also be isolated as by-products. The telomerization with bidentate phosphine complexes of palladium such as $[PdBr_2(diphos)_2]$ or $[Pd(diphos)_2]$ has been found to produce only the 1 : 1-adducts **41** and **42**.

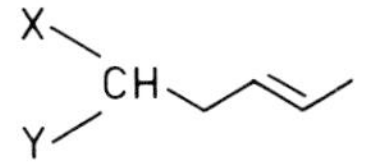

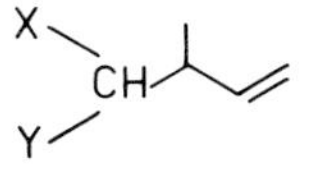

This result shows that the coordination of the ditertiary phosphine hinders the coordination of a second molecule of butadiene to the palladium atom.

The second team which investigated reactions with active methylene carbonyl compounds are Baker and his coworkers [87, 109, 110]. They found that also compounds with only one carbonyl group react with butadiene: the benzyl methyl ketone reacts readily in the presence of catalytic amounts of nickel(0)complexes which are prepared by $[Ni(acac)_2]$, phenyldiisopropoxyphosphine, sodium borohydride and sodium phenoxide.

The telomerization of diethyl malonate and butadiene has been realized with a $NiCl_2$-PPh_3-$NaBH_4$ catalyst. Prior formation of the sodium salt of the diethyl malonate is required for the analogous $CoCl_2$ system. Hydridotris(triphenylphosphine)cobalt may be the active catalyst but it could not be isolated due to its unstableness.

The methylene compounds which are activated only by one electronegative group do not react directly with butadiene. Even in the benzyl methyl ketone already mentioned the methylene group is adjacent to two electron-attracting fragments, the carbonyl and the phenyl group.

It is possible to conduct the telomerization even with aliphatic monoketones or aldehydes via enamines used as intermediates [111]. For instance, cyclohexanone can be converted in the pyrrolidine enamine **43** which reacts with butadiene by using a $[Pd(OAc)_2]/PPh_3$ catalyst (Scheme 8). By hydrolyzing the product **44** with dilute

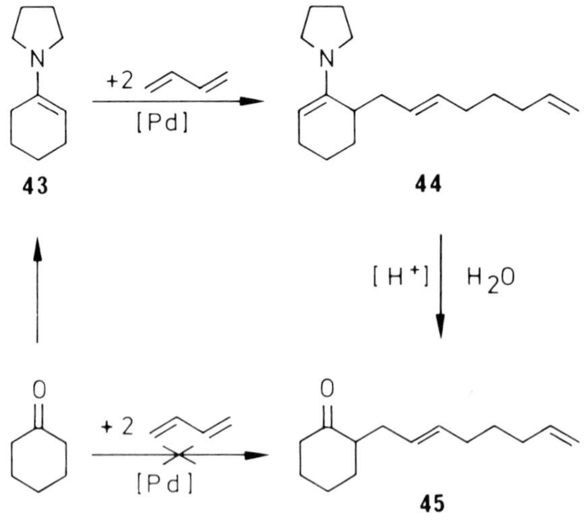

+2 [Pd]

43 **44**

$[H^+]$ H_2O

+2 [Pd]

45

Scheme 8.

acid, the 2-(2,7-octadienyl)cyclohexanone **45** can be obtained in high yield which is accompanied by a small amount of the bis-octadienyl-adduct. The pyrrolidine enamine of cyclopentanone behaves similarly.

The direct catalytic telomerization of butadiene with *p*-quinones is still unknown. Nevertheless, a stoichiometric reaction of these two components can be obtained [112]: bis(triphenylphosphine) (*p*-benzoquinone)palladium(0) reacts with butadiene to yield a complex in which two molecules of butadiene are cyclised across one double bond of the *p*-benzoquinone ligand (Equation 13). With a treatment of $SbPh_3$ this complex liberates a tricyclic adduct with two five-membered

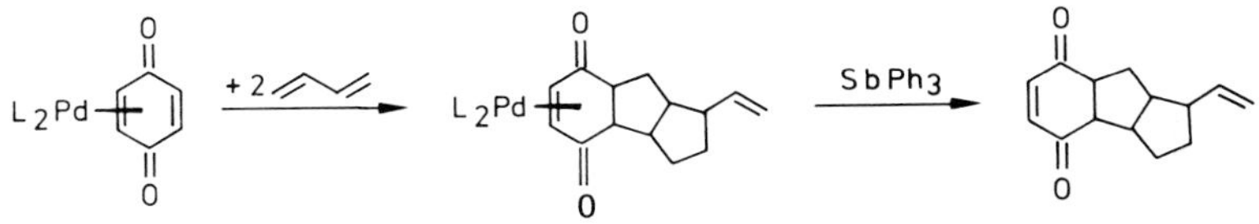

L_2Pd + 2 butadiene → L_2Pd complex —$SbPh_3$→ tricyclic adduct (13)

rings. This reaction has been found to be generally applicable to other quinones, e.g. 1,4-naphthoquinone which are coordinated to a $Pd(PPh_3)_2$ group to yield analogous adducts.

2.4.2. NITROALKANES

As already mentioned, the methylene compounds which contain only one electronegative group are normally inactive in telomerization. One exception is the nitroalkanes which react very smoothly with butadiene [104, 113, 114]. The telomerization of nitromethane in the presence of $[PdCl_2(PPh_3)_2]$ and sodium hydroxide at room temperature produces the nitrocompounds **46**–**48** which are accompanied by a small amount of branched products (Scheme 9). The relative

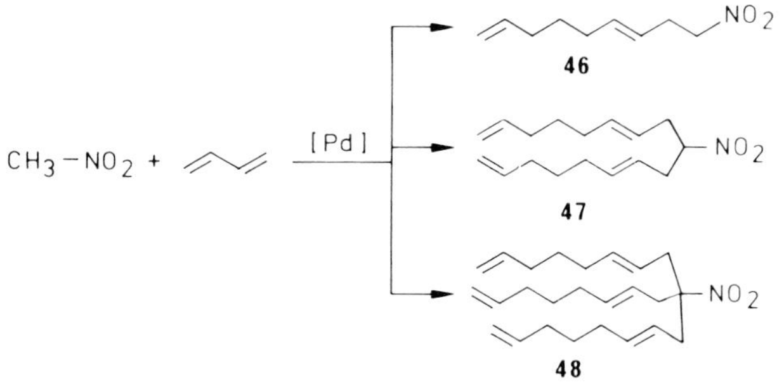

CH_3-NO_2 + butadiene —[Pd]→ **46**, **47**, **48** (NO_2)

Scheme 9.

amounts of these products can be controlled by adjusting the ratio of the reactants and the reaction time.

The catalyst $[PdCl_2(PPh_3)_2]$ used for nitromethane telomerization must be activated by a base such as potassium hydroxide or sodium phenoxide. The reaction proceeds faster and produces higher yields when the ratio of the base and palladium was above 100. It seems likely that the formation of the aci-form of the nitro compounds accelerates the reaction. One mole of the Pd catalyst produces more than 1000 moles of the telomers.

Besides nitromethane, also nitroethane, 1- and 2-nitropropane and nitrocyclohexane can be telomerized. The nitroolefins synthesized with these nucleophiles can be reduced with Raney nickel used as a catalyst to produce saturated long-chain amines **49** with a primary amino group at the middle of the molecule. With $LiAlH_4$ the nitro group can only be reduced to yield the unsaturated amines **50** which are potential precursors for surface active compounds (Scheme 10).

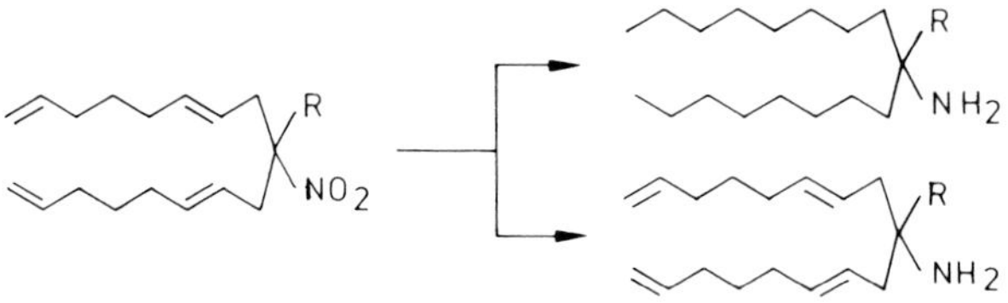

R NO2 → R NH2 **49**; R NH2 **50**

Scheme 10.

2.4.3. NITRILES

Nitriles can also act as active methylene compounds: the cyano-group behaves similarly to the carbonyl group [108]. Ethyl cyanoacetate, cyanoacetamide, benzoylacetonitrile and malononitrile produce the linear mono- and dioctadienyl adducts in good yields by using palladium catalysts. Similar to the benzyl methyl ketone, the benzyl cyanide can react as nucleophile [109, 110]. However, the reactivity of this molecule is rather low: with a nickel catalyst system a yield of only 30% was able to be obtained.

2.4.4. SULFONES

The sulfones have been studied by Hata [108]: ethyl(phenylsulfonyl)acetate reacted with butadiene to produce the telomeric monooctadienyl product in a 91% yield. The butadiene/sulfone telomers are of some interest because with treatment of 6% Na/Hg in MeOH, the $PhSO_2$ group can be substituted by hydrogen [115]. As is shown in Scheme 11, it is thus possible to synthesize linear polyenic esters.

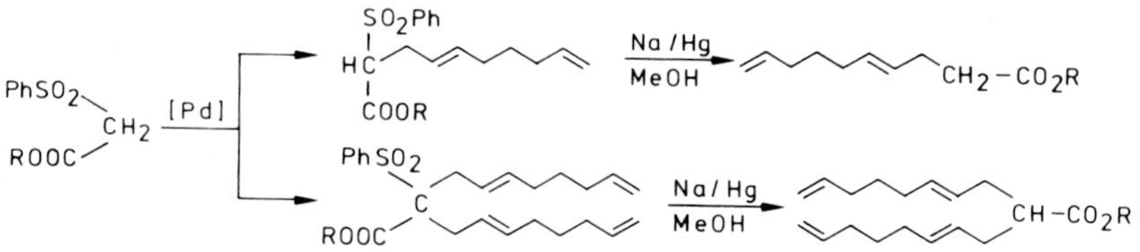

PhSO2 CH2 ROOC —[Pd]→ SO2Ph HC COOR —Na/Hg, MeOH→ CH2–CO2R; PhSO2 C ROOC —Na/Hg, MeOH→ CH–CO2R

Scheme 11.

2.5. Telomerization with Si-H-nucleophiles

Hydrosilanes are very reactive nucleophiles and numerous hydrosilylation reactions with monoenes and conjugated dienes are known. The products of this type of telomerization are somewhat different from the telomers described thus far: the octadienyl chain of the 1 : 2-adducts shows another arrangement of the double bonds. Unlike other telomers which have an 1,6-octadienyl chain, the telomers of silanes have a 2,6-octadienyl chain. In addition to the 2 : 1-adducts, the bissilylbutanes can be formed as by-products. To explain these unusual products a step by step mechanism was proposed in which a methallyl complex is formed in the first step by the insertion of one molecule of butadiene in the palladium–hydrogen bond.

The monosilylbutenes can be derived from complex **51**. On the other hand, a second molecule of butadiene can insert and form the complex **52**. Reductive elimination of this complex produces the 1 : 2-adduct with the 2,6-octadienyl chain (Scheme 12). In normal telomerization the addition of butadiene is not done step

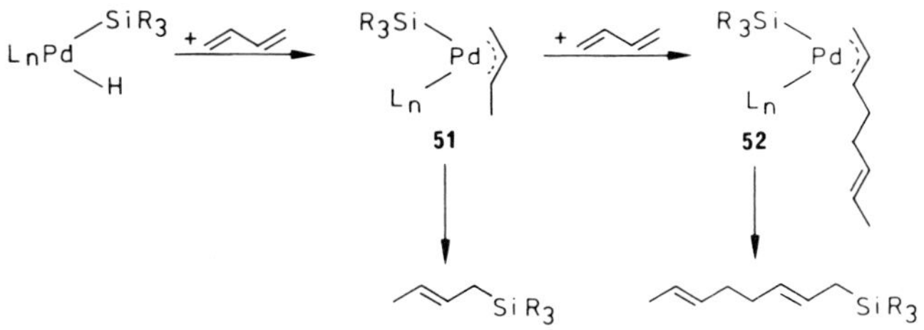

Scheme 12.

by step and a bisallylic complex is formed as an intermediate from which the 1,6-octadienyl chain is liberated. (Compare Section 2.1.)

Palladium complexes are very active hydrosilylation catalysts even at low concentrations. For example $[Pd(PPh_3)_4]$ shows satisfactory activity at a concentration of 10^{-5} mole for one mole of olefin [116]. The order of activity shown in Equation 14 was observed for the ligands.

$$PPh_3 > PEt_3 > PBu_3 > PCy_3 > P(OPh)_3 \quad (14)$$

Unlike all the telomerizations previously mentioned, the hydrosilylation of butadiene is possible even by using metallic palladium when it is combined with various phosphines [117]. The rate of hydrosilylation is also markedly dependent on the kind and the structure of the silanes which are used. Whereas the most reactive silane is trichlorosilane, the reactivity decreases by replacing the chlorine of trichlorosilane by an alkyl group. The order of reactivity shown in Equation 15 has been observed.

$$HSiCl_3 > HSiCl_2Me > HSiClMe_2 > HSiCl_2(OMe) > HSiMe_3 > HSiEt_3 > HSi(OMe)_3 \quad (15)$$

The reaction temperature and the reaction time are of importance for the formation of the various products [118]. With a palladium(II)complex such as $[PdCl_2(PPh_3)_2]$ or $[PdCl_2(PhCN)_2]$ the 1 : 2-adducts are formed in 24 hours at ambient temperature, whereas the 1 : 1-adducts are formed as main products in 4 hours at 100°C. The telomerization of trimethylsilane was found to be barely selective at elevated temperatures: besides the telomeric 1-trimethylsilyl-2,6-octadiene **53**, 4-trimethylsilyl-1-butene **54**, *trans*-1-trimethylsilyl-2-butene **55**, 1-trimethylsilyl-1,3-butadiene **56** and 1,4-bis(trimethylsilyl)-2-butene **57** were also formed.

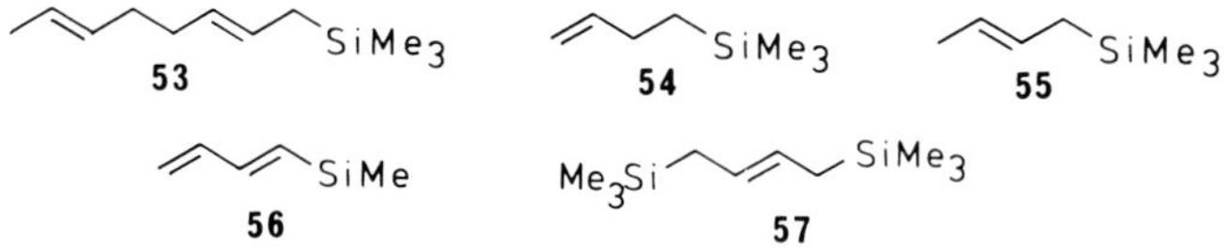

An interesting catalyst system for hydrosilylation is the polymeric chelates of the bis(8-hydroxy-5-chinolyl)-methane with palladium, platinum or rhodium [119]. With the palladium complex **58**, the reaction with trichlorosilane yields only the *cis*-1-trichlorosilyl-2-butene; on the contrary, the reaction with trimethylsilane selectively yields the 2,6-octadienyl derivative.

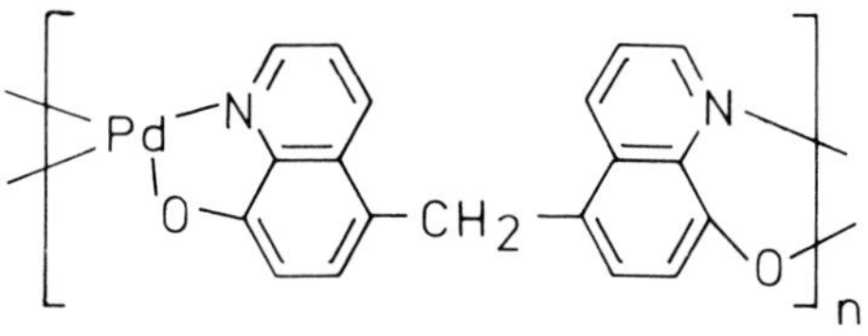

The telomerization with hydrosilanes can also be carried out with compounds of *nickel*(0) and nickel(II). The hydrosilylation of butadiene with alkyl- and alkoxy-silanes occurs with $[Ni(cod)_2]$, nickel-halides, $[Ni(acac)_2]$ and nickel-ligand-systems, but the octadienyl-derivatives are formed only in small amounts [120]. In contrast to the hydrosilylation which is catalyzed by palladium complexes, the ratio of the butenyl to the octadienyl products is only slightly affected by temperature, solvents, and the butadiene to hydrosilane ratio.

Cobalt systems are also active catalysts in hydrosilane telomerization. The systems of a cobalt(II) salt and sodium bis(2-methoxyethoxy)aluminium hydride in particular are very effective. Some important examples for the telomerization of butadiene with silanes are summarized in Table 12.

2.6. Sulfinic acids and sulfur dioxide

The reaction of butadiene with sulfinic acids catalyzed by palladium complexes produces the primary telomers in practically quantitative yields [126]. The

Table 12
TELOMERIZATION OF BUTADIENE WITH SILANES

Silanes	Catalysts	Products	Ref.
$HSiCl_3$ $HSiRCl_2$ $HSiR_3$	(a) Pd-metal + PPh_3 (b) $Pd(PPh_3)_4$ and others	1 : 1-adducts 1 : 2-adducts	116, 117
$HSiCl_3$ $HSiRCl_2$ $HSi(OR)_3$	Pd-compounds	1 : 1-adducts 1 : 2-adducts	121
$HSiR_3$ $HSiR(OR')_2$ $HSi(OR')_3$	Pd-phosphine complexes	1 : 2-adducts	5, 122
$HSiCl_3$ $HSi(OR)_3$	$PdCl_2$-aryl-complexes	1 : 1-adducts 1 : 2-adducts	123
$HSiR_3$ $HSiR_2(OR')$ $HSiR(OR')_2$ $HSi(OR')_3$	Pd-compounds	1 : 1-adducts 1 : 2-adducts	118
$HSiR_3$	Pd- or Ni-compounds	1 : 2-adducts	124
$HSiCl_3$ $HSi(OR)_3$ $HSiR_3$	Polymeric chelates of Pd, Pt, Rh	1 : 1-adducts 2 : 1-adducts	119
$HSiCl_3$ $HSiR_3$ $HSi(OR')_3$	(a) Ni(II) and Ni(O) compounds (b) Ni-Al-systems (c) Co-Al-systems	1 : 1-adducts 2 : 1-adducts 1 : 2-adducts	120
$HSiR_3$ H_2SiR_2 H_3SiR	(a) Ni-compounds (b) Co-compounds	1 : 1-adducts 1 : 2-adducts	125

1 : 1-adducts = monosilylbutenes
1 : 2-adducts = monosilyloctadienes
2 : 1-adducts = bis silylbutanes

unsaturated sulfone **59** is formed with cyclohexane sulfinic acid. Higher branched alkane sulfinic acids produce the 2,7-octadienylsulfones, whereas straight acids such as butane sulfinic acid mainly yield the 1- and 2-butenylsulfones **60** and **61** [127].

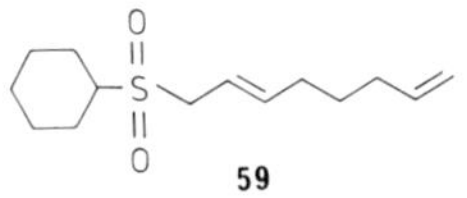

$R-SO_2$ **59** $R-SO_2$ **60** $R-SO_2$ **61**

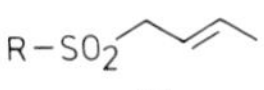

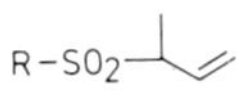

The same 1 : 1-adducts **60** and **61** are formed in the telomerization of toluenesulfinic acid with butadiene which is catalyzed by low-valence nickel complexes [128]. Sulfur dioxide reacts with butadiene to yield the 1 : 1-adducts 3-sulpholene **62** and the 2 : 1-adduct 2,5-divinylsulpholane **63** [129]. It could be shown in a separate reaction that 3-sulpholene and butadiene also yield the adduct **63** [130].

SO_2

62

$CH{=}CH_2$

SO_2

$CH{=}CH_2$

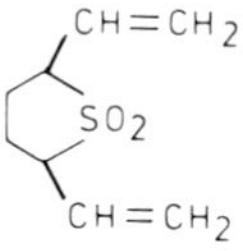

63

2.7. Telomerization with two-centre nucleophiles

The two-centre nucleophiles possess two sites both of which theoretically can become active in telomerization. These nucleophiles and the telomers which were actually able to be obtained are summarized in Table 13.

Table 13

TELOMERIZATION OF BUTADIENE WITH TWO-CENTRE NUCLEOPHILES

Two centres	Nucleophiles	Telomers	Ref.
O–O	diols	mono- and dioctadienyl ethers	51, 131, 132
N–N	diamines	alkylated amines	133
O–N	aminoalcohols	*N*-alkylated amino alcohols and amino ethers	133–135
	oximes	oxime ethers, isoxazolidines	136
N–C	hydrazones	*N*-alkylated hydrazones *C*-alkylated azo compounds	137–140
	imines	*C*-alkylated products	139
	acylaminoketones	*C*-alkylated acylamino-ketones	141
O–C	acyloins	*O*- and *C*-alkylated acyloins	142
Si–Si	disilanes	disilyloctadienes	
	disilacyclo-alkanes	disilamacrocycles	143–145

2.7.1. DIOLS

Two hydroxy groups can act as nucleophiles in diols. Romanelli discovered only the two monooctadienyl derivatives **64** and **65** in the telomerization of butadiene and ethylene glycol in the presence of $[Pd(PPh_3)_4]$ [53].

2 + HO–CH₂CH₂–OH ⟶ **64** + **65** (16)

Dzhemilev, in a recent paper, noted that bis-2,7-octadienyl ethers were also able to be obtained [131]. A yield of 80% for the dioctadienyl derivative was reached by using the catalyst $Pd(acac)_2$-PPh_3-$AlEt_3$ at high temperature (100°C) and with long reaction times (20h). The telomerization of diethylene glycol, 1,2-propylene glycol, 1,3- and 1,4-butylene glycols and glycerol produced similar results. Cyclic diols such as 1,2-cyclohexane-, cyclooctane-, and cyclododecanediol are also active in telomerization [132].

2.7.2. DIAMINES

The reaction of butadiene with tetramethylmethylenediamine on the above-mentioned catalyst system yielded the telomeric product 1,1-dimethyl-amino-3,8-nonadiene **66** and insignificant amounts of the analoguous butenyl derivatives [133].

$$(CH_3)_2N-CH_2-N(CH_3)_2 + 2 \longrightarrow [(CH_3)_2N-]_2CH \quad \mathbf{66} \qquad (17)$$

In this reaction butadiene adds to the methylene group because the two tertiary amine groups are fully blocked. Telomerizations with primary or secondary diamines are not thus far known.

2.7.3. AMINOALCOHOLS

The amino group is more reactive than the hydroxy group in aminoalcohols. Therefore the amino alcohols are exclusively alkylated to form the N-substituted octadienyl compounds. Only when the nitrogen atom of the amino alcohols is fully substituted, does oxygen alkylation take place to produce the unsaturated amino ethers [134].

Only in one case, that of diethanolamine, did any significant ether formation occur (Equation 18). However, only one hydroxy group was able to be etherified [135].

HN(CH₂CH₂OH)₂ + $\xrightarrow{[Pd]}$ (18)

The octadienyl ether **67** is found in the reaction mixture in a quantity of less than 3%, in the telomerization of butadiene with *N*-hydroxymethyldiethylamine.

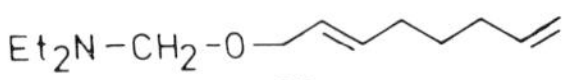

Et$_2$N–CH$_2$–O–

67

The reason for the large number of by-products such as amines and pyranes is the disproportion of the nucleophile and the subsequent reaction of the decomposition products with butadiene [133].

2.7.4. OXIMES

The palladium-catalyzed telomerization of oximes proceeds with either oxygen or nitrogen to yield a series of products [136]. On the one hand, nitrones are fomed which react thermally in a 1,3-dipolar addition reaction with excess butadiene to yield the *N*-alkylated isoxazolidines **68**. On the other hand, the octadienyl oxime ethers **69** and **70**, as well as the butenyl oxime ethers **71** and **72**, are produced. Either [Pd(PPh$_3$)$_4$] or Pd(NO$_3$)$_2$ and PPh$_3$ were typically used as catalysts. With aliphatic aldehyde oximes such as CH$_3$—CH=N—OH or CH$_3$CH$_2$—CH=N—OH the isoxazolidines are exclusively obtained, whereas only oxime ethers are formed with aromatic aldehyde oximes or ketoximes (Scheme 13).

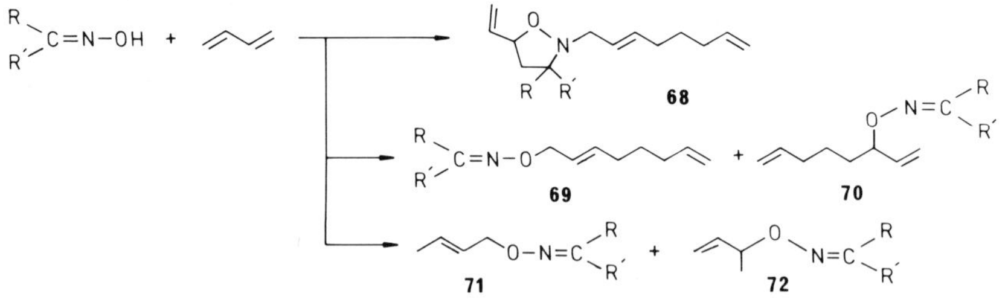

R, R′ C=N–OH + → **68**; R, R′ C=N–O– **69** + **70**; **71** + **72**

Scheme 13.

2.7.5. HYDRAZONES

Aromatic and aliphatic hydrazones produce different results: phenylhydrazones possess both nucleophilic nitrogen and electrophilic carbon. The carbon atom is the active centre in the presence of bis-η^3-allyl nickel and a mixture of the two azo compounds **73** and **74** are able to be obtained in high yields [138]. The same reaction catalyzed by [Pd(PPh$_3$)$_4$] yields the additional *N*-alkylated hydrazone **75**, although the azo compounds are always predominant in the product mixture. The formation of **75** by a Cope rearrangement of **74** was able to be excluded by experiments in which **74** was separately heated in the presence of the palladium catalyst and under identical reaction conditions. The azo compound **74** remained completely unchanged (Scheme 14).

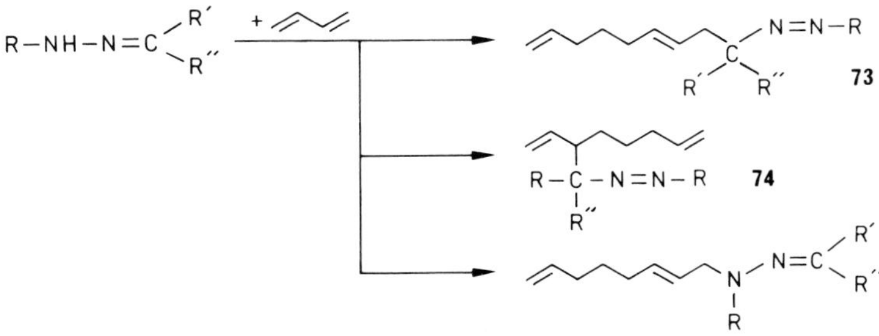

Scheme 14.

The telomerization proceeds more selectively with aliphatic hydrazones, especially with *N*-methylhydrazones: only the *N*-octadienylated methylhydrazones **75** and no azo derivatives were formed in the presence of palladium catalysts [137]. Only the azo compounds were prepared with a [Ni(cod)$_2$]/PPh$_3$ system. This means that the reaction of methylhydrazones with butadiene can be completely controlled by the choice of the complex metal.

Reinehr studied the formation of the azo compounds and found that with extended reaction times the amount of the linear isomer **73** increases substantially which suggests an ensuing rearrangement of **74** to **73** [140].

2.7.6. IMINES

Only the *C*-alkylated compounds **76–78** were able to be obtained with imines when a nickel catalyst was used [139].

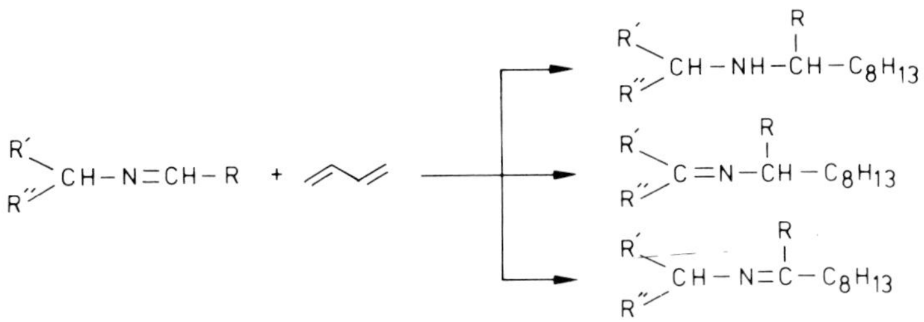

Scheme 15.

2.7.7. ACYLAMINO KETONES

The telomerization of butadiene with the incorporation of α-acylamino ketones selectively yielded the C–C bond formation product, α-acylamino α-(2,7-octadienyl) ketone. No C—N bond formation product was formed. The catalyst was generated *in situ* from palladium dichloride and sodium toluenesulfinate used as a cocatalyst [141].

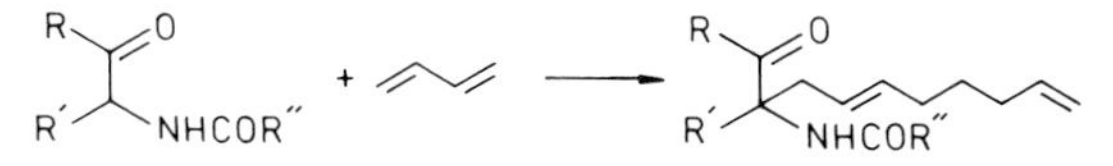

(19)

2.7.8. ACYLOINS

In the telomerization of butadiene with an acyloin catalyzed by palladium the selective formation of α-hydroxy α-octadienyl ketone **79**, the product of C—C bond formation is observed when toluenesulfinate serves as a cocatalyst. The 2-oxy octadienyl ether **80**, the product of C—O bond formation, is produced in high selectivity by using $Pd(PPh_3)_4$ as a catalyst. The tetrameric product **81** was also observed in addition to the 2 : 1 telomers **79** and **80** [142].

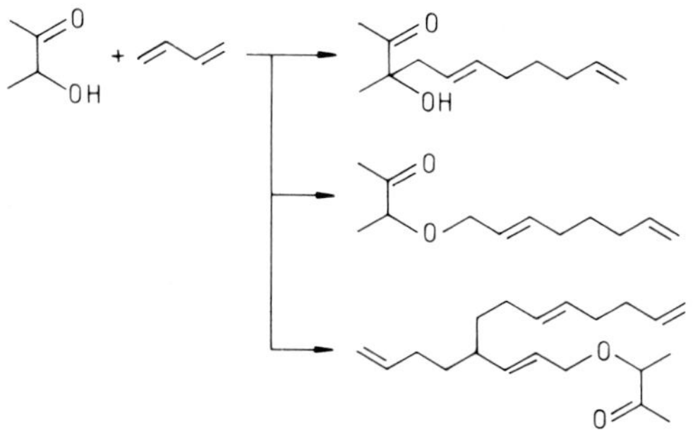

Scheme 16.

2.7.9. ORGANODISILANES

Linear and cyclic organodisilanes can also undergo the telomerization reaction: the two silicon atoms act as nucleophilic centres. This type of two-centre telomerization is a special one because the two active sites are set free only by cleavage of a Si—Si single bond. Cyclic carbon-disilanes such as 1,2-disilacyclopentane or 1,2-disilacyclohexane react with butadiene in the presence of $[PdCl_2(PPh_3)_2]$ to yield the 13-membered or 14-membered macrocycle respectively (Equation 20). Also unstrained organodisilanes such as hexamethyldisilane react by the splitting of

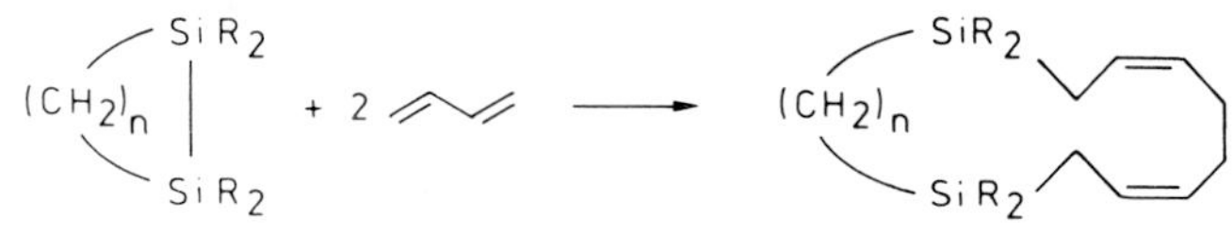

(20)

the Si—Si bond to produce the disilyloctadienes (Equation 21) [143]. The same reaction was able to be observed with 1,2-difluorodisilanes to produce the telomers

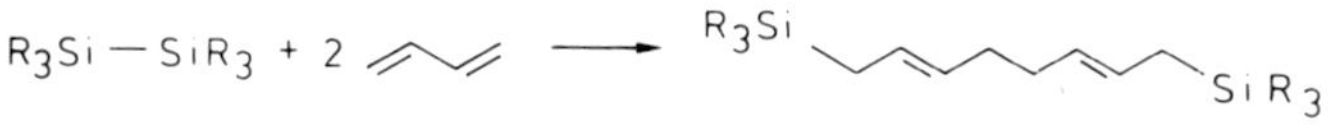

$R_3Si-SiR_3$ + 2 → R_3Si … SiR_3 (21)

in a yield of 29% [144]. The addition of chlorodisilanes to butadiene yields 1 : 1-adducts exclusively [145].

2.8. Telomerization with C=X double bond nucleophiles

An important reaction is the telomerization of butadiene with molecules which possess a one-hetero atom double bound such as C=O and C=N. In this type of telomerization the two sites of the nucleophile react simultaneously with butadiene to form six- or five-membered heterocyclic compounds. The fundamental reaction scheme is shown in the Equations 22 and 23.

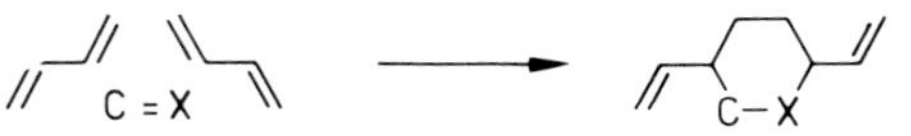

C = X → C–X (22)

C = X → C–X (23)

2.8.1. ALDEHYDES

Butadiene and formaldehyde react under mild conditions in the presence of catalytic amounts of [Pd(PPh_3)$_4$] to produce divinyltetrahydropyrans **82** in good yields (> 65%) and selectivities (> 95%) [146, 148]. Two isomers, the 2,5- and the 3,5-divinyl derivatives, were able to be obtained in a 2 : 1-ratio. The same reaction catalyzed by palladium acetate and tricyclohexylphosphine yielded the C_9-alcohols **83** and **84** [147]. These products are of industrial interest because the 2-ethylheptanol, which is formed by hydrogenation of **83** and **84**, can act as a potential precursor for plasticizers.

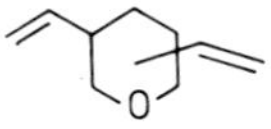

O **82** CH_2OH **83** CH_2OH **84**

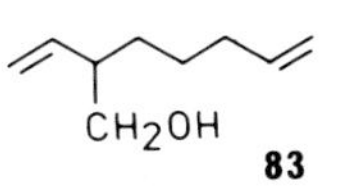

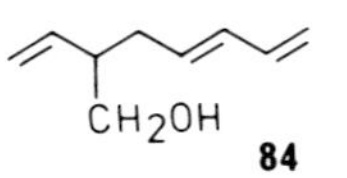

Similar products, such as the substituted divinyltetrahydropyrans **85** and the vinylheptadienols **86**, are able to be obtained with higher aldehydes [18, 42, 149–151].

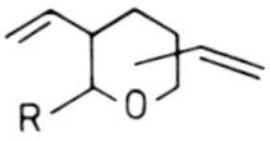

R O **85** R OH **86**

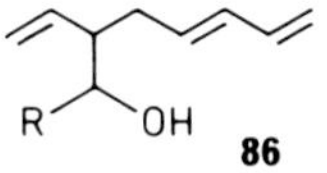

Various palladium–phosphine complexes can be used as catalysts, e.g. $[Pd(OAc)_2]/PPh_3$ or $[Pd(acac)_2]/PPh_3$. The ratio of the two products can be controlled by changing the molar ratio of PPh_3 to palladium: the unsaturated alcohol **86** is obtained as a main product when the ratio is near unity. The pyran is formed almost selectively when the ratio is greater than two.

A large number of both aliphatic and aromatic aldehydes can react with butadiene: benzaldehyde, furfural, acetaldehyde, propionaldehyde and butyraldehyde.

2.8.2. KETONES

The telomerization with ketones is more complicated than the reaction with aldehydes. Tsuji reports that the reactions catalyzed by palladium of butadiene with simple ketones such as acetone are not successful and that only 1,3,7-octatriene is formed [151]. However, Musco claims that acetone and butadiene react in the presence of water and $[Pd(PEt_3)_3]$ to yield the pyran **89** as the main product [152, 169]. The synthesis of C_{11}-alcohols from acetone and butadiene was able to be achieved: the alcohol **87** is formed with $[Pd(PCy_3)_2]$ as the catalyst [152]. The product **88** can be obtained with a nickel-ligand system such as $[Ni(acac)_2]$-$P(O\text{-}o\text{-}MeC_6H_4)_3$-$AlEt_3$ [153].

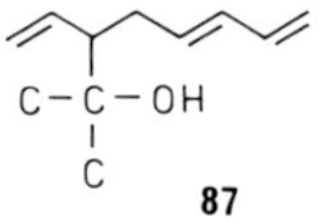

C–C–OH (C) **87**   C–C–OH (C) **88**   O **89**

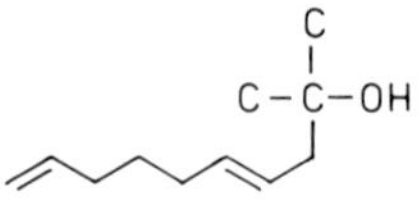

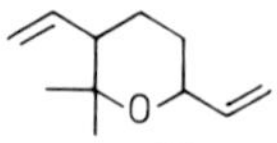

Other specific ketones also undergo the telomerization reaction: perfluoroacetone yields the pyran **90**, and active α-diketones such as benzil and biacetyl yield the corresponding pyranes **91** by the reaction of only one of the carbonyl centres [151].

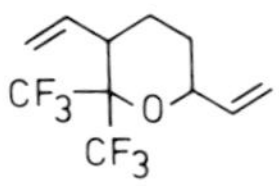

CF_3 O CF_3 **90**   R O R–C=O **91**

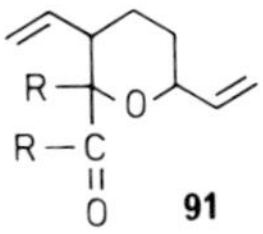

The possible mechanism of the telomerization of butadiene with aldehydes or ketones is shown in Scheme 17. The formation of a bisallylic palladium complex is probably the first step of the reaction. In the presence of aldehydes, an insertion of the carbonyl group into a Pd—C-bond might take place to produce an alkoxide type complex. This intermediate would have an equilibrium between a η^3-allyl and a η^1-alkyl palladium complex, depending on the concentration of the phosphine ligands L. The former complex is then expected to decompose rapidly through a hydrogen transfer from the η^3-allyl moiety to produce the unsaturated alcohol **86**, whereas the latter complex might give rise to the pyran **85**. This mechanism would explain the influence of the ligand concentration already mentioned. The pyran is the main product in the presence of excess PPh_3 and vice versa.

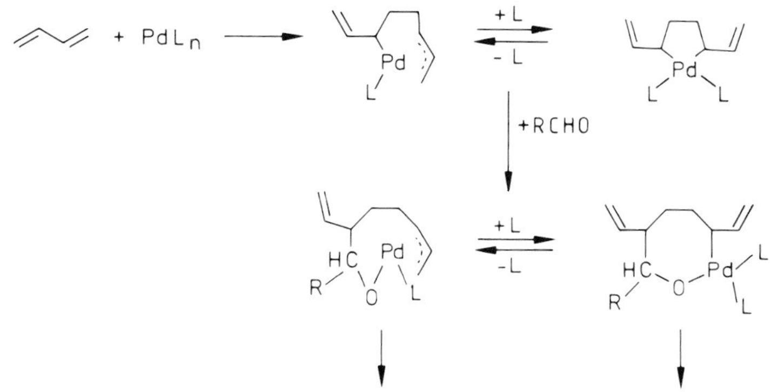

+ PdLn
+L
-L
+RCHO
86
85

Scheme 17.

2.8.3. CARBON DIOXIDE

Y. Inoue and his coworkers discovered the telomerization of butadiene with carbon dioxide [154, 155]. The γ-lactone **95** was formed in small yields when the reaction was carried out in a polar, aprotic solvent, such as dimethylformamide, and with palladium catalysts stabilized by ditertiary phosphines such as diphos. This was the first example of the formation of a new C—C bond between CO_2 and an organic compound catalyzed by a transition metal complex. It is significant that the reaction is very evidently influenced by the complex ligand: no lactone was able to be synthesized with phosphites and arsines. In general the yield of the lactone was very poor when monodentate ligands were used. Basic ligands such as PBu_3 gave slightly higher yields (2.5%) than less basic ligands, e.g. PPh_3 (0.5%). Diphos was the most suitable for the lacton formation (5.4%) among the bidentate ligands. The telomerization with palladium complexes of unidentate tertiary phosphine ligands was also studied by Musco [156, 157]. He obtained the δ-lactone **92** and the two octadienyl esters of the 2-ethylidene-hepta-4,6-dienoic acid **93** and **94** in non-polar solvents such as benzene. He was able to prove that the γ-lactones **95** and **96** were formed through further reaction of the acid in the presence of palladium-phosphine catalysts. The selective syntheses of the δ-lactone **92** and of the γ-lactone **96** have been described recently [158]. Mechanistic studies showed that the formation of the γ-lactone proceeds by a consecutive reaction via the δ-lactone.

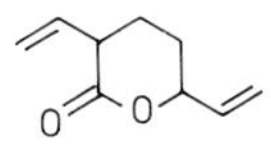

92 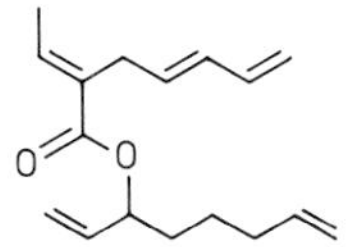 93 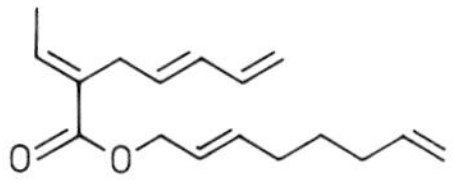 94

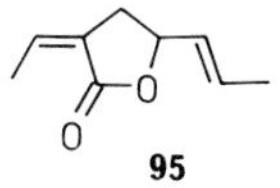

95 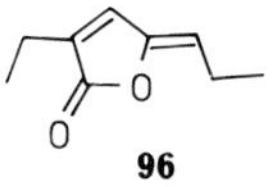 96

2.8.4. SCHIFF BASES

Palladium nitrate catalyzes the reaction of butadiene with Schiff bases in the presence of triphenylphosphine to produce the substituted piperidines **97**. Palladium chloride or acetylacetonate were surprisingly ineffective in this reaction.

This telomerization can be applied to Schiff bases which have aromatic substituents on the carbon atom and alkyl, allyl and aromatic substituents on the nitrogen atom. For example, the piperidine products **97** were able to be obtained in a total yield of 91% with *N*-benzilidene-methyl-amine [159].

Telomerization of Schiff bases is also possible with a nickel catalyst system such as $[Ni(cod)_2]$-PPh_3, but linear products are formed instead of cyclic piperidine compounds [140, 160]. Thus heating butadiene with PhCH=NPr produces 9-phenyl-9-propylidenamino-1,6-nonadiene as main product. If small amounts of morpholine, acetic acid or methanol are added to the reaction mixture, the 9-phenyl-9-propyl-amino-1,3,6 nonatriene **98** becomes the principal telomer.

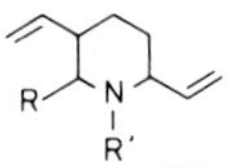

R N R' 97 H N R R' 98

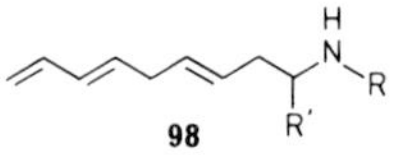

2.8.5. ISOCYANATES

Arylisocyanates are another group of nucleophiles with C=N double bonds [42, 161]. The reaction of butadiene and phenylisocyanates produces the two isomeric piperidone derivatives **99** and **100** which were able to be separated by column chromatography. A double bond migration to the conjugated position takes place in this telomerization. Bis(triphenylphosphine)(maleic anhydride)palladium and a $[Pd(cod)_2]/PPh_3$ system proved to be active catalysts.

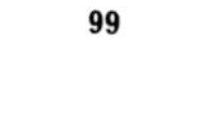

R N O 99 R N O 100

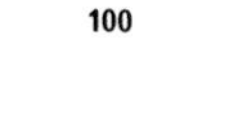

2.9. Telomerization with organomagnesium compounds

Butadiene and phenyl-magnesium bromide and catalytic amounts of $[Ni(acac)_2]/P(OPh)_3/AlEt_3$ yield the telomer 1-phenyl-2,7-octadiene **101** and the 2:2-adduct 1,8-diphenyl-2,6-octadiene **102** in about equal parts. Diphenyl as well as linear and cyclic oligomers are formed as by-products [162].

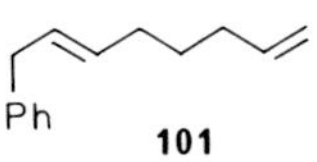

Ph 101 Ph Ph 102

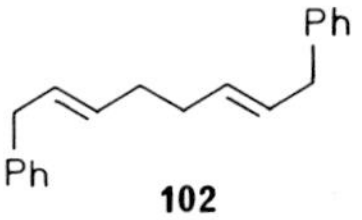

2.10. Carboxy-telomerization

Telomerization combined with the additional incorporation of carbon monoxide is called carboxy-telomerization. The fundamental reaction which yields products which contain a C_9-chain is shown in Equation 24.

2 butadiene + CO + HY ⟶ nonadienoyl–Y (24)

2.10.1 ALCOHOLS

A halide free palladium complex, e.g. one which is generated *in situ* by $[Pd(acac)_2]$ and PPh_3, catalyzes the carboxy-telomerization of butadiene in ethanol to yield the ethyl nona-3,8-dienoate [163]. The ethoxyoctadienes are also formed in a side-reaction. The catalyst described is significantly inactive for the subsequent carbonylation of the ethoxyoctadienes which suggests that the nonadienoate is not produced by a following reaction of the ethoxyoctadienes.

The absence of a halide ion which is coordinated to the palladium is an essential factor in carboxy-telomerizations [164, 165]. With catalysts which contain chloride only the 3-pentenoate **104** is formed by simple carbonylation. Whereas this carbonylation can be obtained even in the absence of phosphine, the carboxy-telomerization does not take place if a ligand is not available. Both triphenylphosphine and tributylphosphine have been used. The highest activity was observed when the relative ratio of palladium and phosphine was 1 : 4.

Other alcohols such as *iso*-propyl or tertiary butyl alcohol can also be used to produce nonadienoates in a yield of about 90%. Methanol shows a low reactivity (9%), but a statisfactory yield can be obtained by diluting the methanol with acetonitrile [66].

Carbon monoxide pressure has an important effect on the reaction; a higher pressure tends to decrease the yield. The best conversions were able to be obtained with a carbon monoxide pressure of about 50 atm.

A probable mechanism of the carboxy-telomerization is shown in Scheme 18. The formation of a palladium-bisallyl complex is followed by carbon monoxide

+ $[PdCl_nL_m]$ ⟶ π-allyl-Pd(Cl)(L) —(+CO)⟶ **103** —(+ROH)⟶ CO_2R **104**

+ $[PdL_m]$ ⟶ Pd-bisallyl(L) —(+CO)⟶ **105** —(+ROH)⟶ CO_2R **106**

Scheme 18.

insertion to give the acyl palladium complex **105**. This acyl complex decomposes to yield the 3,8-nonadienoate **106** with the attack of alcohol. When chloride ions are coordinated to palladium, only a monomeric η^3-allyl complex is formed. Carbon monoxide insertion produces the acyl complex **103** whose reaction with the alcohol yields the pentenoate **104**.

2.10.2. ACIDS

The reaction of butadiene with CO and AcOH in the presence of a palladium catalyst produces nonadienoic acid together with the dimers octatriene and 4-vinyl-1-cyclohexene [167].

2.10.3. AMINES AND AMMONIA

A similar reaction was obtained with secondary or primary amines as well as with ammonia. Diethylamine, benzylamine or cyclohexylamine react with butadiene and carbon monoxide to yield the 3,8-nonadienamides in a one-step synthesis [168].

2.11. Utilization of the butadiene telomers

Many butadiene telomers are useful compounds to synthesize natural products because they have a functional group at one end and a terminal double bond at the other [170–172]. Especially the acetoxyoctadienes (Table 14), but also the products of the telomerization with carbonyl methylene compounds and nitroalkanes are frequently used. Nonadienoates formed by carboxy-telomerization are also important precursors of natural compounds, especially of fragrances and pheromones (Table 15).

3. TELOMERIZATION OF ISOPRENE

In contrast to butadiene, isoprene is an unsymmetrical molecule. The connection of two isoprene molecules can therefore occur on the "head" *h* or on the "tail" *t* of the molecule. Consequently, in the telomerization of isoprene the tail-to-tail (*tt*)-, tail-to-head (*th*)-, head-to-tail (*ht*)- and head-to-head (*hh*)-products are possible. Because of nucleophile HY can attack the chain at positions 1 and 3, eight telomers occur (Table 16). The fact that the primary telomers possess an internal double bond means that these molecules exist as cis and trans isomers and that, on the whole, twelve different molecules can be formed in isoprene telomerization.

The tail-to-tail-linkage can easily be identified by the two methyl groups in the positions 2 and 7; the head-to-head-linkage by methyl groups in positions 3 and 6. It is important to distinguish between the head-to-tail- and tail-to-head-telomers: the molecule must be considered from the side where the nucleophile is added. When one takes this definition into account, the tail-to-head-products are characterized by

Table 14

SYNTHESIS OF NATURAL COMPOUNDS FROM ACETOXYOCTADIENES

	Natural compounds	Ref.
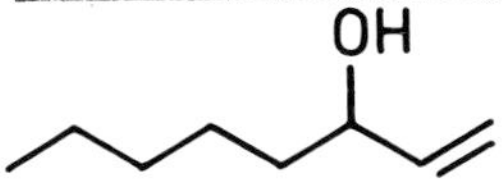 OH	matsutake-alcohol (1-octen-3-ol)	173–175
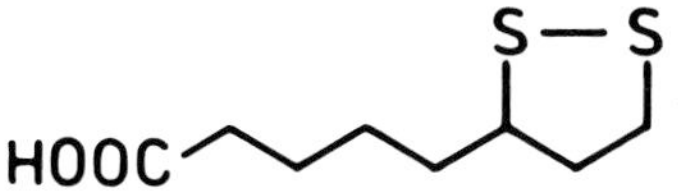 S — S, HOOC	α-lipoic acid (1,2-dithiolane-3-valeric acid	176
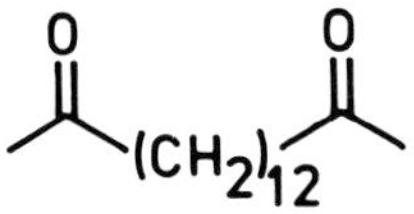 O, O, $(CH_2)_{12}$	2,15-hexadecanedione (precursor of muscone)	177
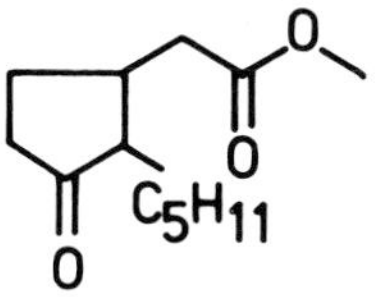 O, O, O, C_5H_{11}	methyl-dihydrojasmonate	178
steroids, e.g.	(+)-19-nortestosterone (±)-D-homo-19-norandrosta-4-en-3-one	179–181

Table 15

SYNTHESIS OF NATURAL COMPOUNDS FROM BUTADIENE TELOMERS

Starting telomer	Natural compound	Ref.
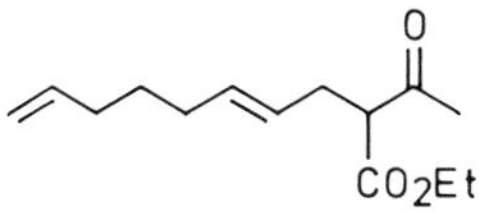 O, CO_2Et	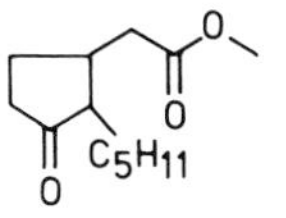 O, O, O, C_5H_{11} methyl-dihydrojasmonate	182
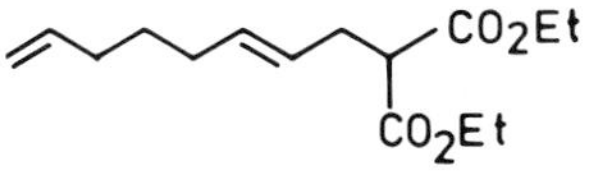 CO_2Et, CO_2Et	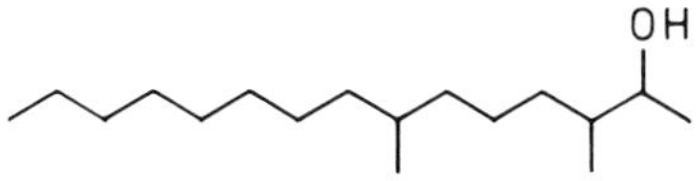 OH 3,7-dimethylpentadecan-2-ol (pheromone)	183
	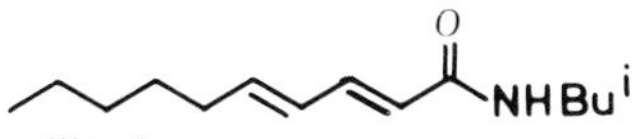 O, $NHBu^i$ pellitorine	184
	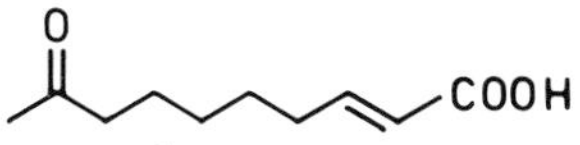 O, COOH queen substance (honey bee pheromone)	185

Table 15 (continued)

Starting telomer	Natural compound	Ref.
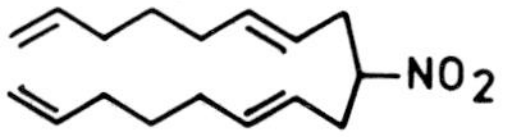 NO_2	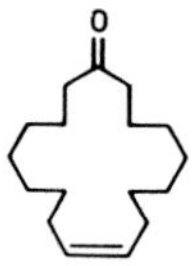 *cis*-civetone	186
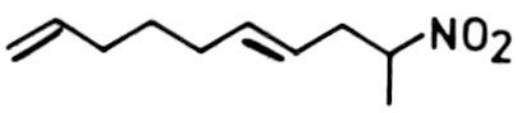 NO_2	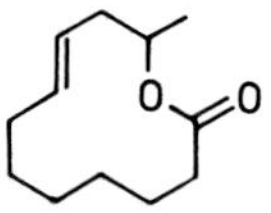 recifeiolide	187
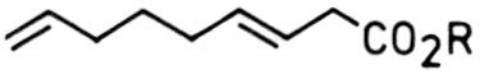 CO_2R	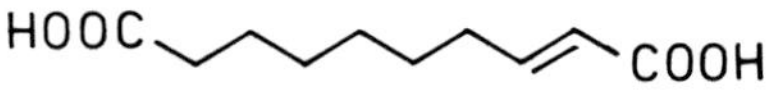 HOOC COOH 2-decenedioic acid (royal jelly acid)	188
	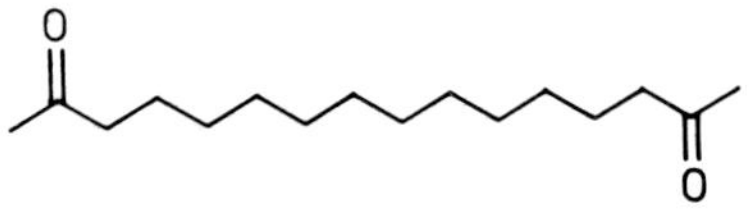 2,15-hexadecanedione (precursor of muscone)	189
	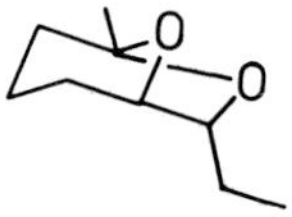 brevicomin	190

Table 16

THEORETICALLY POSSIBLE CONNECTION-TYPES OF ISOPRENE TELOMERIZATION

tt	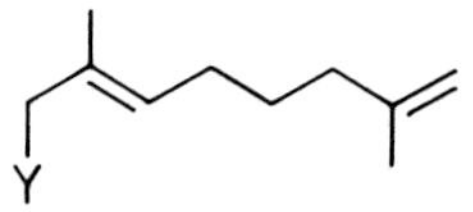 Y	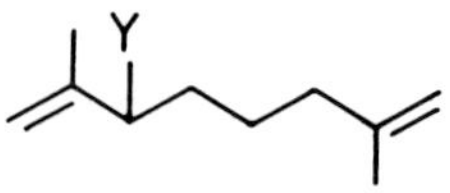 Y
th	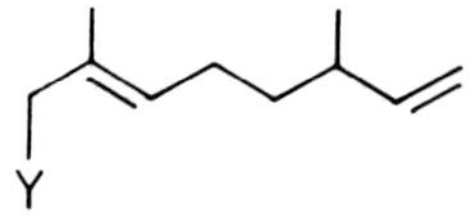 Y	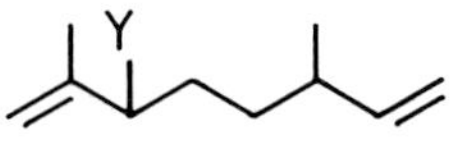 Y
ht	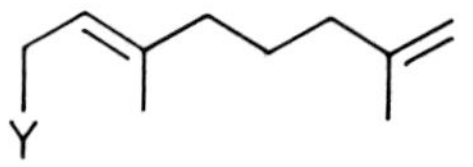 Y	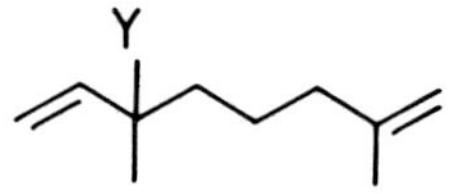 Y
hh	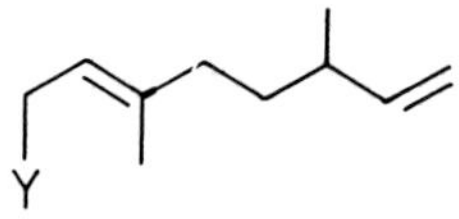 Y	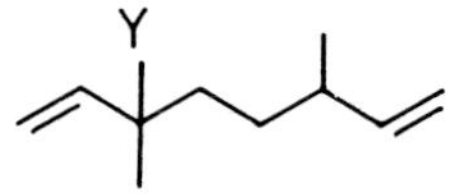 Y

methyl groups in positions 2 and 6, and the head-to-tail-products by methyl groups in positions 3 and 7.

The telomers of isoprene belong to the large class of monoterpenes, a very important group of chemicals which occur in nature. In these natural substances the combination of the isoprene units is normally a head-to-tail-linkage, but some exceptions to this "rule" are well known.

3.1. Telomerization with OH-nucleophiles

3.1.1. ALCOHOLS

Methanol has been thoroughly studied in isoprene telomerization (Table 17). The most usual linkage type is the tail-to-head-combination obtained by a large

Table 17
TELOMERIZATION OF ISOPRENE WITH ALCOHOLS

Nucleophiles	Catalysts	Telomers[a]	Ref.
CH_3OH	$[PdCl_2(PPh_3)_2]$, KOH	*th* (*ht*)	48
	$[Pd(PPh_3)_3]$	*ht* (*tt*)	58
	$[Pd(PPh_3)_4]$, O_2	*th* (*tt*)	191
	$[\{(C_3H_5)PdCl\}_2]$, PR_3, NaOMe	*th*	192
	$[M(PPh_3)_4] + O_2$; $[M(PPh_3)_2O_2]$ $[M(PPh_3)_2CO_3]$ M = Pd, Pt	*th* (*tt*)	193
	$[Pd(OAc)_2]$ + sulfonat. phosphines	*th*	38
	$[Pd(acac)_2]$, PPh_3	*tt*	194
	$[\{(C_3H_5)PdCl\}_2]$, NaOMe, opt. active phosphine	*th*	195
CH_3OH, Pr^iOH	$[\{(C_3H_5)PdCl\}_2]$, PBu^n_3	*th*, *tt*	196
EtOH, Pr^nOH, Bu^iOH and others	$[PdCl(PhCN)_2]$ NaOH, PR_3	*th*	197
EtOH, Pr^nOH, Pr^iOH, Bu^nOH	$[Pd(acac)_2]$ PPh_3	*th*, *tt*	198
CF_3CH_2OH furfuryl alcohol, benzyl alcohol	$[Pd(acac)_2]$ PR_3, $P(OR)_3$	*tt*, *th*	199

[a] By-products in parentheses; *th* = tail-to-head, etc.

number of palladium and platinum catalysts. A catalyst system of $[Pd(OAc)_2]$ or $[Pd(acac)_2]$ with PPh_3 has been used to obtain the tail-to-tail connected telomers in a yield of more than 50% [194].

The primary tail-to-head-telomer can be used to synthesize citronellol **107** by the sequence of reactions shown in Scheme 19. The telomer 1-methoxy-2,6-dimethyl-2,7-octadiene was converted into 2,6-dimethyl-1,3,7-octatriene, which yields the terpenoid citronellol after hydrogenation and hydroboration/oxidation.

Scheme 19.

The telomerization of isoprene with methanol can be carried out with optically active phosphines to yield the telomer in an optically active form [195, 200] which is useful for the synthesis of optically active citronellol. Various menthylphosphorous derivatives have been used as phosphorous ligands, for example, menthyl dialkylphosphines $MenPR_2$, menthyl dialkylphosphonites $MenP(OR)_2$ and menthyl bis(alkylamino) phosphines $MenP(NR_2)_2$. The phosphines produced the (+)-telomer with priority, whereas the phosphonites produced the (–)-telomer. Menthyl diisopropylphosphine produced the (+)-telomer in a maximum optical yield of 35%. On the other hand, menthyl dimethylphosphonite led to the formation of the (–)-telomer in an optical yield of only 8%. Higher alcohols also react with isoprene to yield mixture of tail-to-head- and tail-to-tail-telomers [198]. However, the telomerization of *n*-propanol and *n*-butanol yield only traces of telomers. Interestingly enough furfuryl alcohol and benzyl alcohol react very easily in isoprene telomerization: tail-to-tail-telomers were able to be synthesized in a yield of more than 60%. The electronic properties of the nucleophile have obviously a large influence on the rate of telomerization [199].

This increasing "electronic effect" is also well demonstrated by the high reactivity of trifluoroethanol in comparison with the results obtained by ethanol. The tail-to-head-linkage of the isoprene units was predominant with trifluoroethanol. Palladium complexes with arylphosphanes or alkyl phosphites as ligands produced the best yields of telomers. The reaction with a catalyst which is formed *in situ* by palladium bisacetylacetonate and triethylphosphite produced a yield of 40% telomers containing 70% of tail-to-head-products [199].

3.1.2. CARBOXYLIC ACIDS

In principle the reaction of isoprene and acetic acid can be carried out without any transition metal catalysts. These processes which are catalyzed by inorganic or organic acids normally yield very unselective mixtures of prenyl(C_5)-, monoterpenyl(C_{10})-, sesquiterpenyl(C_{15})- and higher acetates.

The length of the chain and the linkage of the isoprene units can be controlled when palladium catalysts are used (Table 18). Only three prenylacetates were able to

Table 18

TELOMERIZATION OF ISOPRENE WITH ACIDS AND WATER [a]

Nucleophiles	Catalysts	Telomers	Ref.
HOAc	$PdCl_2/PPh_3/NaOAc$ in THF	*th, hh* (*ht*)-esters	201
	$[Pd(OAc)_2]$ in DMSO	*th*-esters	202
HCOOH	$[Pd(OAc)_2]$, PPh_3 in dioxane	*hh*-alcohol	75
	$[Pd(OAc)_2]$ without ligand	3,7-dimethyl-1,6-octadiene	203
	$[\{Pd(C_3H_5)OAc\}_2]$, NEt_3, PR_3 in THF	*ht*-dimers	204
H_2O	(a) $[Pd(acac)_2]$, PR_3 (b) $[Pd(PPh_3)_4]$ (c) $[Pt(PPh_3)_4]$ (d) $[Pt(PPh_3)_2(CH_2{=}CH_2)]$	*tt, th, ht*-alcohols	41
	$[Pd(acac)_2]$, PPh_3	*tt*-alcohols	205

[a] *th* = tail-to-head, etc.

be obtained with the catalyst system $PdCl_2/PPh_3/NaOAc$ when benzene was used as a solvent. Five terpenyl acetates occur in polar solvents such as tetrahydrofuran [201].

The reaction of isoprene and formic acid does not yield any terpenyl formates. Only the dimer 3,7-dimethyl-1,6-octadiene was able to be obtained when the catalyst $[Pd(OAc)_2]$ is used without any further stabilizing ligand [203]. The head-to-head-terpene alcohol, the 3,6-dimethyl-2,7-octadienol was produced by adding the ligand PPh_3 [75].

3.1.3. WATER

Though the nucleophile water is a very inactive one, isoprene and water also undergo telomerization when a palladium or a platinum catalyst and the cocatalyst carbon dioxide are used. The monoterpenic alcohols which are important fragrances are formed in one reaction step [41]. Seven terpenols have been synthesized in a yield up to 50% in which the isoprene units are combined in tail-to-tail (**108–110**), tail-to-head (**111–113**) and head-to-tail (**114**) manners. The highest conversions are obtained with palladium catalysts such as [Pd$(PPh_3)_4$], but the largest amount of terpenols was found with platinum catalysts such as [Pt$(PPh_3)_4$] or [Pt$(PPh_3)_2$ $(CH_2{=}CH_2)$]. Palladium catalysts favour the tail-to-tail-products whereas platinum gives rise to more tail-to-head-telomers.

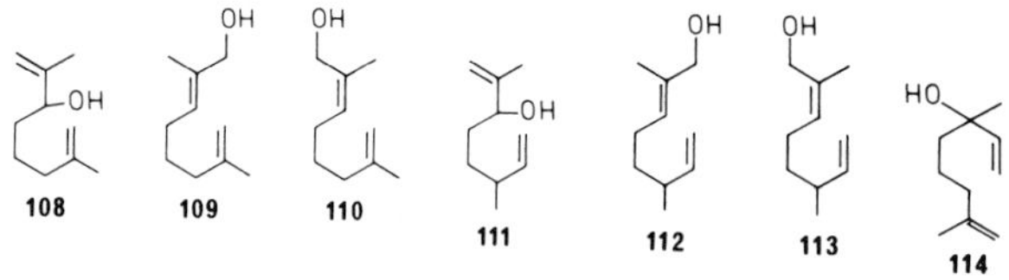

108 **109** **110** **111** **112** **113** **114**

The mechanism of this reaction is analogous to that one proposed for butadiene telomerization. However, with regard to the bimetallic mechanism instead of the one bis-allylic complex **11**, the three intermediates **115–117** are able to be formed.

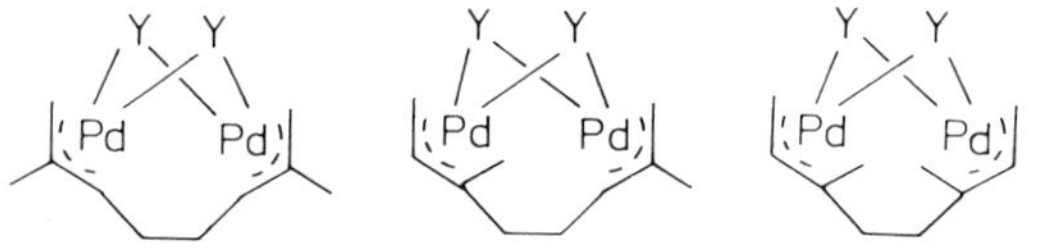

115 **116** **117**

The fact that no head-to-head-isomers were able to be obtained, shows that the complex **117** is very unstable on account of the two internal methyl groups.

3.2. Telomerization with NH-nucleophiles

Both nickel- and palladium catalysts were studied in the telomerization of isoprene with amines (Table 19). In the reaction of isoprene and morpholine catalyzed by nickel the primary tail-to-tail-amine was isolated besides four 1 : 1-adducts. Tail-to-head-connected isomers were observed in the same reaction when [Pd$(cod)_2$] was used as a catalyst [42].

Telomerization of isoprene with diethylamine in the presence of palladium-dichloride, triphenylphosphine and carbon dioxide as a cocatalyst produced the tail-to-head-product **118**, the tail-to-tail product **119** and the head-to-tail telomer **120** [208]. In addition Röper found the head-to-head-isomer **121** which is of special interest because reports of head-to-head-products of isoprene are extremely rare in the literature [207].

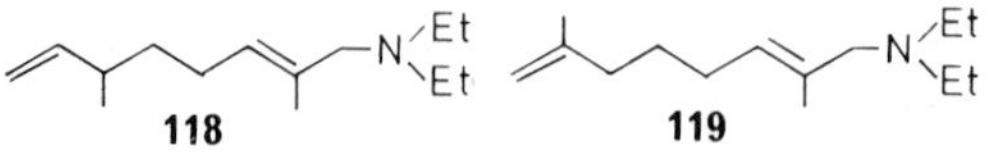

118 **119**

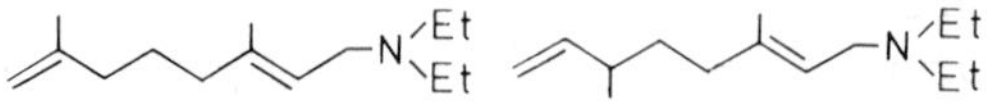

120 **121**

Table 19
TELOMERIZATION OF ISOPRENE WITH NH-NUCLEOPHILES [a]

Nucleophiles	Catalysts	Products	Ref.
NH_3	$[Pd(acac)_2]$, $P(OR)_3$	primary, secondary and tertiary *tt*, (*ht*)-amines	206
NHR_2, NH_2R	$[Pd(acac)_2]$, $P(OR)_3$	*tt, ht, th, hh*	207, 235
	$NiCl_2 - PPh_3$ $NaBH_4$	*tt, th*-telomers and 1 : 1-adducts	87
	$[Ni(acac)_2]$, $PhP(OPr^i)_2$	*tt*-telomer and 1 : 1-adducts	33
	$[Pd(cod)_2]$	*tt* (*th*)	42
	$[PdBr_2(diphos)_2]$ PhONa	only 1 : 1-adducts	84
	$PdCl_2$, NaOPh	*ht*	93
	$PdCl_2$, PPh_3, CO_2/CH_3OH	*th, tt, ht*	208

[a] *th* = tail-to-head, etc.

Other secondary amines also react very smoothly with isoprene [235]. The yield of terpene amines is strongly influenced by the structure of the nucleophile applied. Nucleophiles with linear substituents such as di-*n*-propyl- or di-*n*-butyl-amines produce high yields (~70–80%); branching in β-position of the substituent as in diisobutylamine leads to medium yields (~50%), whereas branching in α-position as in diisopropylamine results in very low yields (~3%).

Acidic cocatalysts have a surprisingly great influence on the isoprene/amine telomerization. Significant amounts of the head-to-head-telomer are only observed in the reaction with diethylamine when strong acids such as trifluoroacetic acid or $BF_3 \cdot OEt_2$ were added to the palladium complex. By using a catalyst of $Pd(acac)_2$, tricyclohexylphosphine and BF_3OEt_2 in a ratio of 1 : 1 : 100, the selectivity towards the telomer **121** is increased to almost 80%. Cationic palladium catalysts are also highly selective towards the head-to-head terpene amine. They are active even at room temperature.

Another method to synthesize terpene amines is the telomerization of isoprene with ammonia [206]. Seven amines whose structures are shown in Scheme 20 were isolated as the main products by using a catalyst formed by palladium acetylacetonate and tributyl phosphite. Tail-to-tail-coupling of the isoprene units was predominant but a head-to-tail-derivative, the α-linalylamine, was also able to be isolated. The formation of the primary terpene amines was favoured by short reaction times and high ammonia/isoprene ratios.

Scheme 20.

3.3. Telomerization with CH-, SiH- and SiSi-nucleophiles

In 1969 Hata and Takahashi discovered the telomerization of isoprene with CH-nucleophiles [108]. In the reaction of isoprene with ethyl acetoacetate or acetylacetone under catalysis of $[PdCl_2(PPh_3)_2]$/NaOPh or $[Pd(PPh_3)_4]$ the tail-to-tail-products were able to be isolated (Table 20). Only 1 : 1-adducts are formed

Table 20

TELOMERIZATION OF ISOPRENE WITH CH-, SiH- AND SiSi-NUCLEOPHILES[a]

Nucleophiles	Catalysts	Products	Ref.
$R-CH_2-R'$	$NiCl_2/PPh_3/NaBH_4$	*th, tt*	87
	$[Ni(acac)_2]$, $PhPPr^i_2$, $NaBH_4$, NaOPh	*th, tt*	109 110
	$[Pd(diphos)_2]$	1 : 1-adducts	84
	$NiCl_2$, PPh_3, $NaBH_4$	1 : 1-adducts and *th, tt, ht*-telomers	209
	$[PdCl_2(PPh_3)_2]$, PhONa	*tt*	108
	$[Ni(cod)]_2$, PR_3	C_{13}-alcohols (from isoprene and acetone)	153
	$PdCl_2$, P-ligand	1 : 1-adducts	210
R_3SiH	$[Pd(PPh_3)_4]$		116
	$[PdCl_2(PhCN)_2]$, PPh_3	1 : 1-adducts	211
	$[RhCl(PPh_3)_3]$		212
$R_3Si-SiR_3$	$[PdCl_2(PPh_3)_2]$ or $[Pd(PPh_3)_4]$	1 : 1- and 2 : 1-adducts	144
	$[PdCl_2(PPh_3)_2]$	2 : 1-adducts	143

[a] *th* = tail-to-head, etc.

by using palladium catalysts with bidentate phosphine ligands such as [$PdBr_2(diphos)_2$] [84]. Tail-to-head-connected telomers were also able to be identified with nickel catalysts [109, 110, 209]. Only 1 : 1-adducts and no telomers were able to be obtained with silanes [116, 211, 212]. The telomerization of isoprene was also possible with strained electron-rich or strain-free but electron-deficient Si—Si compounds [143, 144].

3.4. Telomerization with C=X double bond nucleophiles

Only a few examples are known in which isoprene is telomerized with C=X double bond nucleophiles. The reaction of isoprene with an aqueous solution of formaldehyde catalyzed by palladium acetate and tricyclohexylphosphine yielded a mixture containing 54% of unsaturated C_{11}-alcohols [147]. In the presence of [$Ni(cod)_2$] and PPh_3 isoprene and acetaldehyde produced mainly telomers in which the isoprene is dimerized in a head-to-tail-manner to yield the products **122** and **123** [209, 213].

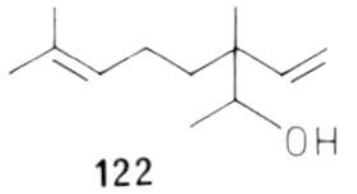

OH OH
122 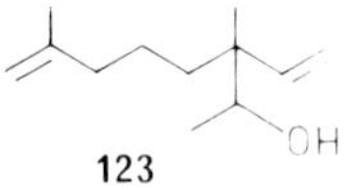 123

The telomerization of isoprene and acetone yielded the open-chain products **124** and **125** [153].

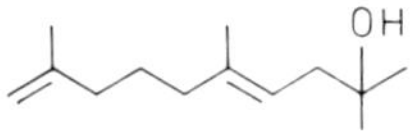

OH OH
124 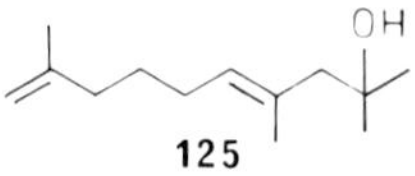 125

The cyclization reaction of isoprene with phenyl isocyanate catalyzed by bis(triphenylphosphine) (maleic anhydride)palladium produced piperidone derivatives **126** with a tail-to-tail-linkage [161]. The telomerization of isoprene with carbon dioxide yields the five-membered lactones **127**–**129** in small amounts [214].

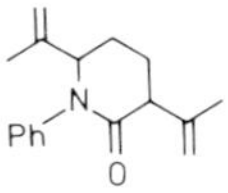

N Ph O O O O O O O
126 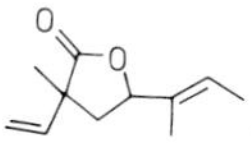 127 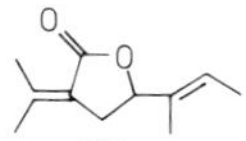 128 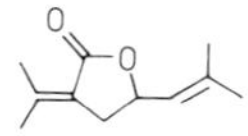 129

[$Pd(diphos)_2$] was the most active catalyst; however the yield of lactones does not exceed 1% under the best reaction conditions.

3.5. Carboxy-telomerization

The reaction of isoprene with carbon monoxide and acetic acid yields the dimethylnonadienoic acid [167]. Palladium acetate and triphenylphosphine form an active *in situ* catalyst. When alcohols such as methanol, ethanol or *t*-butanol are used instead of the acid, usually only 1 : 1-adducts were obtained [215, 216]. Knifton found [164] that under special conditions the telomeric product isopropyl dimethylnonadienoate can be formed from isoprene and isopropanol, however only in small amounts (1.3%).

3.6. Utilization of the isoprene telomers

Numerous terpene esters, -ethers, -alcohols and -amines can be formed by telomerization of isoprene. The terpene alcohols are especially important for fragrancies and perfumes. They are also useful in conserving food or used as insecticides. Terpene amines became well known as drugs.

The terpene alcohols are important precursors of vitamines. It is shown in Scheme 21 that vitamine E can be synthesized by coming from the head-to-tail-terpenol β-linalool **130**. Vitamine K can also be produced with β-linalool as an educt.

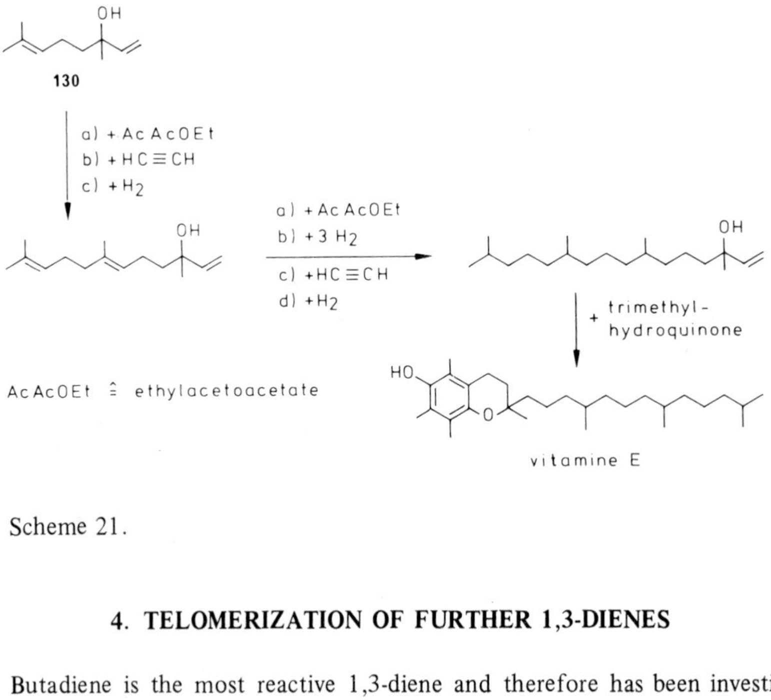

OH

130

a) + AcAcOEt
b) + HC≡CH
c) + H_2

OH

a) + AcAcOEt
b) + 3 H_2
c) + HC≡CH
d) + H_2

OH

+ trimethyl-hydroquinone

AcAcOEt ≙ ethylacetoacetate

HO

O

vitamine E

Scheme 21.

4. TELOMERIZATION OF FURTHER 1,3-DIENES

Butadiene is the most reactive 1,3-diene and therefore has been investigated the best in telomerization. The unsymmetric molecule isoprene shows a lower reactivity. Hence the structure of the 1,3-diene has an important influence on the telomerization reaction. Several other dienes have been investigated, especially 1,3-pentadiene, 2,3-dimethylbutadiene, 2,4-hexadiene and myrcene.

4.1. 1,3-Pentadiene

Like isoprene 1,3-pentadiene or piperylene is an asymmetrical molecule. Therefore it is possible to distinguish between head "*h*" and tail "*t*" of the monomeric unit. (See Formula **131**.) The telomerization of piperylene theoretically

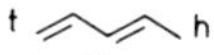

t h

131

also yields eight isomeric products. These different types of linkage are shown in Table 21.

Table 21

THEORETICALLY POSSIBLE LINKAGE-TYPES OF PIPERYLENE TELOMERIZATION

tt	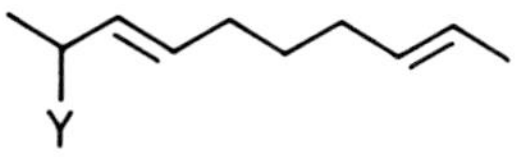 Y	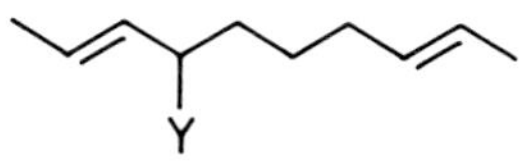 Y
th	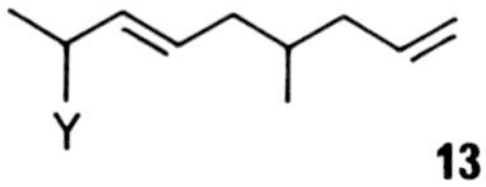 Y **132**	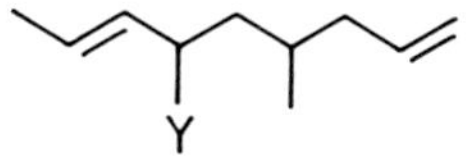 Y
ht	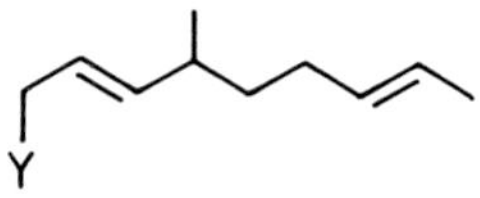 Y	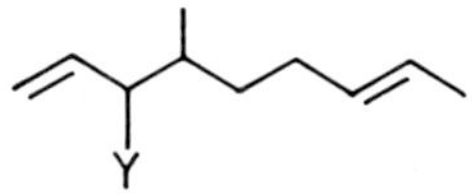 Y
hh	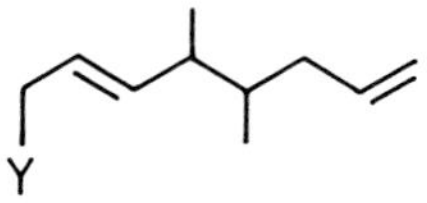 Y **133**	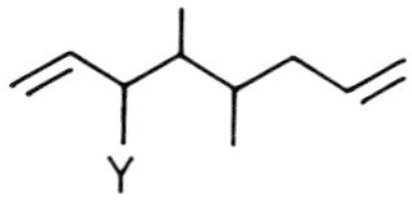 Y

The possible product number is nevertheless much higher than in the case of isoprene because certain telomers possess two internal double bonds: as a total 21 isomers should be formed by piperylene telomerization. It is surprising that only the two structures **132** and **133** have been able to be obtained up till now.

Products of these structures were synthesized by telomerization of piperylene and short-chained alcohols [198]. A mixture of [Pd(acac)$_2$] and PPh$_3$ is used as a catalyst system which is able to isomerize piperylene until the equilibrium of 33% *cis*- and 67% *trans*-piperylene is obtained. By using methanol as a nucleophile, the head-to-head-telomer **133** is formed in a yield of 50%. The conversion of piperylene decreases to 41% with ethanol. The product mixture consists of only small traces of the head-to-head-telomer (4%), what remains are the tail-to-head-telomer (34%) and piperylene dimers (62%). The mechanism shown in Scheme 22 explains the synthesis of the two telomers obtained.

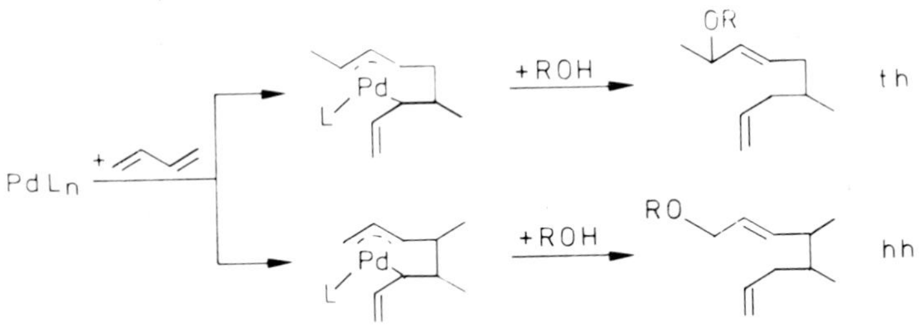

PdLn + piperylene → Pd intermediates; +ROH → th, hh

Scheme 22.

The telomerization of piperylene and propylene glycol has been described, but the structures or yield are not given [53]. Furfuryl-, benzyl- and trifluoroethyl alcohol also undergo telomerization with piperylene [218]. Ethers of structure **134** are formed in these reactions.

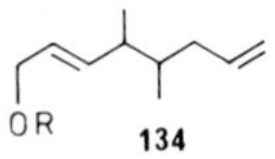

OR **134**

The reaction of piperylene and water yielded the telomer 6-methyl-3,8-nonadien-2-ol [217]. This product was formed in a high selectivity when a $Pd(acac)_2/PPh_3$ catalyst was used at room temperature and with long reaction times.

Telomerization does not take place with amines, neither with palladium nor with nickel catalysts [33, 84]. Only the 1 : 1-adducts **135** and **136** are formed.

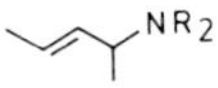

NR_2 **135** NR_2 **136**

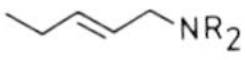

The telomerization with active methylene compounds yields the 1 : 1-adducts in predominance [58, 84]. The telomer **137** with a tail-to-head-structure was formed

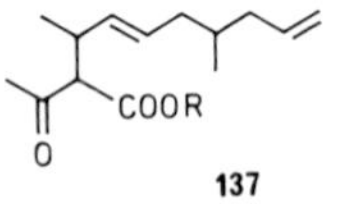

COOR O **137**

from piperylene and acetoacetate in a yield of 32% only under catalysis of $[PdCl_2(PPh_3)_2]$/PhONa [108].

Whereas Tsuji isolated only 1 : 1-adducts by hydrosilylation of piperylene [116], the reaction with trimethylsilane which was carried out by Yur'ev also yields the telomeric product **138**. As it has already been described in the case of butadiene (Scheme 12) the double bond has "migrated" along the chain in this tail-to-head-telomer [219].

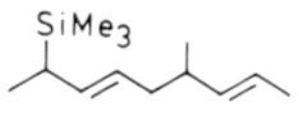

$SiMe_3$

138

The carboxy-telomerization of piperylene with carbon monoxide and ethanol in the presence of palladium catalysts produces the 2-methyl-3-pentenoate in yields of up to 55% and telomers are not formed [216, 220].

4.2. 2,3-Dimethyl-1,3-butadiene

2,3-Dimethylbutadiene is a symmetrical molecule such as butadiene and only three isomeric telomers can be expected. However, this diene is a very inactive one and only one case of telomerization has been reported. The reaction of 2,3-dimethyl-1,3-butadiene with ethyl acetoacetate in the presence of a $[PdCl_2(PPh_3)_2]$/NaOPh catalyst results in the formation of ethyl-2-acetyl-2,3,6,7-tetramethyl-2,7-decadienoate **139** with a yield of 7% [108].

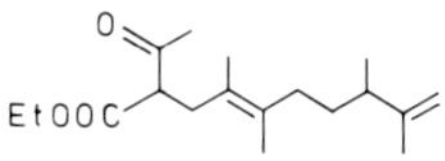

O

EtOOC

139

All other experiments using alcohols [198], water [218], methylene compounds [109, 110, 210] or Si—Si-compounds [144] led to 1 : 1-adducts or dimers.

4.3. 2,4-Hexadiene

No telomerization reactions are known with 2,4-hexadiene. The reactions with ethyl acetoacetate and aniline selectively yielded the 1 : 1-adducts **140** in the presence of palladium catalysts [84].

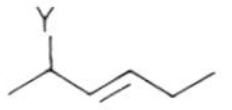

Y

$Y = CH_3-CO-\overset{|}{C}H-CO_2Et$

$= PhNH-$

140

4.4. Myrcene

Myrcene **141** is a very reactive 1,3-diene. It has been reported to react with alcohols [221, 222], acetic acid [221, 223] and silanes [224]. Active methylene compounds such as diethyl malonate, ethyl acetoacetate or acetylacetone also react with myrcene to yield the products **142**–**144** [225].

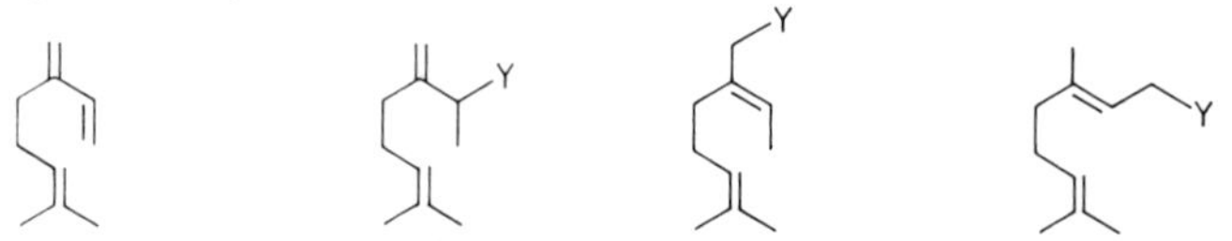

Y Y Y

141 **142** **143** **144**

Product **145** can be obtained from the reaction of myrcene with acetaldehyde in presence of [Ni(cod)$_2$] in a yield of 25%. With a 25% conversion the reaction with acetone produces a mixture of the 1 : 1-adduct **146** and the 2 : 1-adduct **147** [225].

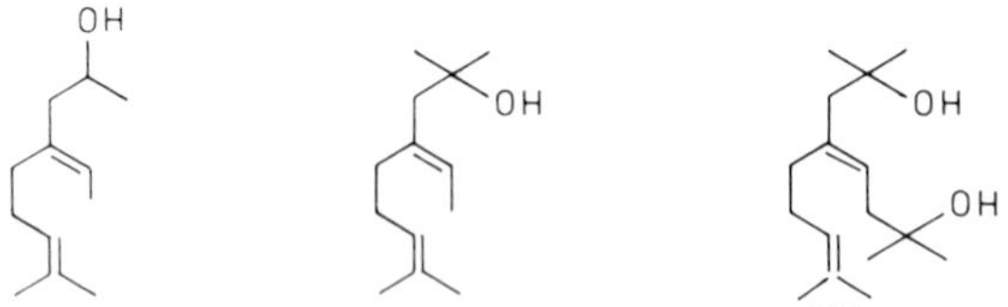

OH OH OH OH

145 **146** **147**

The addition of diethylamine to myrcene yields the (*Z*)- and (*E*)-isomers of the products **148** and **149**. High selectivity (98%) and good yields (65%) have been obtained with the catalyst system Pd(acac)$_2$/dppe/BF$_3$ · OEt$_2$ in a molar ratio of 1 : 2 : 2 [226].

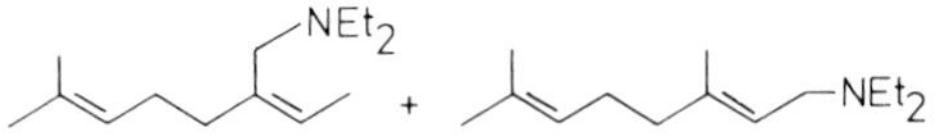

NEt_2 + NEt_2

148 **149**

A real telomer, that is a 1 : 2-adduct consisting of one molecule of nucleophile and two molecules of myrcene, was not able to be observed. It is worth noting that the 1 : 1 addition with the diene myrcene yields similar products like the 1 : 2 telomerization with isoprene.

5. TELOMERIZATION OF 1,2-DIENES

In glacial acetic acid at 50° and 1 atm 1,2-propadiene (allene) is converted in the presence of a palladium acetate catalyst to a mixture of allyl acetate **150**, 2,3-dimethylbuta-1,3-diene **151**, the main products 3-methyl-2-hydroxymethylbuta-1,3-diene acetate **152** and 2,3-dihydroxymethylbuta-1,3-diene diacetate **153**.

$CH_2{=}CH{-}CH_2{-}OAc$

150

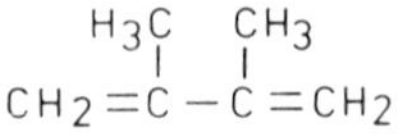

$CH_2{=}C(CH_3){-}C(CH_3){=}CH_2$

151

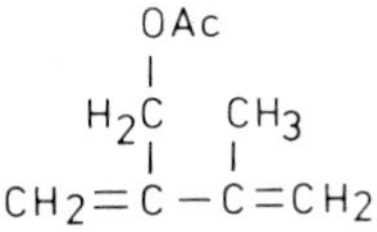

$CH_2{=}C(CH_2OAc){-}C(CH_3){=}CH_2$

152

$CH_2{=}C(CH_2OAc){-}C(CH_2OAc){=}CH_2$

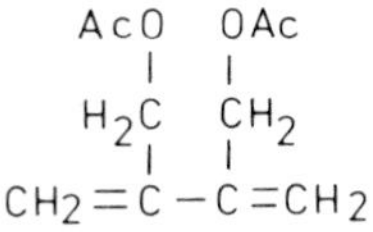

153

Carbon—carbon bonds are formed between internal carbon atoms of allene and carbon—oxygen bonds are formed at the terminal carbon atoms [227].

A variety of palladium and rhodium compounds were found to be effective catalysts for the reactions of allene with amines. This telomerization can be carried out by bubbling allene into a solution of the amine and the catalyst in hexamethylphosphorictriamide (HMPT) at 70–90°. Reactions are reported for a great number of amines and yields up to 80% are obtained [6, 228]. Telomeric 1 : 2 products **154** or 1 : 4 products **155** are formed depending on reaction conditions and amines applied.

When the amines and an excess of allene are heated up to 100° in the presence of the catalytic system nickel bromide and $PhPPr^i_2$ in ethanol solvent, the 1 : 3 adducts **156** and **157** occur, whereas the monoadduct **158** is formed only in traces (Scheme 23).

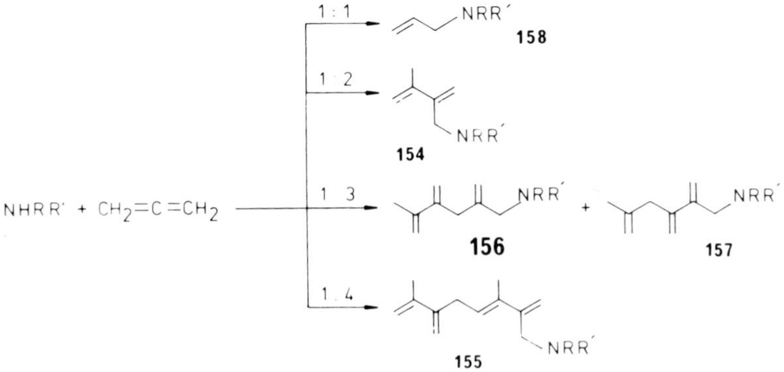

NHRR' + $CH_2=C=CH_2$ → 1 : 1 **158** (NRR'); 1 : 2 **154** (NRR'); 1 : 3 **156** (NRR') + **157** (NRR'); 1 : 4 **155** (NRR')

Scheme 23.

Reaction of allene and acetyl chloride at –30° in the presence of $[Ni(cod)_2]$ in ether produces a mixture of **159**–**161**. Similar results are obtained by using the nucleophile acetaldehyde.

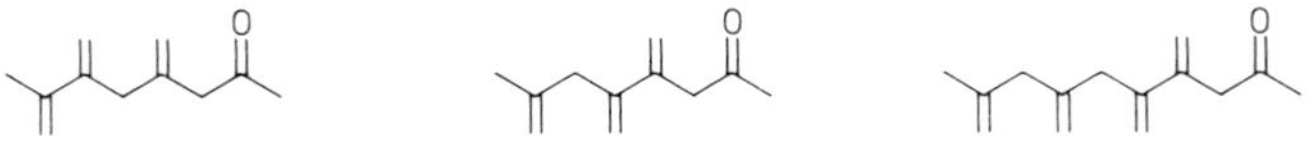

159 **160** **161**

To inquire the mechanism of this reactions Baker carried out some stoichiometric work [209]. The addition of allene to $[Ni(cod)_2]$ in ether yields a red solution containing the bis η^3-allylnickel complex **162**. By treating this solution with an organophosphorous compound L, for example PPh_3 or PCy_3, the η^1-η^3-allyl nickel species **163** occurs. By later addition of acetaldehyde the two isomeric alcohols **164** and **165** are obtained in a ratio of 7 : 1, whereas, in contrast to this formation, the reaction with amines produces the adduct **166** in high selectivity. It appears that the addition of the aldehyde occurs with priority at the η^1-allyl-nickel bond and that amines react at the η^3-allyl sites (Scheme 24).

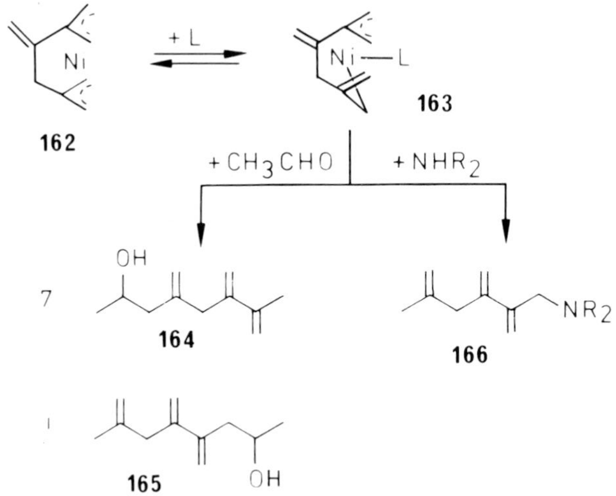

Scheme 24.

The hydrosilylation reaction of allene is described by Tsuji [116]. The reaction with trichlorosilane at 120° for 5h in the presence of $[Pd(PPh_3)_4]$ produces 1-trichlorosilyl-2-propene **167** in a 70% yield. Further reaction for 10h yields the 2 : 1-adduct 1,3-di(trichlorosilyl)propane **168**. Telomers are not formed under these conditions.

The telomerization of allene with carbon dioxide catalyzed by palladium has been examined [229]. The most effective catalyst has been prepared by treating $[Pd(\eta^3\text{-}C_3H_5)_2]$ with bis-dicyclohexylphosphinoethane. The highest yield of the telomeric products **169**–**171** are obtained at a reaction temperature of 110°. This telomerization is always accompanied by the formation of numerous oligomers.

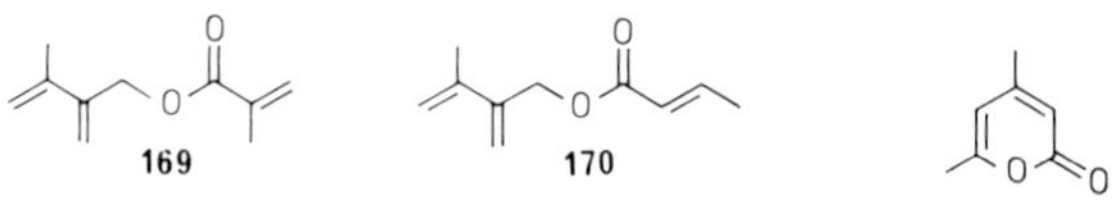

Product **171** is reminiscent of the lactones formed in the reaction of carbon dioxide with alkynes [230, 231] and suggests that the allene might be initially isomerized to propyne. This presumption was able to be ruled out by an experiment involving propyne in which only traces of telomers were formed.

The carboxy-telomerization of allene yields a great variety of products depending on the reaction conditions [6]. The reaction of allene, carbon monoxide and methanol yields methyl methacrylate **172** and the glutarate **173** as main products in the presence of ruthenium catalysts. The lactones **174** and **175** are formed by using water instead of methanol.

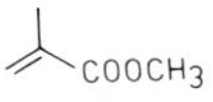

$COOCH_3$ **172** H_3COOC $COOCH_3$ **173** 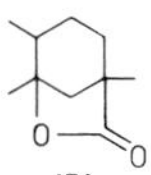 **174** 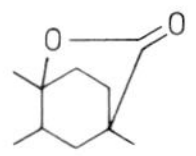 **175**

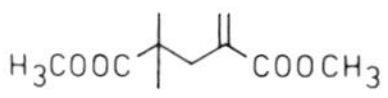

The stoichiometric reaction of palladium chloride with allene yields the two complexes **176** and **177**. With the step by step addition of carbon monoxide the latter produces the three products **178**–**180**.

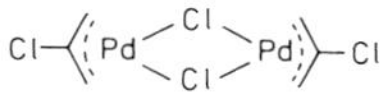

176 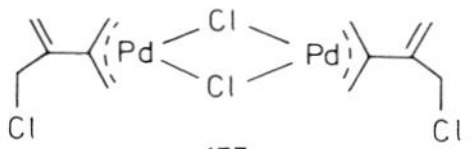 **177**

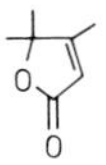

CO_2Et CO_2Et EtO_2C **178** 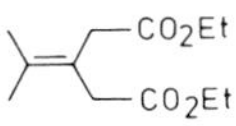 **179** 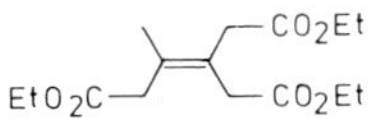 **180**

6. COTELOMERIZATIONS

The telomerization with different dienes has been called cotelomerization or cross-telomerization. An interesting example is the cotelomerization of butadiene, 1,3,7-octatriene and a nucleophile HY as shown in Equation 25.

+ + HY ⟶ Y (25)

When phenol is inserted as the nucelophile the phenoxydodecatriene is synthesized [3, 4]. This reaction has been carried out at ambient temperature with η^3-allyl palladium chloride–sodium phenoxide as the catalyst. An important outcome of this reaction was the discovery that the same catalyst also enables a highly selective degradation reaction. By further heating of the phenoxydodecatriene/catalyst-mixture, the added nucelophile leaves the molecule and 1,3,7,11-dodecatetraene **181** is set free in a good conversion and selectivity. The next higher linear telomer, 1-phenoxy-2,7,11,15-hexadecatetraene **182** was synthesized in an analogous manner. Phenol, butadiene and the dodecatetraene **181** yield the telomer **182** whose ensuing degradation leads to 1,3,7,11,15-hexadecapentaene **183**. Hence, this reaction provides a convenient path to straight-chain polyolefins with a high regioselectivity (Scheme 25). The cotelomerization has been realized with numerous other nucleophiles. Especially the reaction of butadiene, 1,3,7-octatriene and carbonic acids such as

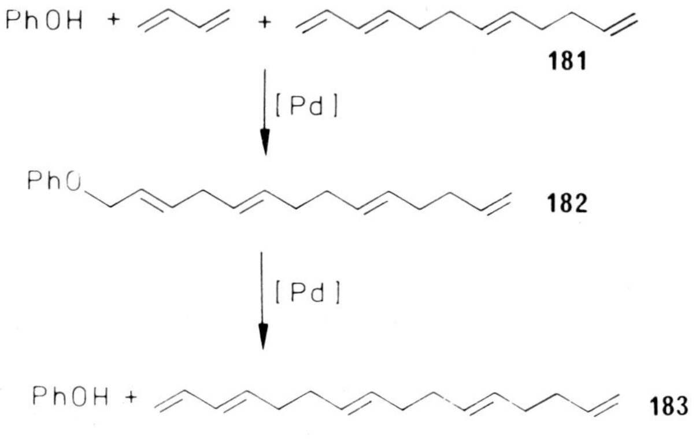

PhOH + + 181

[Pd]

PhO 182

[Pd]

PhOH + 183

Scheme 25.

acetic-, propionic- or benzoic acid has been well studied [232]. Ammonia and amines are also suitable nucleophiles [233].

The reaction of butadiene and isoprene with the nucelophile water proved to be another significant cotelomerization of 1,3-dienes [234]. In the presence of palladium acetylacetonate, triphenyl phosphine and carbon dioxide, isoprene reacts exclusively with its tail to produce the six nonadienols **184**–**189** in a yield of 40% besides a mixture of octadienols and terpene alcohols.

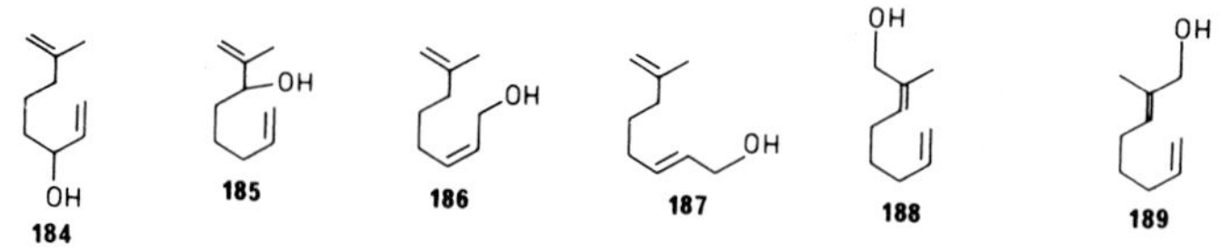

OH 184 | OH 185 | OH 186 | OH 187 | OH 188 | OH 189

The 1,2-diene allene can also enter into cotelomerization [9]. For example, the interaction of allene with propyne and acetic acid in the presence of a palladium salt yields the co-oligomer 2-methyl-1-pentene-3-yne **190** and the telomer 2-acetoxymethyl-1-pentene-3-yne **191**.

190 | OAc 191

7. ACKNOWLEDGEMENT

The author wishes to thank Prof. Dr W. Keim for his helpful discussions and extensive assistance.

8. REFERENCES

1. E. J. Smutny: *J. Am. Chem. Soc.* **89**, 6793 (1967).
2. E. J. Smutny, H. Chung, K. C. Dewhirst, W. Keim, T. M. Shryne and H. E. Thyret:
   (a) *Chem. Eng. News*, 11. Dez., 21 (1967);
   (b) *Prog. Coord. Chem.*, 824 (1968);
   (c) *Am. Chem. Soc., Div. Petrol. Chem., Prepr.* **14** (1969), B 100; *C. A.* **74**, 42059 j (1971);
   (d) *Chem. Eng. News*, 21. Apr., 48 (1969);
   (e) *Erdöl Kohle, Erdgas, Petrochem. Brennst.-Chem.* **23**, 44 (1970).
3. W. Keim: in *Transition Metals in Homogeneous Catalysis* (ed. G. N. Schrauzer), p. 59, Marcel Dekker, New York (1971).
4. E. J. Smutny: *Ann. N. Y. Acad. Sci.* **214**, 125 (1973).
5. (a) S. Takahashi, T. Shibano and N. Hagihara: *Tetrahedron Lett.*, 2451 (1967); *Bull. Chem. Soc. Jpn.* **41**, 454 (1968). *J. Chem. Soc., Chem. Commun.*, 161 (1969).
   (b) S. Takahashi, H. Yamazaki and N. Hagihara, *Bull. Chem. Soc. Jpn.* **41**, 254 (1968).
6. R. Baker: *Chem. Rev.* **73**, 487 (1973).
7. J. Tsuji: *Acc. Chem. Res.* **6**, 8 (1973).
8. J. Tsuji: *Adv. Organomet. Chem.* **17**, 141 (1979).
9. P. N. Rylander: *Organic Chemistry*, vol. 28, p. 175, Academic Press, New York (1973).
10. D. Rose and H. Lepper: *J. Organomet. Chem.* **49**, 473 (1973).
11. C. U. Pittman, Jr., S. K. Wuu and S. E. Jacobson: *J. Catal.* **44**, 87 (1976).
12. C. U. Pittman, Jr. and S. E. Jacobson: *J. Mol. Catal.* **3**, 293 (1977/78).
13. C. U. Pittman, Jr. and Q. Ng: *J. Organomet. Chem.* **153**, 85 (1978).
14. K. Kaneda, H. Kurosaki, M. Terasawa, T. Imanaka and S. Teranishi: *J. Org. Chem.* **46**, 2356 (1981).
15. A. Döhring, P. W. Jolly, R. Mynott, K.-P. Schick and G. Wilke: *Z. Naturforsch.* **36b**, 1198 (1981).
16. P. M. Maitlis: *The Organic Chemistry of Palladium*, vol. 2, p. 46, Academic Press, New York (1971).
17. M. L. H. Green and H. Munakata: *J. Chem. Soc., Chem. Commun.*, 549 (1971).
18. W. Keim, A. Behr and G. v. Ilsemann: unpublished results.
19. H. Felkin and G. K. Turner: *J. Organomet. Chem.* **129**, 429 (1977)..
20. H. Werner and A. Kühn: *Angew. Chem.* **89**, 427 (1977); *Angew Chem., Int. Ed. Engl.* **16**, 412 (1977).
21. G. Holloway, B. R. Penfold, R. Colton and M. J. McCormick: *J. Chem. Soc., Chem. Commun.*, 485 (1976).
22. D. Medema and R. van Helden: *Recl. Trav. Chim. Pays-Bas* **90**, 324 (1971).
23. B. Åkermark, G. Åkermark, C. Moberg, C. Björklund and K. Siirala-Hansén: *J. Organomet. Chem.* **164**, 97 (1979).
24. P. Heimbach: *Angew. Chem.* **80**, 967 (1968); *Angew. Chem. Int. Ed. Engl.* **7**, 882 (1968).
25. (a) H. Chung, W. Keim and E. J. Smutny, reported: Gordon Res. Conf. on Hydrocarbon (1969).
    (b) H. Chung and W. Keim: U.S. Pat. 3.636.162 (1972); *C. A.* **76**, 85545 y (1972).
26. T. C. Shields and W. E. Walker: *J. Chem. Soc., Chem. Commun.*, 193 (1971).
27. H. Yashida and S. Yuguchi: Japan. Pat. 7.009.729 (1970); *C. A.* **73**, 55612 p (1970).
28. R. Baker, D. E. Halliday and T. N. Smith: *J. Chem. Soc., Chem. Commun.*, 1583 (1971).
29. D. Rose: *Tetrahedron Lett.*, 4197 (1972).
30. J. Beger, C. Duschek, H. Füllbier and W. Gaube: *J. Prakt. Chem.* **316**, 26 (1974).
31. D. Commereuc and Y. Chauvin: *Bull. Soc. Chim. Fr.*, 652 (1974).
32. F. J. Weigert and W. C. Drinkard: *J. Org. Chem.* **38**, 335 (1973).

33. R. Baker, A. H. Cook, D. E. Halliday and T. N. Smith: *J. Chem. Soc.*, Perkin Trans. 2, 1511 (1974).
34. G. Wilke, B. Bogdanovič, P. Borner, H. Breil, P. Hardt, P. Heimbach, G. Herrmann, H.-J. Kaminsky, W. Keim, M. Kröner, H. Müller, E. W. Müller, W. Oberkirch, J. Schneider, J. Stedefeder, K. Tanaka and K. Weyer: *Angew. Chem.* 75, 10 (1963); *Angew. Chem. Int. Ed. Engl.* **2**, 105 (1963).
35. K. C. Dewhirst: *J. Org. Chem.* **32**, 1297 (1967).
36. K. C. Dewhirst, W. Keim and H. E. Thyret:
    (a) U.S. Pat. 3.502.725 (1970); *C. A.* **72**, 111023 r (1970);
    (b) U.S. Pat. 3.652.614 (1972); *C. A.* 77, 34707 r (1972).
37. R. Baker and D. E. Halliday: *Tetrahedron Lett.*, 2773 (1972).
38. E. Kuntz, Ger. Offen. 2.733.516 (1978); *C. A.* **88**, 152026 t (1978).
39. E. J. Smutny: U.S. Pat. 3.407.224 (1968); *C. A.* **70**, 3300 d (1969).
40. A. M. Lazutkin, A. I. Lazutkina and Yu. I. Ermakov: *Kinet. Katal.* **14**, 1593 (1973).
41. W. Keim, A. Behr and H. Rzehak: *Tenside Detergents* **16**, 113 (1979).
42. M. Green, G. Scholes and F. G. A. Stone: *J. Chem. Soc.*, Dalton Trans., 309 (1978).
43. A. M. Lazutkin and A. I. Lazutkina: *React. Kinet. Catal. Lett.* 8, 263 (1978).
44. A. M. Lazutkin, V. M. Mastikhin and A. I. Lazutkina: *Kinet. Katal.* **19**, 1061 (1978).
45. A. I. Lazutkina, A. M. Lazutkin and Yu. I. Ermakov: *Katalizatory, Soderzhashchie Nanesen. Kompleksy.*, 115 (1977); *C. A.* **88**, 104603 g (1978).
46. A. Kasahara, T. Izumi, K. Sato and M. Watanabe: *Yamagata Daigaku Kiyo*, Kogaku, **12**, 145 (1973); *C. A.* **79**, 41602 x (1973).
47. S. Hattori, H. Munakata, T. Suzuki, Y. Nishikawa and N. Imaki: Ger. Offen. 1.807.491 (1969); *C. A.* **71**, 70098 c (1969).
48. W. T. Dent:
    (a) Brit. Pat. 1.354.507 (1970);
    (b) Ger. Offen. 2.148.156 (1972); *C. A.* 77, P 100777 s (1972).
49. Y. Tamaru, M. Kagotani, R. Suzuki and Z. Yoshida: *Chem. Lett.*, 1329 (1978).
50. B. Castro, P. Grenouillet, D. Neibecker and I. Tkatchenko: Fr. Demande 2.459.075 (1981); *C. A.* **95**, 168543 q (1981).
51. P. Grenouillet, D. Neibecker, J. Poirier and I. Tkatchenko: 2nd Int. Symp. Homogeneous Catal., Düsseldorf 1980, preprints p. 98.
52. M. G. Romanelli: U.S. Pat. 3.670.029 (1972).
53. M. G. Romanelli and R. J. Kelly:
    (a) Ger. Offen. 2.011.163 (1970); *C. A.* **74**, 53040 x (1971).
    (b) U.S. Pat. 3.670.032 (1972).
54. H. Jadamus and K. Diebel, Ger. Offen. 2.635.250 (1978); *C. A.* **88**, 153231 t (1978).
55. Japan Synthetic Rubber Co. Ltd., Brit. Pat. 1.178.812 (1970); *C. A.* **72**, 78361 z (1970).
56. J. Beger and H. Reichel, *J. Prakt. Chem.* **315**, 1067 (1973).
57. M. Perree-Fauvet and Y. Chauvin, *Tetrahedron Lett.*, 4559 (1975).
58. T. M. Shryne, U.S. Pat. 3.530.187 (1970); *C. A.* **73**, 130615 x (1970).
59. K. C. Dewhirst, U.S. Pat. 3.489.813 (1970); *C. A.* **72**, 66603 s (1970).
60. C. Duschek, H. Füllbier and J. Beger, Ger. Pat. (East) 104.968 (1974); *C. A.* **81**, 151497 r (1974).
61. M. Čapka, P. Svoboda and J. Hetflejš, *Collect. Czech. Chem. Commun.* **38**, 1242 (1973).
62. N. Hagiwara, S. Takahashi and T. Shibano: Japan. Pat. 7.700.006 (1977); *Chem. Abstr.* **86**, 189198 k (1977).
63. S. Takahashi, T. Shibano and N. Hagihara: *Kogyo Kagaku Zasshi* 72, 1798 (1969).
64. S. Gardner and D. Wright: *Tetrahedron Lett.*, 163 (1972).
65. P. Roffia, G. Gregorio, F. Conti, G. F. Pregaglia and R. Ugo: *J. Organomet. Chem.* **55**, 405 (1973).
66. D. Wright: Ger. Offen. 2.240.719 (1973); *C. A.* **78**, 135608 k (1973).

67. C. U. Pittman, Jr., R. M. Hanes and J. J. Yang: *J. Mol. Catal.* **15**, 377 (1982).
68. W. E. Walker, R. M. Manyik, K. E. Atkins and M. L. Farmer: *Tetrahedron Lett.*, 3817 (1970).
69. W. Keim, A. Durocher and P. Voncken:
   (a) *Erdöl Kohle, Erdgas, Petrochem. Brennst.-Chem.* **29**, 31 (1976);
   (b) *Erdöl Kohle, Erdgas, Petrochem. Brennst.-Chem.*, Compendium 75/76, 347.
70. R. Fletcher: Brit. Pat. 1.505.317 (1978); *C. A.* **89**, 90137 q (1978).
71. R. Klüter, M. Bernd and H. Singer: *J. Organomet. Chem.* **137**, 309 (1977).
72. U. M. Dzehmilev, R. V. Kunakova, D. L. Minsker, E. V. Vasileva and G. A. Tolsti!.ov: *Izv. Akad. Nauk SSSR*, Ser. Khim., 1466 (1978).
73. K. E. Atkins, W. E. Walker and R. M. Manyik: *J. Chem. Soc., Chem. Commun.*, 330 (1971).
74. J. Tsuji and T. Mitsuyasu, Japan. Pat. 7.435.603 (1974); *C. A.* **82**, 124743 g (1975).
75. T. Mitsuyasu and J. Tsuji: Japan. Kokai 7.378.107 (1973); *C. A.* **80**, 59446 e (1974).
76. K. E. Atkins, R. M. Manyik and G. L. O'Connor:
   (a) Ger. Offen. 2.018.054 (1970); *C. A.* **74**, 41892 p (1971);
   (b) Brit. Pat. 1.307.101 (1970);
   (c) French. Pat. 2.045.369 (1970);
   (d) U.S. Pat. 816.792 (1969).
77. J. F. Kohnle, L. H. Slaugh and K. L. Nakamaye: *J. Am. Chem. Soc.* **91**, 5904 (1969).
78. J. F. Kohnle and L. H. Slaugh: U.S. Pat. 3.444.258 (1969); *C. A.* **71**, 12505 x (1969).
79. A. Musco and A. Silvani: *J. Organomet. Chem.* **88**, C 41 (1975).
80. S. Enomoto, H. Takita, H. Wada, Y. Mukaida and M. Yanaka: Japan. Kokai 7.626.809 (1976); *C. A.* **85**, 77643 z (1976).
81. N. Yoshimura and M. Tamura: Brit. U.K. Pat. Appl. GB 2.074.156 (1981); *C. A.* **96**, 122195 v (1982).
82. N. Yoshimura and M. Tamura: Fr. Demande FR 2.479.187 (1981); *C. A.* **96**, 103630 s (1982).
83. D.-L. Deng, K.-W. Liu, T.-B. Pang, H.-F. Zhou, C.-Q. Ye and T.-L. Wang: *Yu Chi Hua Hsueh* **3**, 184 (1981); *C. A.* **95**, 149868 h (1981).
84. K. Takahashi, A. Miyake and G. Hata:
   (a) *Chem. Ind.* (London), 488 (1971);
   (b) *Bull. Chem. Soc. Jpn.* **45**, 1183 (1972).
85. J. Beger and F. Meier: Ger. Pat. (East) 129.779 (1978); *C. A.* **89**, 108099 r (1978).
86. J. Beger and F. Meier: *J. Prakt. Chem.* **322**, 69 (1980).
87. R. Baker, A. Onions, R. J. Popplestone and T. N. Smith: *J. Chem. Soc.*, Perkin Trans. 2, 1133 (1975).
88. Yu. I. Ermakov, A. M. Lazutkin, A. I. Lazutkina and V. I. Prozorova: *Kinet. Katal.* **19**, 911 (1978).
89. A. M. Lazutkin, A. I. Lazutkina and Yu. I. Yermakov: *React. Kinet. Catal. Lett.* **8**, 353 (1978).
90. D. White: *J. Chem. Res.* (S), 266 (1977); *J. Chem. Res.* (M), 2401 (1977).
91. S. Miyamoto: Japan. Kokai 7.329.706 (1973); *C. A.* **79**, 78084 f (1973).
92. U. M. Dzhemilev, A. Z. Yakupova and G. A. Tolstikov: *Izv. Akad. Nauk SSSR*, Ser. Khim., 1068 (1978).
93. E. J. Smutny: U.S. Pat. 3.350.451 (1967); *C. A.* **68**, 21674 s (1968).
94. U. M. Dzhemilev, R. N. Fakhretdinov, A. G. Telin, M. Yu. Dolomatov, E. G. Galkin and G. A. Tolstikov: *Izv. Akad. Nauk SSSR*, Ser. Khim., 163 (1980).
95. C. Moberg and B. Åkermark: *J. Organomet. Chem.* **209**, 101 (1981).
96. C. Moberg: *Tetrahedron Lett.* **22**, 4827 (1981).
97. U. M. Dzhemilev, L. Yu. Gubaidullin and G. A. Tolstikov: *Izv. Akad. Nauk SSSR*, Ser. Khim., 2557 (1978).

98. U. M. Dzhemilev, R. J. Khusnutdinov, Z. S. Muslimov, G. A. Tolstikov and O. M. Nefedov: *Izv. Akad. Nauk SSSR*, Ser. Khim., 220 (1980).
99. U. M. Dzhemilev, F. A. Selimov, G. A. Tolstikov, E. A. Galkin and V. I. Khvostenko: *Izv. Akad. Nauk SSSR*, Ser. Khim., 652 (1980).
100. U. M. Dzhemilev, F. A. Selimov, A. Z. Yakupova and G. A. Tolstikov: *Izv. Akad. Nauk SSSR*, Ser. Khim., 1412 (1978).
101. U. M. Dzhemilev, F. A. Selimov and G. A. Tolstikov: *Izv. Akad. Nauk SSSR*, Ser. Khim., 2652 (1979).
102. U. M. Dzhemilev, R. N. Fakhretdinov, A. G. Telin and G. A. Tolstikov: *Izv. Akad. Nauk SSSR*, Ser. Khim., 1324 (1980).
103. U. M. Dzhemilev, R. N. Fakhretdinov, A. G. Telin, G. A. Tolstikov, A. A. Panasenko and E. V. Vasil'eva: *Izv. Akad. Nauk SSSR*, Ser. Khim., 2771 (1980).
104. T. Mitsuyasu, M. Hara and J. Tsuji: *J. Chem. Soc., Chem. Commun.*, 345 (1971).
105. J. Tsuji: Japan. Pat. 7.522.014 (1975); *C. A.* 84, 16736 q (1976).
106. J. Tsuji and M. Takahashi: *J. Molec. Catal.* **10**, 107 (1981).
107. R. N. Fakhretdinov, G. A. Tolstikov and U. M. Dzhemilev: *Neftekhimiya* **19**, 468 (1979).
108. G. Hata, K. Takahashi and A. Miyake:
   (a) *Chem. Ind.* (London), 1836 (1969).
   (b) *J. Org. Chem.* **36**, 2116 (1971).
109. R. Baker, A. H. Cook and T. N. Smith: *J. Chem. Soc.*, Perkin Trans. 2, 1517 (1974).
110. R. Baker, D. E. Halliday and T. N. Smith: *J. Organomet. Chem.* **35**, C 61 (1972).
111. J. Tsuji: *Bull. Chem. Soc. Jpn.* **46**, 1896 (1973).
112. H. Minematsu, S. Takahashi and N. Hagihara: *J. Chem. Soc., Chem. Commun.*, 466 (1975).
113. T. Mitsuyasu and J. Tsuji: *Tetrahedron* **30**, 831 (1974).
114. J. Tsuji and T. Mitsuyasu: Japan Pat. 7.424.887 (1974); *C. A.* **82**, 30951 w (1975).
115. R. V. Kunakova, G. A. Tolstikov, U. M. Dzhemilev, F. V. Sharipova and D. L. Sazikova: *Izv. Akad. Nauk SSSR*, Ser. Khim., 931 (1978).
116. J. Tsuji, M. Hara and K. Ohno: *Tetrahedron* **30**, 2143 (1974).
117. M. Hara, K. Ohno and J. Tsuji: *J. Chem. Soc., Chem. Commun.*, 247 (1971).
118. J. Langová and J. Hetflejš: *Collect. Czech. Chem. Commun.* **40**, pp. 420 and 432 (1975).
119. V. Vaisarová, J. Hetflejš, H.-W. Krause and H. Pracejus: *Z. Chem.* **14**, 105 (1974).
120. M. Čapka and P. Hetflejš:
   (a) *Collect. Czech. Chem. Commun.* **40**, 2073 (1975); **40**, 3020 (1975); **40**, 3186 (1975); **41**, 1024 (1976);
   (b) Czech. Pat. 162.588 (1976); *C. A.* **85**, 177609 m (1976).
121. P. Svoboda, V. Vaisarova, M. Čapka, J. Hetflejš, M. Kraus and V. Bazant: Ger. Offen. 2.260.260 (1973); *C. A.* **79**, 78953 p (1973).
122. N. Hagihara, S. Takahashi and T. Shibano: Japan. Pat. 7.307.416 (1973); *C. A.* **79**, 53532 n (1973).
123. V. Vaisarová, J. Langová, J. Hetflejš, G. Oehme and H. Pracejus: *Z. Chem.* **14**, 64 (1974).
124. N. Hagiwara and S. Takahashi: Japan. Pat. 7.524.948 (1975); *C. A.* **84**, 105749 p (1976).
125. V. P. Yur'ev, J. M. Salimgareeva, O. Zh. Zhebarov, G. A. Tolstikov and S. R. Rafikov: *Dokl. Akad. Nauk SSSR*, Ser. Khim. **224**, 1092 (1975).
126. U. M. Dzhemilev, R. V. Kunakova, R. L. Gaisin and G. A. Tolstikov: *Izv. Akad. Nauk SSSR*, Ser. Khim., 2702 (1979).
127. U. M. Dzhemilev, R. V. Kunakova, R. L. Gaisin, G. A. Tolstikov, R. F. Talipov and S. I. Lomakina: *Zh. Org. Khim.* **17**, 763 (1981).
128. R. V. Kunakova, R. L. Gaisin, G. A. Tolstikov, L. M. Zelenova and U. M. Dzhemilev: *Izv. Akad. Nauk SSSR*, Ser. Khim., 2610 (1980).
129. U. M. Dzhemilev, R. V. Kunakova, Yu. T. Struchkov, G. A. Tolstikov, F. V. Sharipova, L. G. Kuz'mina and S. R. Rafikov: *Dokl. Akad. Nauk SSSR*, Ser. Khim., **250**, 105 (1980).

130. U. M. Dzhemilev, R. V. Kunakova, F. V. Sharipova, L. V. Spirikhin, L. M. Khalilov, E. V. Vasil'eva and G. A. Tolstikov: *Izv. Akad. Nauk SSSR*, Ser. Khim., 1822 (1979).
131. U. M. Dzhemilev, R. V. Kunakova, N. Z. Baibulatova, G. A. Tolstikov and A. A. Panasenko: *Zh. Org. Khim.* **16**, 1157 (1980).
132. U. M. Dzhemilev, R. V. Kunakova, N. Z. Baibulatova, G. A. Tolstikov and L. M. Zelenova: *Izv. Akad. Nauk SSSR*, Ser. Khim., 1837 (1981).
133. U. M. Dzhemilev, F. A. Selimov and G. A. Tolstikov: *Izv. Akad. Nauk SSSR*, Ser. Khim., 348 (1980).
134. L. W. Watts, Jr. and W. H. Brader, Jr.: U.S. Pat. 4.054.605 (1977); *C. A.* 88, 61964 f (1978).
135. J. D. Umbleby: *Helv. Chim. Acta* **61**, 2243 (1978).
136. R. Baker and M. S. Nobbs: *Tetrahedron Lett.*, 3759 (1977).
137. R. Baker, M. S. Nobbs and P. M. Winton: *J. Organomet. Chem.* **137**, C 43 (1977).
138. R. Baker, M. S. Nobbs and D. T. Robinson: *J. Chem. Soc., Chem. Commun.*, 723 (1976).
139. H.-U. Blaser and D. Reinehr: *Helv. Chim. Acta* **60**, 208 (1977).
140. D. Reinehr: *Pure Appl. Chem.* **52**, 2417 (1980).
141. Y. Tamaru, R. Suzuki, M. Kagotani and Z. Yoshida: *Tetrahedron Lett.* **21**, 3791 (1980).
142. Y. Tamaru, R. Suzuki, M. Kagotani and Z. Yoshida: *Tetrahedron Lett.* **21**, 3787 (1980).
143. H. Sakurai, Y. Kamiyama and Y. Nakadaira: *Chem. Lett.* 8, 887 (1975).
144. K. Tamao, S. Okazaki and M. Kumada: *J. Organomet. Chem.* **146**, 87 (1978).
145. H. Matsumoto, K. Shono, A. Wada, I. Matsubara, H. Watanabe and Y. Nagai: *J. Organomet. Chem.* **199**, 185 (1980).
146. P. Haynes: *Tetrahedron Lett.*, 3687 (1970).
147. H. A. Jung: Ger. Offen. 2.141.186 (1972).
148. H. Watanabe, K. Matsuzaki, R. Tsumura and T. Mazaki: Japan. Pat. 7.036.463 (1973); *C. A.* 79, 42345 c (1973).
149. R. M. Manyik, W. E. Walker, K. E. Atkins and E. S. Hammack: *Tetrahedron Lett.*, 3813 (1970).
150. K. Ohno, T. Mitsuyasu and J. Tsuji: *Tetrahedron Lett.*, 67 (1971).
151. K. Ohno, T. Mitsuyasu and J. Tsuji: *Tetrahedron* **28**, 3705 (1972).
152. A. Musco: *Inorg. Chim. Acta* **11**, L 11 (1974).
153. S. Akutagawa: *Bull. Chem. Soc. Jpn.* **49**, 3646 (1976).
154. Y. Inoue, Y. Sasaki and H. Hashimoto: *Bull. Chem. Soc. Jpn.* **51**, 2375 (1978).
155. Y. Sasaki, Y. Inoue and H. Hashimoto: *J. Chem. Soc., Chem. Commun.*, 605 (1976).
156. A. Musco, C. Perego and V. Tartiari: *Inorg. Chim. Acta* **28**, L 147 (1978).
157. A. Musco: *J. Chem. Soc.*, Perkin Trans. 1, 693 (1980).
158. (a) A. Behr, K.-D. Juszak and W. Keim: presented at the 2nd IUPAC Symposium on Organometallic Chemistry Directed Toward Organic Syntheses, Dijon, France, Aug. 28–Sept. 1, 1983;
(b) A. Behr, K.-D. Juszak and W. Keim: *Syntheses*, 1983 (7), 574;
(c) A. Behr and K.-D. Juszak: *J. Organomet. Chem.* **255**, 263 (1983).
159. J. Kiji, K. Yamamoto, H. Tomita and J. Furukawa: *J. Chem. Soc., Chem. Commun.*, 506 (1974).
160. P. Heimbach, A. Roloff and E. F. Nabbefeld: Ger. Offen. 2.638.430 (1978); *C. A.* 88, 169568 u (1978).
161. K. Ohno and J. Tsuji: *J. Chem. Soc., Chem. Commun.*, 247 (1971).
162. U. M. Dzhemilev, L. Yu. Gubaidullin and G. A. Tolstikov: *Izv. Akad. Nauk SSSR*, Ser. Khim., 915 (1979).
163. W. E. Billups, W. E. Walker and T. C. Shields: *J. Chem. Soc., Chem. Commun.*, 1067 (1971).
164. J. F. Knifton: *J. Catal.* **60**, 27 (1979).
165. J. F. Knifton: *Ann. N. Y. Acad. Sciences* **333**, 264 (1980).

166. J. Tsuji, Y. Mori and M. Hara: *Tetrahedron* **28**, 3721 (1972).
167. H. Fukutani, M. Tokizawa and H. Okada: Japan. Pat. 7.325.169 (1973); *C. A.* **79**, 104765 a (1973).
168. J. Tsuji and M. Hara: Japan. Pat. 7.311.086 (1973); *C. A.* **85**, 123386 t (1976).
169. R. Bortolin, G. Gatti and A. Musco: *J. Mol. Catal.* **14**, 95 (1982).
170. J. Tsuji: *Pure Appl. Chem.* **51**, 1235 (1979).
171. J. Tsuji: *Top. Curr. Chem.* **91**, 29 (1980).
172. J. Tsuji: *Ann. N.Y. Acad. Sciences* **333**, 250 (1980).
173. J. Tsuji and T. Mandai: *Chem. Lett.*, 975 (1977).
174. J. Tsuji, K. Tsuruoka and K. Yamamoto: *Bull. Chem. Soc. Jpn.* **49**, 1701 (1976).
175. H. Munakata and N. Imaki: Japan. Kokai 7.596.511 (1975); *C. A.* **84**, 4461 h (1976).
176. J. Tsuji, H. Yasuda and T. Mandai: *J. Org. Chem.* **43**, 3606 (1978).
177. J. Tsuji, K. Mizutani, I. Shimizu and K. Yamamoto: *Chem. Lett.*, 773 (1976).
178. J. Tsuji, Y. Kobayashi and I. Shimizu: *Tetrahedron Lett.*, 39 (1979).
179. J. Tsuji, I. Shimizu, H. Suzuki and Y. Naito: *J. Am. Chem. Soc.* **101**, 5070 (1979).
180. J. Tsuji, Y. Kobayashi and T. Takahashi: *Tetrahedron Lett.*, 483 (1980).
181. I. Shimizu, Y. Naito and J. Tsuji: *Tetrahedron Lett.*, 487 (1980).
182. J. Tsuji, K. Kasuga and T. Takahashi: *Bull. Chem. Soc. Jpn.* **52**, 216 (1979).
183. R. Baker, P. M. Winton and R. W. Turner: *Tetrahedron Lett.*, 1175 (1980).
184. J. Tsuji, N. Nagashima, T. Takahashi and K. Masaoka: *Tetrahedron Lett.*, 1917 (1977).
185. J. Tsuji, K. Masaoka and T. Takahashi: *Tetrahedron Lett.*, 2267 (1977).
186. J. Tsuji and T. Mandai: *Tetrahedron Lett.*, 3285 (1977).
187. J. Tsuji, T. Yamakawa and T. Mandai: *Tetrahedron Lett.*, 565 (1978).
188. J. Tsuji and H. Yasuda: *J. Organomet. Chem.* **131**, 133 (1977).
189. J. Tsuji, M. Kaito, T. Yamada and T. Mandai: *Bull. Chem. Soc. Jpn.* **51**, 1915 (1978).
190. N. T. Byrom, R. Grigg and B. Kongkathip: *J. Chem. Soc., Chem. Commun.*, 216 (1976).
191. H. Yamazaki: Japan. Kokai 7.323.707 (1973); *C. A.* **79**, 52781 f (1973).
192. Y. Uchida: *Asahi Garasu Kogyo Gijutsu Shoreikai Kenkyu Hokoku* **23**, 103 (1973); *C. A.* **82**, 98467 y (1975).
193. H. Yamazaki: Japan. Kokai 7.448.613 (1974); *C. A.* **81**, 119953 a (1974).
194. L. I. Zakharkin and S. A. Babich: *Izv. Akad. Nauk SSSR*, Ser. Khim., 2099 (1976).
195. M. Hidai, H. Ishiwatari, H. Yagi, E. Tanaka, K. Onozawa and Y. Uchida: *J. Chem. Soc., Chem. Commun.*, 170 (1975).
196. H. Yagi, E. Tanaka, H. Ishiwatari, M. Hidai and Y. Uchida: *Synthesis*, 334 (1977).
197. W. Hoffmann, F.-J. Müller and K. von Fraunberg: Ger. Offen. 2.154.370 (1973); *C. A.* **79**, 18077 w (1973).
198. J. Beger, C. Duschek and H. Reichel: *J. Prakt. Chem.* **315**, 1077 (1973).
199. A. Behr and W. Keim: *Chem. Berichte* **116**, 862 (1983).
200. M. Hidai, H. Mizuta, H. Yagi, Y. Nagai, K. Hata and Y. Uchida: *J. Organomet. Chem.* **232**, 89 (1982).
201. K. Suga, S. Watanabe and K. Hijikata: *Aust. J. Chem.* **24**, 197 (1971).
202. S. Hattori, Y. Nishikawa and T. Imaki: Japan. Pat. 7.534.002 (1975); *C. A.* **86**, 4946 m (1977).
203. D. Wright: Ger. Offen. 2.050.774 (1971); *C. A.* **75**, 37618 g (1971).
204. J. P. Neilan, R. M. Laine, N. Cortese and R. F. Heck: *J. Org. Chem.* **41**, 3455 (1976).
205. L. I. Zakharkin, S. A. Babich and I. V. Pisareva: *Izv. Akad. Nauk SSSR*, Ser. Khim., 1616 (1976).
206. W. Keim and M. Röper: *J. Org. Chem.* **46**, 3702 (1981).
207. W. Keim and M. Röper: presented at the 10th Int. Conf. on Organomet. Chem., Toronto, Canada, Aug. 9–14, 1981.
208. M. Hidai, H. Mizuta, K. Hirai and Y. Uchida: *Bull. Chem. Soc. Jpn.* **53**, 2091 (1980).
209. R. Baker, A. H. Cook and M. J. Crimmin: *J. Chem. Soc., Chem. Commun.*, 727 (1975).

210. S. Watanabe, K. Suga and T. Fujita: *Can. J. Chem.* **51**, 848 (1973).
211. I. Ojima: *J. Organomet. Chem.* **134**, C 1 (1977).
212. I. Ojima and M. Kumagai: *J. Organomet. Chem.* **134**, C 6 (1977).
213. R. Baker and M. J. Crimmin: *J. Chem. Soc.*, Perkin Trans. 1, 1264 (1979).
214. Y. Inoue, S. Sekiya, Y. Sasaki and H. Hashimoto: *Yuki Gosei Kagaku Kyokai Shi* **36**, 328 (1978); *C. A.* 89, 42280 j (1978).
215. J. Tsuji and H. Yasuda: *Bull. Chem. Soc. Jpn.* **50**, 553 (1977).
216. S. Hosaka and J. Tsuji: *Tetrahedron* **27**, 3821 (1971).
217. W. Keim, A. Behr and H. Rzehak: unpublished results.
218. A. Behr, V. Falbe and W. Keim: presented at the 4th Int. Symp. on Homogeneous Catalysis, Leningrad, U.S.S.R., Sept. 24–28, 1984.
219. V. P. Yur'ev, I. M. Salimgareeva, O. Zh. Zhebarov and G. V. Tolstikov: *Izv. Akad. Nauk SSSR*, Ser. Khim., 2135 (1975).
220. C. Bordenca and W. E. Marsico: *Tetrahedron Lett.*, 1541 (1967).
221. R. J. H. Duprey, W. D. Fordham, J. F. Janes, D. V. Banthorpe and M. R. Young: *Chem. Ind.* (London), 847 (1973).
222. K. Dunne and F. J. McQuillin: *J. Chem. Soc.* (C), 2196 (1970).
223. K. Suga, S. Watanabe and K. Hijikata: *Chem. Ind.* (London), 33 (1971).
224. I. Ojima and M. Kumagai: *J. Organomet. Chem.* **157**, 359 (1978).
225. R. Baker and R. J. Popplestone: *Tetrahedron Lett.*, 3575 (1978).
226. W. Keim, M. Röper and N. Finke: unpublished results.
227. G. D. Shier: *J. Organomet. Chem.* **10**, P 15 (1967).
228. R. Baker and A. H. Cook: *J. Chem. Soc.*, Perkin Trans. 2, 443 (1976).
229. A. Döhring and P. W. Jolly: *Tetrahedron Lett.*, 3021 (1980).
230. Y. Inoue, Y. Itoh and H. Hashimoto: *Chem. Lett.*, 855 (1977).
231. Y. Inoue, Y. Itoh and H. Hashimoto: *Chem. Lett.*, 633 (1978).
232. S. Hattori and N. Imaki: Ger. Offen. 2.162.990 (1972); *C. A.* 77, 126020 q (1972).
233. W. Keim, M. Röper and S. Osanai: unpublished results.
234. J.-P. Bianchini, B. Waegell, E. M. Gaydou, H. Rzehak and W. Keim: *J. Mol. Catal.* **10**, 247 (1981).
235. W. Keim, K.-R. Kurtz and M. Röper: *J. Mol. Catal.* **20**, 129 (1983).

The Cobalt-Catalyzed Synthesis of Pyridine and Its Derivatives

H. BÖNNEMANN AND W. BRIJOUX

Max-Planck-Institut für Kohlenforschung, Postfach 01 13 25
D–4330 Mülheim a.d. Ruhr 1

R. Ugo (ed.), Aspects of Homogeneous Catalysis, Vol. 5, 75–196.

1. INTRODUCTION

Although quite spectacular results have been obtained in the last few decades in the field of homogeneous transition metal catalyzed transformations of olefins and alkynes [1], reactions which could lead to heterocycles have been partly neglected. An obvious reason for this is that substrates containing heteroatoms such as N, O or S could coordinate the metal and suppress the catalytic activity. Nevertheless, some interesting early examples of transition-metal-catalyzed syntheses of heterocyclic compounds have been reported and these have been reviewed by C. W. Bird [2].

More recently the incorporation of CO_2, which enables esters and lactones to be synthesized from olefinic starting materials, has begun to attract attention (see, for example, ref. [3]). The dominant role of palladium as the catalyst for the formation of O-containing heterocycles has been suggested to be associated with the relatively low strength of the Pd—O bond.

Among the first examples of a nitrogen-containing heterocycle to be formed by homogeneous catalysis is the triazine shown in Equation 1 which is the product of the trimerization of benzonitrile in the presence of iron pentacarbonyl or Raney-nickel [4].

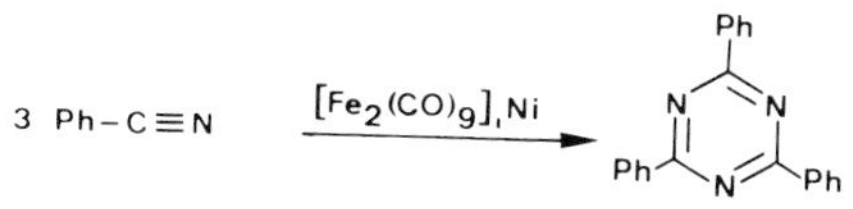

$$3\ \mathrm{Ph{-}C{\equiv}N} \xrightarrow{[\mathrm{Fe_2(CO)_9}],\ \mathrm{Ni}} \text{(2,4,6-triphenyl-1,3,5-triazine)} \qquad (1)$$

Further developments have shown that homogeneous transition metal catalysts may also be used to synthesize heterocycles by co-cyclization of simple unsaturated organic molecules in a manner familiar from carbocyclic chemistry. J. Tsuji and K. Ohno [5], for example, expanded the access to *N*-heterocycles by developing a 2 : 1 cycloaddition of conjugated dienes with isocyanate: in the presence of palladium-triphenylphosphine complexes, piperidone derivatives are formed in excellent yields (Equation 2).

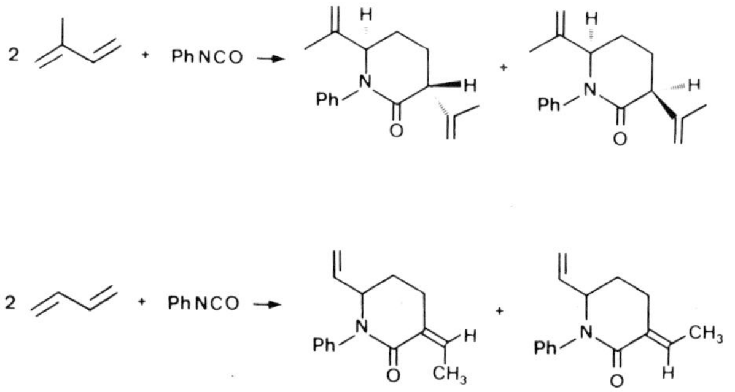

2 (isoprene) + PhNCO → (piperidone isomers) + (piperidone isomers)

2 (butadiene) + PhNCO → (piperidone isomers) + (piperidone isomers) (2)

In 1974 J. Kiji *et al.* [6] reported that a similar reaction between 1,3-dienes and Schiff's bases gives piperidine derivatives (Equation 3).

$$2\ CH_2{=}CH{-}CH{=}CH_2 + CH_3N{=}CHPh \xrightarrow[DMF/80^{\circ}]{[Pd(NO_3)_2\ (PPh_3)]} \quad (3)$$

At about this time, P. Heimbach and his coworkers [7] developed a synthesis of nitrogen-containing 12-membered ring systems by reacting butadiene with azines using nickel(0)-catalysts. A 3–10 butadiene to azine molar ratio has to be maintained in the reaction mixture and, depending upon the nature of the substituents, the yields vary between 70% and 90% (Equation 4).

$$\xrightarrow{\text{Ni or Ni-L}} \quad (4)$$

R^1 and R^3 = Me, Et, Ph ; R^2 and R^4 = H, Me, Et

The reaction of Schiff's bases with vinyl ethers led, in the presence of cobalt- or nickel-carbonyl catalysts, N. Hagihara *et al.* [8] to a synthesis of quinoline derivatives. The cobalt catalyzed reaction between benzilidene-aniline and ethyl-vinyl ether proceeds smoothly at room temperature (Equation 5). Oxygen is apparently

$$\xrightarrow[THF]{[Co_2(CO)_8][Ni(CO)_4]} \qquad \xrightarrow[-\,EtOH,\ -\,H_2O]{+\,\tfrac{1}{2}O_2} \quad (5)$$

necessary to promote the dehydrogenation. In recent years, Y. Watanabe *et al.* [9] have obtained quinoline derivatives by reacting aniline with 2,3-unsaturated alcohols such as allyl alcohol and crotyl alcohol at 180° in the presence of $[RuCl_2(PPh_3)_3]$. The reaction proceeds equally well using propanal or butanal (Equation 6).

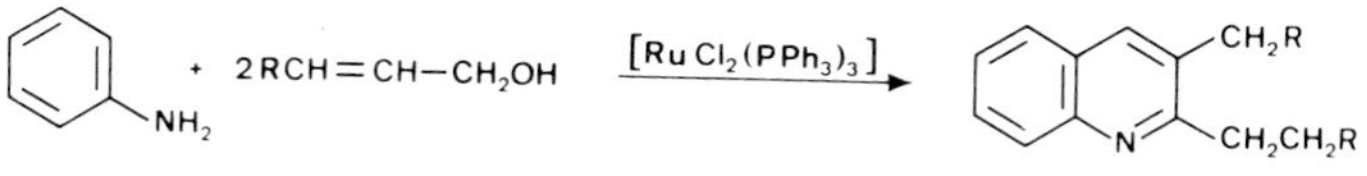

$$C_6H_5NH_2 + 2RCH{=}CH{-}CH_2OH \xrightarrow{[RuCl_2(PPh_3)_3]}$$ (6)

R = H (33%), CH_3 (45%)

Some synthetically attractive transformations of azirines induced by molybdenum(0)- and palladium(0)-complexes have been described by H. Alper [10]. For example, the intramolecular cycloaddition of 3-phenyl-2-substituted-2-H azirines to give five-membered ring heterocycles occurs at room temperature in high yields (Equation 7, 8).

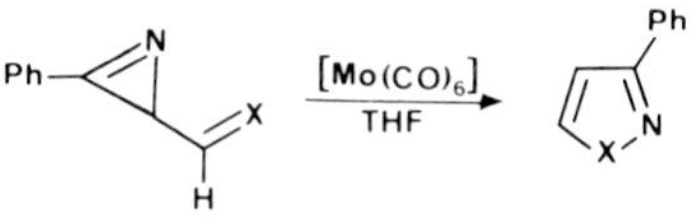

$[Mo(CO)_6]$, THF (7)

X = O, NR

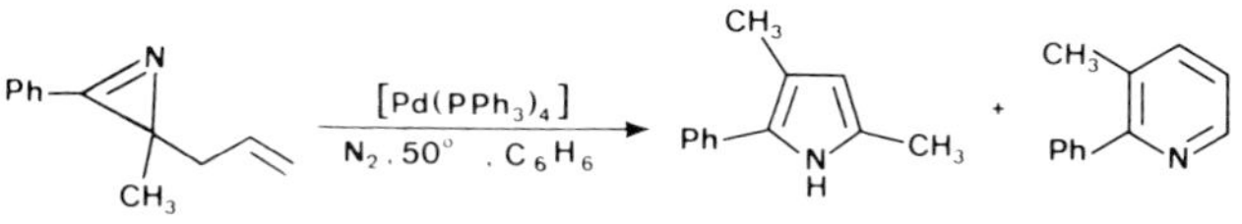

$[Pd(PPh_3)_4]$, N_2, 50°, C_6H_6 (8)

Finally, the recently reported palladium-catalyzed synthesis of pyrroles (Equations 9–11) should be mentioned [11], [12], [13].

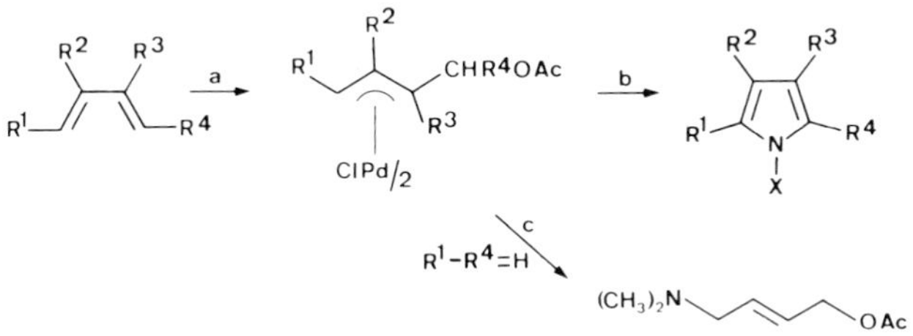

(9)

a: $PdCl_2$, $CuCl_2$, AcOH

b: $AgBF_4$, PPh_3, $Cu(BF_4)_2$, XNH_2

c: $AgBF_4$, $(CH_3)_2NH$

R^1–R^4 : H, CH_3

$$HOCH_2CH{=}CHCH_2OH + RNH_2 \xrightarrow[120°]{Pd} \text{N-R pyrrole} \quad (10)$$

R: CH_2CH_2OH, c-C_6H_{11}, n-C_6H_{13}, Ph

$$R^1{-}C(OH)(CH_2NH_2){-}C{\equiv}C{-}R^2 \xrightarrow{Pd(II)} \text{3-}R^1\text{-5-}R^2\text{-pyrrole (NH)} \quad (11)$$

R^1 = H, Et, $(CH_3)_3C$

R^2 = n-C_6H_{13}, Ph

We were first induced to investigate the use of soluble organocobalt complexes as catalysts for the formation of heterocyclic compounds during attempts to influence the selectivity of the dimerization of butadiene catalyzed by "naked cobalt(I)" [14]. All our efforts to change the ratio of the branched to linear products in Equation 12

(12) [scheme: Co complex ⇌ (20°) [Co–H] + branched diene (92%) + linear triene (8%); recycled with butadiene]

by adding donor ligands containing polar heteroatoms such as P, As, N or even S proved unsuccessful: the rate of reaction was hardly altered even in the presence of a large excess of ligand in polar solvents such as alcohols. (This is in contrast to the analogous nickel(0)-catalyzed reaction [15].) In view of this we decided to extend our work to alkynes and to introduce heterofunctional substrates. The transition-metal-catalyzed cyclotrimerization of acetylene (Equation 13) was discovered by

$$3\ HC{\equiv}CH \xrightarrow{\text{Transition metal system}} C_6H_6 \quad (13)$$

M. Berthelot [16] way back in the last century. He used heterogeneous systems. The merits of homogeneous catalysts in this field were then demonstrated most convincingly by W. Reppe [17], and recently K. P. C. Vollhardt [18] has developed a number of elegant applications in synthetic organic chemistry.

Our first experiments were carried out in 1973 [19] when we reacted 1 mole propionitrile with 0.5 moles propyne at 80° using 8 mmoles of methylheptadienyl-cobalt-butadiene as catalyst. It is this catalyst which is active for the dimerization reaction shown in Equation 12. To our satisfaction, one of the main reaction products was a mixture of substituted pyridines, i.e., the use of the R′-Co(diene) moiety as the catalyst enables one alkyne molecule in the cyclotrimerization reaction to be replaced by a C≡N triple bond. A turnover number (TON) of 15 moles of pyridine derivatives per g-atom cobalt was achieved and the yield was 35% based on reacted nitrile. The product distribution is shown in Equation 14.

$R^1C{\equiv}N + 2\,R^2C{\equiv}CH \longrightarrow$ (Co catalyst) 52% + 20% + 10% + 18% (14)

$R^1 = C_2H_5$

$R^2 = CH_3$

While our work was in progress, H. Yamazaki and Y. Wakatsuki reported that phosphane-substituted cobaltacyclopentadiene complexes, in which the two alkyne molecules are already linked together at the cobalt atom, also react with nitriles (Equation 15) [20a]. The complex shown in Equation 15 was the first organocobalt

[CpCoPPh$_3$]; $R^2{-}C{\equiv}C{-}R^2$; $R^1{-}C{\equiv}N$; CpCo; PPh$_3$; $R^2{-}C{\equiv}C{-}R^2$ (15)

system to be published which was active for the catalytic conversion of alkynes and nitriles to pyridine derivatives [20b]. However, the multistep synthesis necessary to prepare Yamazaki's complex made it unattractive for general preparative work and it was later shown that the stabilizing phosphine ligand is superfluous and actually depresses the catalytic activity. The Japanese authors themselves later turned to cobaltocene as a readily accessible catalyst-precursor [21].

Since the cyclotrimerization of alkynes (Equation 13) is catalyzed by a large variety of transition metal complexes [22], in addition to cobalt, we first of all tested the activity of a whole spectrum of transition metals, e.g. Zr, Hf, Cr, Fe, Ni, Pd, Rh. With the exception of the acetylacetonato-rhodium system for which slight activity was detected, no pyridine derivatives could be detected – even by thin-layer chromatography. Recently H. Hogeveen [23] has reported the use of aluminium halides in the stoichiometric transformation of propyne and ethylcyanoformate to the corresponding pyridine derivatives. Soluble organo*cobalt* complexes generally proved to be the most active catalysts (Equation 14). A typical selection of our data from this screening programme is shown in Table 1.

Table 1

SCREENING (TLC) OF TRANSITION METAL CATALYSTS FOR THE REACTION SHOWN IN EQUATION 14

Catalyst	Result	Catalyst	Result
$[Ti(cot)_2]$	–	$[FeCO(C_4H_6)_2]$	–
$[Ti_2(cot)_3]$	–	$[Ni(cod)_2]$	–
$[Zr(allyl)_2(cot)]$	–	$Pdacac_2$	–
$[Hf(allyl)_2(cot)]$	–	$[Rh(C_8H_{13})(C_8H_{12})]$	–
$[Cr_2(cot)_3]$	–	$[Rhacac(C_2H_4)_2]$	traces
$Cracac_3/AlEt_3/$	–		
Butadiene	–		

On the basis of these results, we decided to concentrate on developing an easily accessible, soluble organocobalt catalyst and to explore the scope and limitations of the reaction [24]. The result was the development of a highly selective method for the catalytic synthesis of 2-substituted pyridine derivatives (Equation 16).

$$2\,HC{\equiv}CH + N{\equiv}C{-}R \xrightarrow{R'-Co^{I}} \text{2-R-pyridine} \qquad (16)$$

R' = Organylgroup e.g. η^3-Allyl, η^5-Cp

Variation of the organic substituent R′ associated with the cobalt enabled us to tailor the catalyst in order to optimize the turnover number for the various reactants. In the case of monosubstituted alkynes (Equation 14), the selectivity control was studied and a continuous-flow apparatus has been devised to optimize the reaction parameters for those reactions which are of technical interest. Work is currently in progress to extend the cobalt-alkyne reactions to the synthesis of heterocycles other than pyridine and its derivatives.

2. SURVEY OF THE ORGANOCOBALT CATALYSTS

Two types of organocobalt complexes proved effective as catalysts for the co-cyclization of alkynes with cyanocompounds: the allyl-cobalt type, where the organic group R′ is η^3-bonded to the metal, and the η^5-cyclopentadienyl-cobalt-halfsandwich compounds (Figure 1). Our investigations have shown that the organic

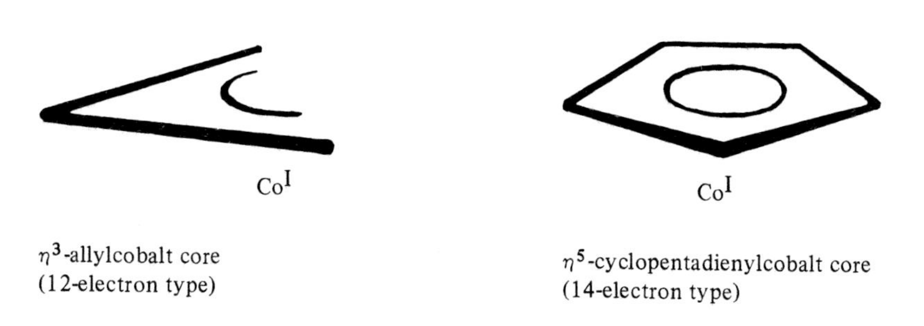

Fig. 1. Effective organocobalt systems.

group R′ remains bonded to the cobalt throughout the catalytic cycle, whereas additional ligands bonded to the metal are removed in the initial stages of the catalysis. In the case of the η^3-allyl-cobalt catalyst, a 12-electron system is regenerated during the catalytic cycle, whereas in the case of the η^5-cyclopentadienyl-cobalt complexes the catalytic reaction involves a 14-electron moiety.

As can be seen from Equation 15, Yamazaki's cobalt(III) complex is best regard as a special case of the Cp—Co type catalyst. Cobaltocene is quite efficient [21], [25], since it acts as a precursor for the true catalyst – a Co—Co(diene) complex. In an elegant model reaction (Equation 17), it was demonstrated that

$$2\ \mathrm{Co} \xrightarrow{CH_3OOC-C\equiv C-COOCH_3} \quad \mathrm{Co} \quad \mathrm{H} \quad \mathrm{H} \quad \mathrm{Co} \quad CH_3OOC \quad COOCH_3 \qquad (17)$$

cobaltocene reacts in an initial step with the alkyne to give a dimeric diene complex in which one cyclopentadienyl group has been removed and which contains the 14-electron Cp—Co fragment (Equation 17). However, there is no need to synthesize sophisticated organocobalt complexes since the catalysis can be carried out in a very simple one-pot reaction in which the active species are generated *in situ* [19].

2.1. The *in situ* catalyst

The *in situ* method proved to be a convenient procedure for many laboratory scale experiments whose main purpose was to test rapidly and qualitatively the scope of the reaction shown in Equation 18, rather than to obtain maximum efficiency.

$$R^1C{\equiv}N + 2\ R^2C{\equiv}CH \longrightarrow \text{(4,6-}R^2_2\text{-2-}R^1\text{-pyridine)} + \text{(3,6-}R^2_2\text{-2-}R^1\text{-pyridine)} \quad (18)$$

R^1, R^2 = Me, Et, Ph etc.

5–10 mol% (based on the nitrile) of any readily available cobalt salt (hydrated or anhydrous) is added to the alkyne-nitrile mixture; where necessary in the presence of a solvent. The active species is then formed by adding a stochiometric amount of a reducing agent, e.g., a main group metal, a commercially available metal-hydride or a metal-alkyl, in order to remove the anions from the metal. The cobalt salt is converted into the homogeneous organocobalt catalyst (which is generally brown) and exothermic pyridine formation occurs immediately (Equation 18). Only those reducing agents which are spontaneously inflammable in air, e.g. alkylaluminium compounds, need to be handled in an inert atmosphere. The catalytic reaction itself does not have to be carried out in an inert atmosphere nor do the substrates have to be of high purity. Examples of the use of *in situ* catalysts are shown in Table 2 and two are discussed in more detail below. The $CoCl_2/Li$ system **1** is fairly active and has been used, among others, to synthesize a mixture of diphenyl, methyl-pyridine isomers: cobaltous chloride (10 mmoles) in 40 ml THF (see Section 10 for abbreviations) and 1 mole acetonitrile were treated with lithium (20 mmoles) and heated for

Table 2

IN SITU CATALYSTS

No.	CoX_n/Metal	No.	CoX_n/MH	No.	CoX_n/MR
1	$CoCl_2/Li$ [a]	7	$CoCl_2/LiH$	15	$CoCl_2/n\text{-}C_4H_9Li$
2	$CoCl_2/Na$ [b]	8	$CoCl_2/LiAlH_4$	16	$CoCl_2/C_2H_5{-}Mg{-}X$
3	$CoCl_2/Mg$ [c]	9	$CoCl_2/NaBH_4$	17	$CoCl_2/Al(C_2H_5)_3$
4	$Coacac_2/Mg$ [c]	10	$CoCl_2{*}6H_2O/NaBH_4$	18	$Coacac_3/Al(C_2H_5)_3$
5	$Coacac_3/Mg$ [c]	11	$Coacac_2/NaBH_4$	19	$Coacac_2/(C_2H_5)_2Al{-}OC_2H_5$
6	$Co(HCO_2)_2/Mg$ [c]	12	$Coacac_3/NaBH_4$		
		13	$Co(CH_3CO_2)_2/NaBH_4$		
		14	$Co(HCO_2)_2/NaBH_4$		

[a] Reduction is carried out in the presence of the nitrile, an alkyne is then added.
[b] Reduction is carried out in the presence of the alkyne and the nitrile is then added.
[c] Activated with iodine.

30 min at 60° before adding phenylacetylene (0.5 moles). It is our experience that the combination cobalt chloride hexahydrate with sodium borohydride is the most versatile *in situ* catalyst (Equation 19). A typical example of its use is in the preparation of 2,3,6- and 2,4,6-collidine from acetonitrile and propyne:

$$\boxed{CoCl_2 \cdot 6\,H_2O\,/\,2\,NaBH_4 + \text{Nitrile/Alkyne}} \tag{19}$$

2.4 g (10 mmoles) $CoCl_2 \cdot 6H_2O$ are dissolved in 82 g (2 moles) of acetonitrile. The resulting blue solution is cooled to –40° and 40 g (1 mole) of liquid propyne is added. 0.76 g (20 mmoles) of $NaBH_4$ is added to the solution which is then allowed to warm up to –20°. The colour changing through pink to brown. This solution, which contains the active catalyst, is sucked at –30° into an evacuated and precooled 500 ml stainless steel autoclave and heated over 1 h to 88°. The pressure rises to 12 bar. After 4 h, the autoclave is vented and the reaction mixture fractionally distilled.
Fraction 1: bp 45–85° (1013 mbar), 69.4 g
Fraction 2: bp 20–56° (20 mbar), 17.7 g
Residue: 5.2 g
A sample of fractions 1 and 2 was analyzed by GC (Instrument: Varian 1400; conditions: column: 100 m FFAP; detector: FID; temperature: 200°/110°/250°; carrier gas: He 2.2 bar)

Conversion (nitrile):	20.4%
yield (based on converted nitrile):	25.5% 2,4,6- and 2,3,6-collidine, (60 : 40)
TON:	10.4 moles pyridines per g atom Co.

In those cases where both substrates are liquids, the catalyst may be prepared at 20°. Mixing the reagents in accompanied by a colour change from blue through dark green to brown while the temperature rises to give a gentle reflux, indicating that an exothermic catalytic co-cyclization reaction is taking place. Thin layer chromatography (TLC) on silica gel with, for example, benzene as eluent proved to be a rapid and convenient tool for the qualitative analysis of the reaction mixture providing direct information without a lengthy work up.

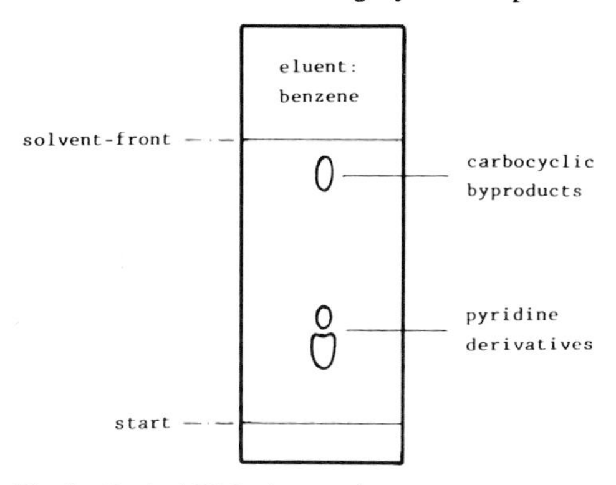

Fig. 2. Typical TLC of a reaction mixture.

Three spots are formed on the TLC-plate, the lower two generally correspond to the pyridine isomers (comparison with known products using 366 nm UV light) whereas the carbocyclic by-products are found near the eluent front. The spots may be removed, extracted with ether and subsequently analyzed by spectroscopic methods (MS and NMR).

2.2. η^3-Allyl-cobalt-olefin complexes (12e$^-$-type catalysts)

The η^3-allyl-cobalt(I) systems depicted in Figure 3 were among the first organocobalt complexes for which activity was established [19]. 5-Methyl-heptadienyl-

Fig. 3. η^3-Allylcobalt complexes: (a) Natta's complex, **20** (dec. –30°);
(b) Otsuka's complex, **21** (mp 60–65°, dec. 70°).

cobalt-butadiene (**20**, Table 3) was obtained as early as 1965 by G. Natta, P. Pino and coworkers [26a] from the reaction of cobalt(II) chloride with butadiene and sodium borohydride in ethanol at –30°. Since no details for the preparation exist in the literature we have included a procedure in Section 7.1.1.1. which allows Natta's complex to be prepared in yields as high as 70%. An X-ray structural determination [26b] shows that two butadiene molecules are linked together and that the resulting 7-membered chain is bonded to the cobalt through an η^3-allyl system and a terminal vinyl group. In an initial step the "single cis" butadiene molecule is removed from the cobalt. Natta's complex is rather labile at –30° but may be stabilized by dissolving it in the alkyne-nitrile substrate mixture. The η^3-allyl group in **20** can be replaced by other substituents such as the indenyl group giving the η^5-indenyl complex.

Cyclooctenyl-cobalt-cyclooctadiene complexes were first prepared by G. Wilke and Ch. Grard in 1967 [27a]. S. Otsuka and M. Rossi [27b] were able to prepare the pure η^3-cyclooctenyl-cobalt complex **21** by reacting $CoCl_2$ with cod and sodium metal in a THF/pyridine mixture and they subsequently established the structure [27c]. An electrochemical synthesis has also been developed [29]. In our laboratory **21** is now known as "Otsuka's complex" and a new one-pot procedure

Table 3

PRE-PREPARED η^3-ALLYLCOBALT(I) CATALYSTS

No.	Complex	Ref.
20 [a]	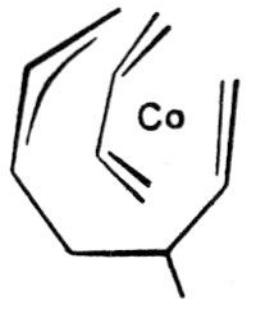 Co	[26]
21 [a]	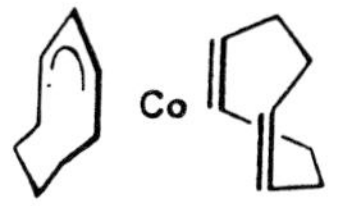 Co	[27], [29]
22	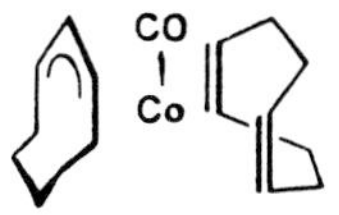 CO Co	[28]

[a] Preparative procedures are presented in Section 7.1.1.

for the production of pure **21** in 70% yield from $Coacac_3$, cod and triethylaluminium is reported in Section 7.1.1.2.

The coordinatively unsaturated 16-electron complex **21** readily adds a 2-electron-donor ligand. With CO, for example, [Co(CO) (η^4-cod) (η^3-cyclooctenyl)] **22** is formed [28]. However, the blocking of a coordination site at the cobalt is associated with a decrease in catalytic activity.

2.3. η^5-Cyclopentadienyl-cobalt complexes and related systems ($14e^-$-type catalysts)

The remarkable thermal and chemical stability of the Cp—Co moiety (14-electrons) [46] and the possibility of modifying the system by introducing substituents into the ring has led to considerable interest being shown in the catalytic properties of cobalt complexes containing a Cp-group. A large variety of olefins and diolefins stabilize the Cp—Co moiety and typical examples are shown in Table 4. The resulting [Co(η^2-olefin)$_2$(η^5-Cp)] and [Co(η^4-diolefin) (η^5-Cp)] complexes are 18-electron systems and are easy to handle. Since the olefin molecules are displaced by the substrate prior to the catalytic reaction, their only function is to stabilize the

Table 4

CATALYSTS CONTAINING CYCLOPENTADIENYL- AND RELATED GROUPS

No.	Complex	Ref.
23 [a]	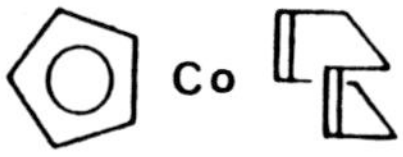 Co	[47], [50b]
24[a]	 CH_3- Co	[47c]
25 [a]	 CH_3, CH_3, CH_3 Co	[47c]
26	 Co	[29]
27 [a]	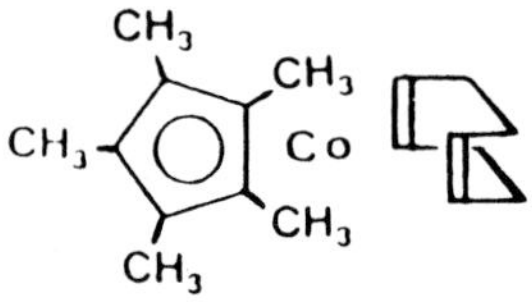 CH_3, CH_3, CH_3, CH_3, CH_3 Co	
28 [a]	 Co	[47c]
29	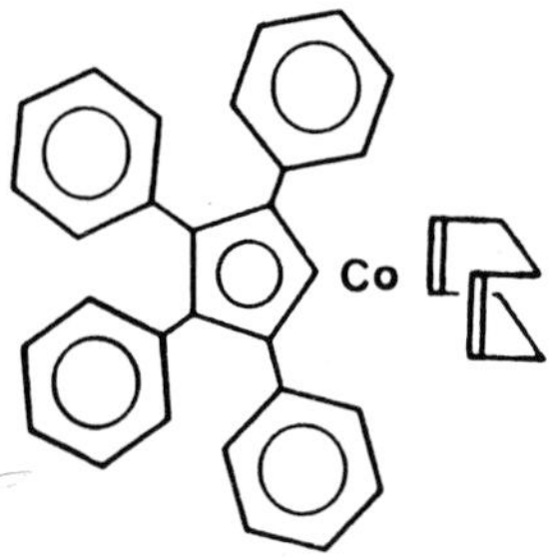 Co	[47c]
30	 O ‖ CH_3-C- Co	[56]

Table 4 (continued)

No.	Complex	Ref.
31 [a]	 $(CH_3)_3Si$ Co	[41], [47c]
32 [a]	 Co	[41], [47c]
33 [a]	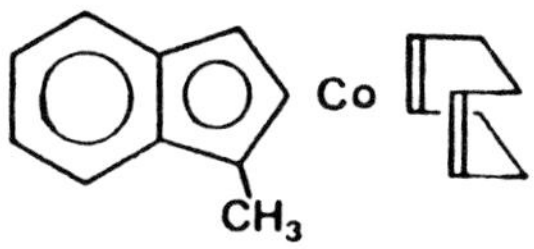 Co CH_3	[47c]
34 [a]	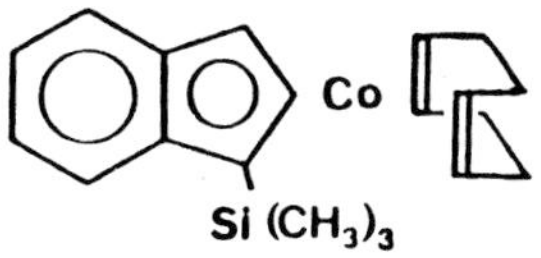 Co $Si(CH_3)_3$	[47c]
35	 CH_3CH_2–B Co	cf. [30]
36 [a]	 B Co	[30], [47c]
37	 Co	[31]
38	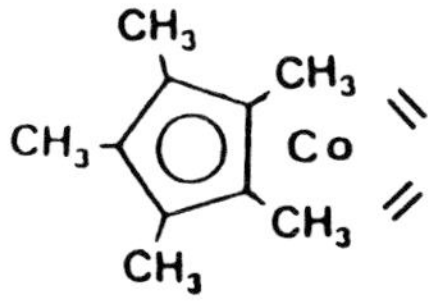 CH_3 CH_3 CH_3 Co CH_3 CH_3	[32]
39	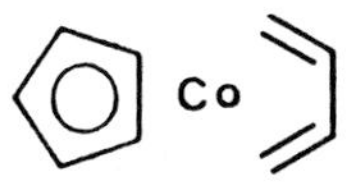 Co	[33]

Table 4 (continued)

No.	Complex	Ref.
40 [a]	Co	[47c], [57]
41	Co CH_3	[34]
42	Co CH_3 CH_3 CH_3	cf. [34]
43	Co	[35]
44	Co	[36]
45	Co CH_2CN	[37]
46	Co CF_3	[38]
47	Co CCl_3	[34]
48	Co $O-CH_3$	[36]
49	Co CH_3 CH_3 CH_3 CH_3 CH_3 CH_3	[39]
50	$Co(CO)_2$	[40]

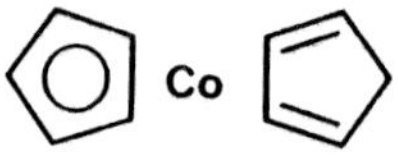

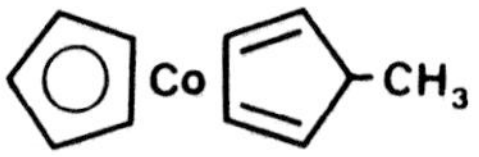

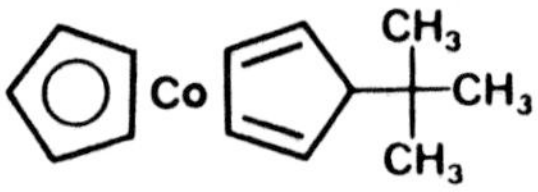

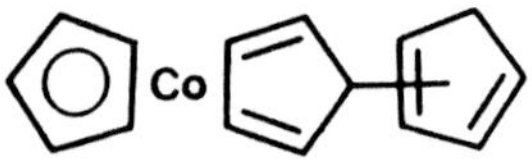

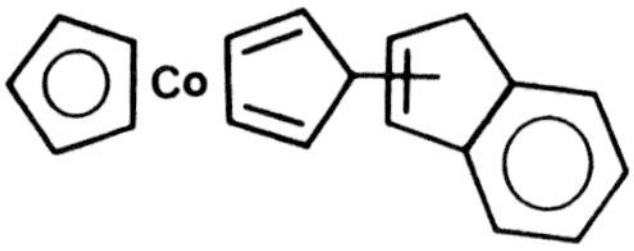

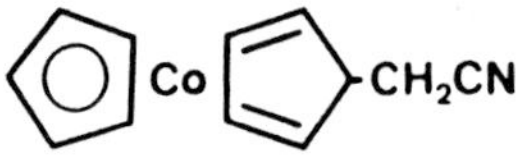

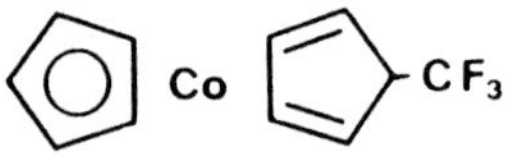

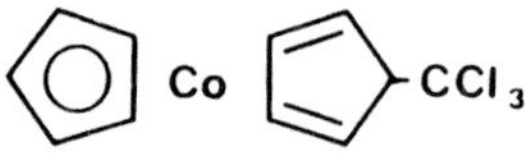

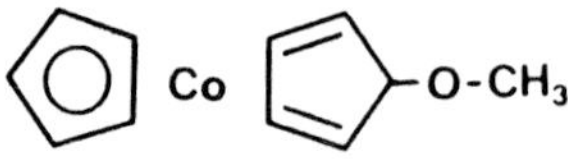

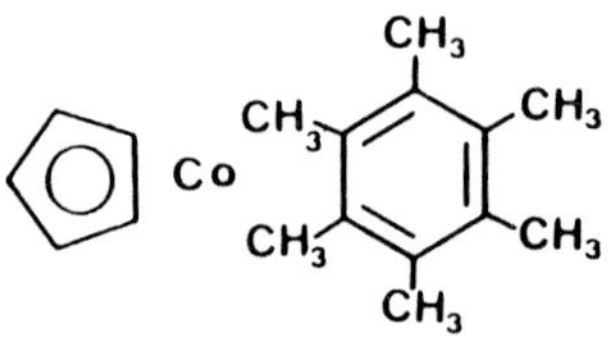

Table 4 (continued)

No.	Complex	Ref.
51	$(CH_3)_3Si$ Co(CO)$_2$	cf. [40]
52	Co N N	[41]
53 [a]	Co	[47c]
Cobaltocene		
54	Co	[42]
55	CH_3—C(=O)— Co —C(=O)—CH_3	cf. [42]
56	B Co B	[43]
Binuclear systems		
57	Co Co	[44]
58	Co Co CH_3 CH_3 CH_3 CH_3	[45]

[a] Preparative procedures are discussed in Section 7.1.2.

Cp—Co moiety. It should be mentioned that the presence of the η^4-cyclobutadiene system completely suppresses the catalytic activity. Cod is the diolefin of choice since it is both readily available and its chelating properties enable pure, crystalline cobalt complexes to be obtained in satisfactory yields (**23**–**36**, Table 4). The orange $[Co(\eta^4\text{-cod})(\eta^5\text{-Cp})]$ [47] complex **23** is not only thermally stable (mp 102–104°) but may also be exposed to air without noticeable decomposition and is thus a useful all-round catalyst of moderate activity. The corresponding bis-ethylene compound **37** is easily available from $[Co(\eta^5\text{-Cp})_2]$ by reaction with alkali metals [31]. This complex is less stable and eliminates ethylene above ambient temperature. As a result the coordination sites at the cobalt are more readily accessible for alkyne substrates and the initial dissociation step, which leads to the formation of the propagating catalytic species, generally occurs at lower temperatures e.g. Equation 20.

$$\left.\begin{array}{l} CpCo(cod) + 2\,HC{\equiv}CH \ (\text{in } C_2H_5CN) \xrightarrow{T:125^\circ} \\ CpCo(C_2H_4)_2 + 2\,HC{\equiv}CH \ (\text{in } C_2H_5CN) \xrightarrow{T:25^\circ} \end{array}\right\} \longrightarrow CpCo(HC{\equiv}CH)_2 + cod,\ 2\,C_2H_4 \qquad (20)$$

The most important catalysts listed in Table 4 are **23**–**34** and **50**–**53** and three main pathways (Method A, B, C) are now available for their preparation and these are compared in Section 7.2.

Method A: The "Carbonyl Method". In a first step dicobalt octacarbonyl is refluxed with cyclopentadiene in CH_2Cl_2. This exothermal reaction leads to formation of $[Co(CO)_2(\eta^5\text{-Cp})]$ **50** [47a], presumably through the intermediate formation of a hydride $[CoH(CO)_2(\eta\text{-Cp})]$ which reacts further with transfer of hydrogen to excess C_5H_6 to give cyclopentene and cyclopentane (Equation 21) [49]. Substituted

$$2\,C_5H_6 + [Co_2(CO)_8] \xrightarrow{60^\circ} 2\,CpCo(CO)_2\ (\mathbf{50}) + 4\,CO + [H_2] \qquad (21)$$

cyclopentadienyl derivatives, $[Co(CO)_2(\eta^5\text{-RCp})]$ [40], may be prepared similarly in high yield [48]. The second step, a ligand exchange reaction (Equation 22), is generally performed in refluxing xylene [47b], [50]. The evolution of CO may be

$$CpCo(CO)_2 + \text{cod} \xrightarrow{130^\circ} CpCo(cod)\ (\mathbf{23}) + 2\,CO \qquad (22)$$

accelerated by photolysis [51]. In this way it is possible to prepare the pentamethyl-cyclopentadienyl-cobalt-cod complex **27** from the dicarbonyl complex [41], [52] (Equation 23). Full experimental details are given in Section 7.1.2.4.

$$[(\eta^5\text{-}C_5(CH_3)_5)Co(CO)_2] + \text{cod} \xrightarrow[20^\circ]{h\nu} [(\eta^5\text{-}C_5(CH_3)_5)Co(\text{cod})]\ \mathbf{27} \quad (23)$$

Method B: Protolytic substitution of Natta's complex [41]. Reaction of Natta's complex **20** with cyclopentadiene derivatives containing active C—H bonds in pentane leads to formation of [Co(η^4-cod) (η-RCp)] complexes (Equation 24) (for

$$\mathbf{20} + \text{indene} \xrightarrow[\text{step 1}]{-20^\circ} [\ldots] \rightarrow [(\text{indenyl})Co(\ldots)] + \ldots \quad (24)$$

$$[(\text{indenyl})Co(\ldots)] \xrightarrow[\text{step 2}]{\text{cod}} [(\text{indenyl})Co(\text{cod})]\ \mathbf{32}$$

experimental details see Section 7.1.2.7.). In this way it proved possible to prepare the [Co(η^4-cod) (η^5-indenyl)] catalyst **32** [41]. The protolytic splitting of an enyl–cobalt system has precedence in the reaction reported by F. G. A. Stone [53] in which a cyclooctatrienyl–cobalt system was reacted with indene. [Co(η^4-cod) (η^5-indenyl)] **32** crystallizes from pentane as red-brown prisms. The bonding situation was established by C. Krüger and L.-K. Liu [54] using X-ray crystallography and compared with that in [Co(η^4-cod) (η^5-Cp)] (Figure 4).

As can be seen, the 5-membered ring of the indenyl group is η^5-bonded to the cobalt atom as in the parent complex **23**. As a consequence, the π-bonds in the benzene ring are localized. The average Co—C distance in the indenyl molecule is 2.120(59) Å, i.e., slightly longer than in **23** (Co—C = 2.096(19) Å). The C—C-bonds in **32** fall into two groups: C7—C8, C8—C9, (av. 1.408(10) Å) and C9—C1, C1—C6, C6—C7 (av. 1.435(10) Å), indicating that charge accumulates in the C7, C8, C9 fragment which would lead to a stronger Co—C bonding interaction than in the parent complex **23**. We have observed that an increase in π-bonding as the result of the introduction of substituents into the η^5-Cp moiety is usually associated with enhanced catalytic activity (see Section 2.4.).

Cod, in its role as a bidentate π-ligand, was found in **32** to have mean Co—C distances of 2.026(08) Å and mean C=C lengths of 1.390(10) Å. Comparison with **23**

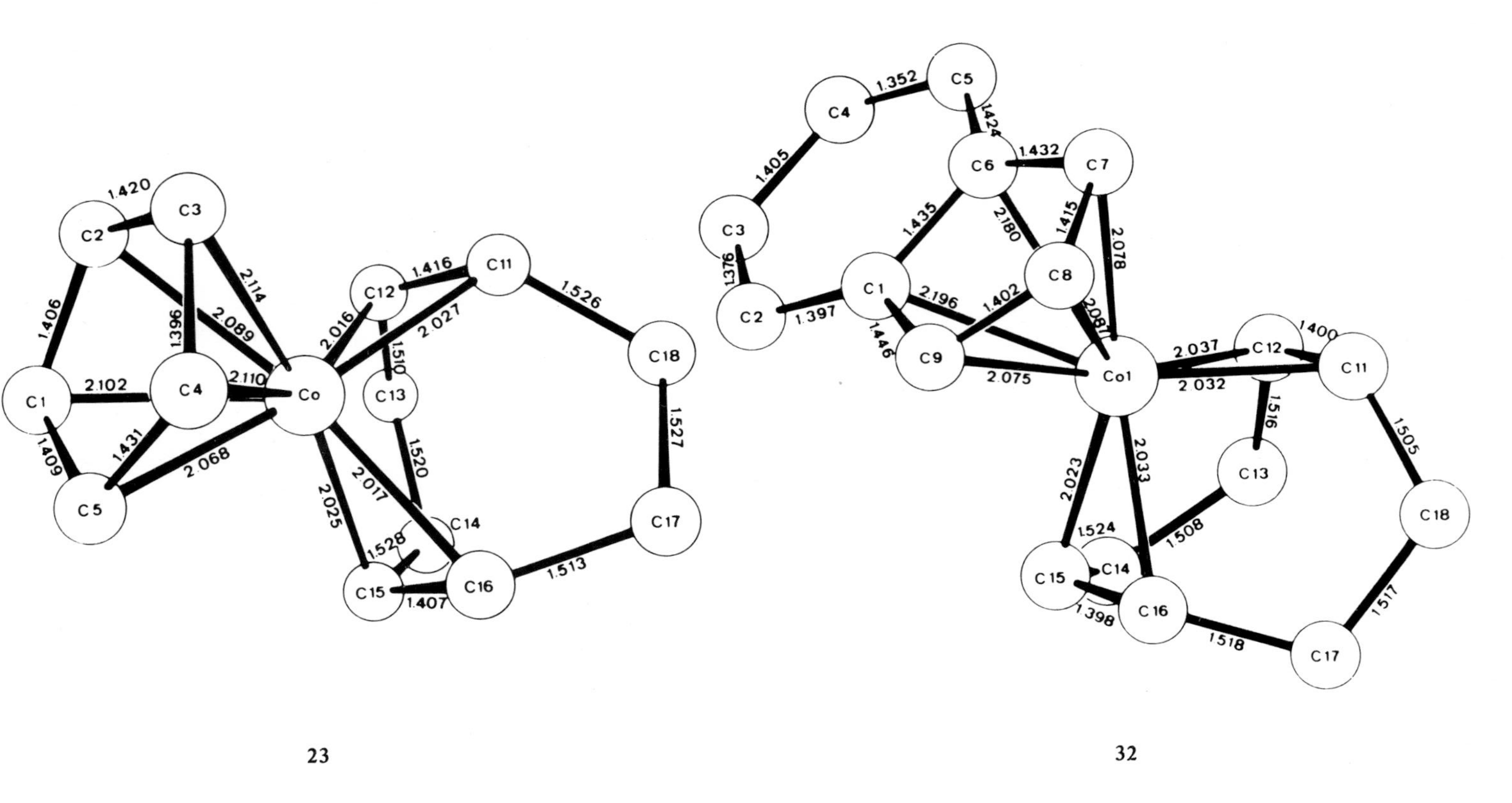

Fig. 4. The crystal structure of Co(η^4-cod) (η-Cp) **23** and Co(η^4-cod) (η-indenyl) **32** [54].

indicates that the Co–C distance is elongated which, together with the short C=C distance, suggests weaker complexation of the cod moiety to Co. This is in agreement with the fact that the initial olefin-dissociation step in the reaction with propyne and propionitrile occurs at 125° for **23** and 105° for **32**.

Method C: One-pot reduction of Co-salts in the presence of olefins. In 1979 our efforts to provide a simple, large-scale access to $[Co(\eta^4\text{-diolefin})(\eta^5\text{-Cp})]$ compounds led to the development of a one-step procedure for converting any organic or inorganic cobalt salt into the desired complex: the various anions are removed from the cobalt by the addition of a reducing agent in the presence of the corresponding RC_5H_5 compounds and the stabilizing diolefin (Equation 25). Preferred reducing agents are an aluminium trialkyl (Variant C1) a Group II metal (Variant C2) or an alkali metal (Variant C3). For a comparative evaluation, see Section 7.

$$CoX_{2/3} + RCpH + \text{diolefin} + \text{reducing agent} \longrightarrow [Co(\eta^4\text{-diolefin})(\eta^5\text{-RCp})] \qquad (25)$$

All three pathways are associated with a certain experimental know-how in order to achieve maximum yield and purity of the desired complex. Of particular interest is a new catalytic activation of metallic magnesium using anthracene (Variant C2), which we have developed together with B. Bogdanović [47c], [116]. η^5-Cyclopentadienyl-cobalt-bisethylene **37**, is readily accessible from cobaltocene using a method developed by K. Jonas (Equation 26) [31]. This complex is, however, rather difficult to handle and has been largely neglected as a catalyst.

$$[Co(\eta^5\text{-Cp})_2] + K + 2C_2H_4 \xrightarrow[-20\text{ to }0°]{Et_2O,\ 85\%} \underset{\mathbf{37}}{[Co(\eta^2\text{-}C_2H_4)_2(\eta^5\text{-Cp})]} + KCp \qquad (26)$$

The $[Co(\eta^4\text{-cod})(\eta^6\text{-phenylborinato})]$ Complex **36**, prepared by G. E. Herberich [30] by elimination of an η^6-ligand from **56** in a manner similar to that shown in Equation 26, turned out to have unique catalytic activity for the reaction of acrylonitrile with acetylene and in the synthesis of pyridine itself. An alternative one-step synthesis of **36** using the Variant C2 is shown in Equation 27.

B–Ph / Co / B–Ph (**56**) $\xrightarrow{Li}$ [B–Ph / Co / B–Ph] Li $\xrightarrow{cod}$ B–Ph / Co / cod (**36**) (27)

$Coacac_3$ + H,H-C$_5$H$_4$B–Ph + Mg + cod ⟶ **36**

Cobaltocene **54**, which is itself active [21a], [25], [55], is also a versatile starting material for the preparation of a number of derivatives. For example, the complexes **40–48** which contain the η^4-cyclopentadiene group can be prepared either by nucleophilic addition to $[Co(\eta^5\text{-Cp})_2]^+$ or by reactions of $[Co(\eta^5\text{-Cp})_2]$ with organic molecules containing active C—H bonds, e.g. $CHCl_3$, in the presence of oxygen. A variation of the latter pathway [37] is the exposure of cobaltocene to air in the presence of excess cyclopentadiene (Figure 5) [35], [36]. The resulting

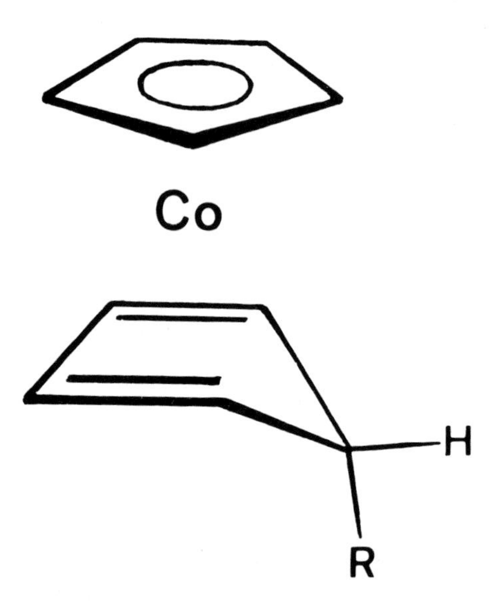

Fig. 5. The structure of $Co(\eta^4\text{-}C_2H_5R)\ (\eta^5\text{-Cp})$; R = cyclopentadienyl, indenyl.

η^4-cyclopentadiene complexes react already at lower temperatures in the pyridine synthesis than the parent sandwich compound: **44** becomes active for the 2-ethylpyridine synthesis at 90° whereas cobaltocene is not active below 185°.

An exotic, but effective catalyst, is the 20-electron system **49** which has been prepared by K. Jonas and D. Habermann [39] (Equation 28). The borabenzene ligand, a η^6-bonded five-electron donor, may be derived from cobaltocene by stepwise

$$\underset{\underline{\underline{37}}}{CpCo(C_2H_4)_2} + 3\,CH_3C{\equiv}CCH_3 \xrightarrow[\text{-10°, 46\%}]{\text{5 days, pentane}} \underset{\underline{\underline{49}}}{CpCo(C_6(CH_3)_6)} \qquad (28)$$

ring expansion (Equation 29) [43]. The voltammetric redox potentials for **56** were shown to be more anodic than for cobaltocene indicating that the borabenzene

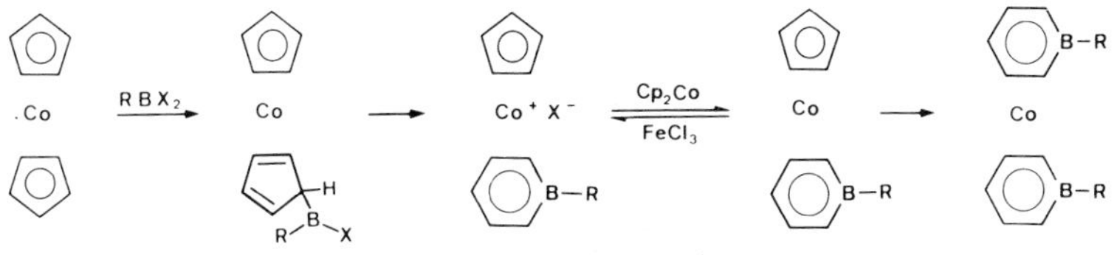

(29)

$R = C_2H_5, C_6H_5$

ligand attracts electrons more strongly than the η^5-Cp moiety. The binuclear system **57** is obtained photochemically from reaction of α-photopyrone with $[Co(CO)_2(\eta^5\text{-}Cp)]$ and may be purified by chromatography over alumina [44]. The analogous complex **58** may also be obtained by reacting $[CoI_2(\eta^5\text{-}Cp)]_2$ with but-2-yne [45].

2.4. The effect of the catalyst on the extent of alkyne-nitrile conversion

The influence of the organic group in the R′Co complexes employed as the catalysts makes itself apparent in the *chemoselectivity* of the reaction. In general catalysts containing a 14-electron core favour pyridine formation whereas those containing a 12-electron core favour carbocyclic formation. For example, in the reaction of propyne and propionitrile at 140° in the presence of Otsuka's complex **21** (a 12-electron system) a 2 : 1 ratio of benzene to pyridine derivatives is formed whereas, under identical conditions, a catalyst containing the Cp—Co moiety (a 14-electron system) leads to the predominant formation of pyridine derivatives (Table 5). In order to obtain some insight into the role of the organic ligands bonded to cobalt upon the catalytic behaviour, we have determined the temperature at which the catalytic reaction of acetylene with propionitrile (Equation 30) begins (using constant molar concentrations of cobalt and substrates) [36].

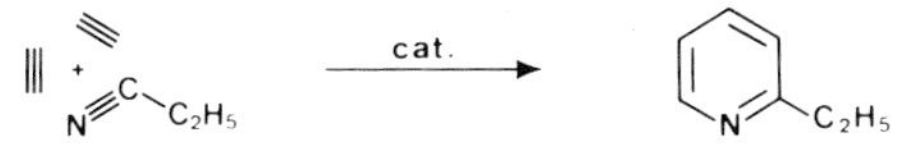

(30)

Inspection of the data shown in Table 6 shows that the dissociation of the various ligands from the Cp—Co core controls the start of the catalysis. This is clearly seen by comparing the reactivity of the bisethylene complex **37** with its cod analogue **23**: a 100° higher temperature is needed for the latter to induce activity. The bis-carbonyl complex **50** is even less active and a temperature of 180° is needed. It is our experience that phosphanes, arsanes, isocyanides as well as the presence of

Table 5

CHEMOSELECTIVITY OF THE η^3-ALLYLCOBALT (12e) AND η^5-CYCLOPENTADIENYLCOBALT (14e) CORES

Catalyst	methyl-substituted 2-methyl-6-ethylpyridine (CH_3, CH_3, N, C_2H_5)	:	trimethylbenzene (CH_3, CH_3, CH_3)
Co **21**	1	:	2
Co **23**	1.8	:	1

$2CH_3C{\equiv}CH + C_2H_5C{\equiv}N$; 140°; 82 min.; cont. flow apparatus.

excess carbon monoxide in the solution suppress the catalytic reaction – presumably as the result of competition with the alkyne molecules for the coordination sites at the cobalt atom. η^4-Cyclobutadiene derivatives are not active since the organic group cannot be displaced from the cobalt atom by the substrates even at temperatures in excess of 200°.

Comparison of the different complexes within the cod-series, i.e., **23**, **32**, **24**, to **27** (Table 6), clearly indicates that substituents on the Cp-ring may also influence the initial equilibrium (Equation 20) which allows entrance into the catalytic cycle. As mentioned above, the reluctance of cobaltocene to catalyze the pyridine formation can easily be overcome by transformation of the Cp-ring into the η^4-cyclopentadiene derivates **43–48** and is accompanied by a decrease in the starting-temperature of the catalytic reaction by ca. 100° [35], [36].

The marked effect of substituents in the RC_5H_4-ring upon the catalytic *activity* first became apparent during experiments in which the cod-complexes **23–36** were tested as catalysts for the co-cyclization of propyne with propionitrile (Equation 31).

$$2\,CH_3C{\equiv}CH + N{\equiv}C{-}C_2H_5 \xrightarrow{\text{cat.}} CH_3(CH_3)C_5H_2N(C_2H_5) \qquad (31)$$

Table 6

THE TEMPERATURE DEPENDENCE OF THE REACTION OF HC≡CH WITH $C_2H_5C{\equiv}N$ UPON THE CATALYST (EQUATION 30)

No.	Catalyst	Temperature
37	 Co	25°
44	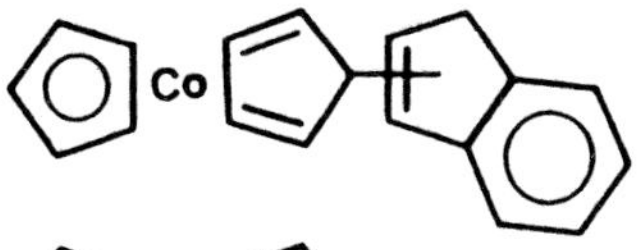 Co	90°
45	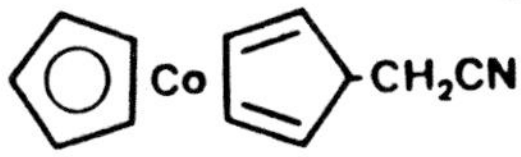 Co · CH_2CN	95°
43	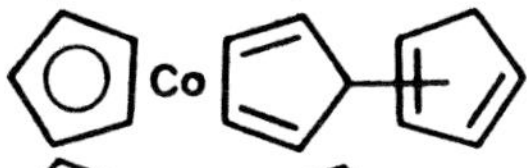 Co	95°
48	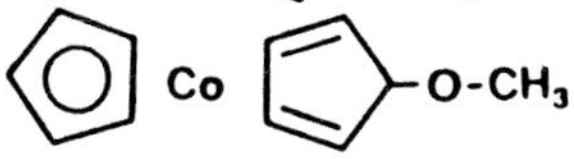 Co · $O-CH_3$	115°
23	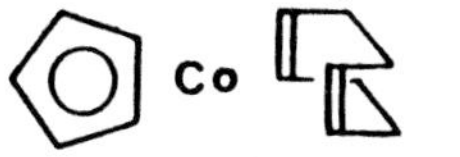 Co	125°
31	 $(CH_3)_3Si$ · Co	136°
24	 CH_3 · Co	125°
25	 CH_3, CH_3, CH_3 · Co	143°
26	 Co	166°
50	 $Co(CO)_2$	180°
27	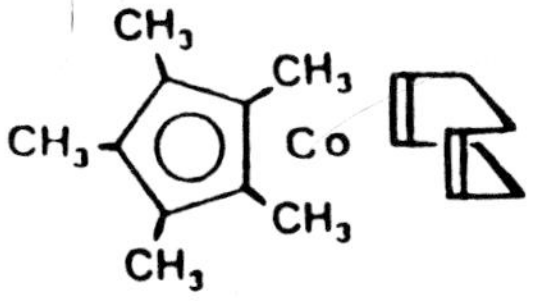 CH_3, CH_3, CH_3, CH_3, CH_3 · Co	185°
54	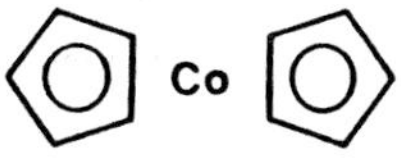 Co	185°

In order to obtain comparable results for the different catalysts, we have run the reaction shown in Equation 31 in a continuous-flow reactor under standard conditions. The reaction temperature which the various catalysts need to reach a 65% propyne-conversion may be regarded as a measure of the catalyst activity. Table 7 summarizes some of our results.

Table 7

CATALYST ACTIVITY IN THE REACTION OF $CH_3C{\equiv}CH$ AND $C_2H_5C{\equiv}N$ (EQUATION 31)

No.	Complex	Required temperature for 65% propyne-conversion
26	 Co	180°
24	 CH_3 Co	162°
31	 $(CH_3)_3Si$ Co	144°
23	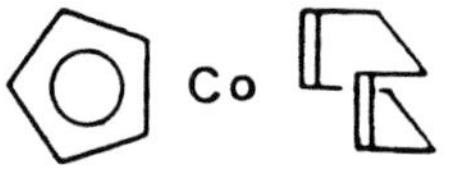 Co	147°
32	 Co	126°

From these data it is evident that the catalytic activity is greatly enhanced if **32** is used instead of **26**. Two isomeric pyridine derivatives are formed; a symmetrically 2,4,6-substituted isomer and the asymmetric 2,3,6-isomer. The symmetrical isomer predominates (Figure 6).

These isomers result from the head-tail or tail-tail coupling at the cobalt. Formal insertion of the nitrile occurs preferentially into the non-substituted Co—C bond to give a system containing a Co—N bond. As can be seen from the data given in Figure 6, the proportion of the isomers formed is dependent upon the catalyst and can vary considerably. A systematic examination of numerous organocobalt systems indicates that the *regioselectivity* of the process can be controlled within certain limits by varying the organic group associated with the cobalt.

Table 8, in which the different cobalt systems are listed in the order of

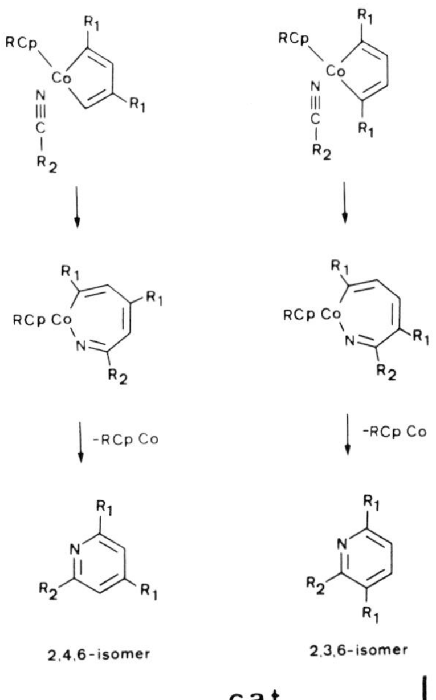

	cat.	2,4,6- : 2,3,6-ratio
26	Co	2.50
24	CH_3- Co	2.02
31	$(CH_3)_3Si$- Co	1.67
23	Co	1.71
32	Co	1.48

Fig. 6. Regioselectivity of collidine formation ($R^1 = CH_3$, $R^2 = C_2H_5$) and its dependence on the catalyst.

Table 8

EFFICIENCY OF VARIOUS TYPES OF CATALYSTS IN THE REACTION OF HC≡CH AND $C_2H_5C{\equiv}N$ (TIME: 120 MIN.) (EQUATION 30)

No.	Catalyst (0.13 mmole complex in 2.1 mole C_2H_5CN)	Pressure at 25°C (bar)	Maximum pressure (bar)	Temp.	Conversion (nitrile) (%)	Yield (%)	Turn-over number (mole product/ g-atom Co)
1	$CoCl_2$/Lithium 11 mmoles/23 mmoles '*in situ*'	16	31.5	167°	37.2	33.1	32
27	CH_3, CH_3, CH_3, CH_3, CH_3 Co	10	50.0	185°	1.9	15.4	49
21	Co	15	43.5	135°	5.7	27.4	268
28	Co	14	42.0	154°	9.2	86.7	1300
30	CH_3-C(=O)- Co	11	27.3	141°	8.4	100.0	1423
26	Co	10	38.5	176°	13.6	67.8	1527

Table 8 (continued)

36	 B Co	16	40.5	122°	8.8	91.9	1604
50	 $Co(CO)_2$	14	56.5	196°	13.3	75.1	1646
24	 CH_3– Co	15	49.0	176°	18.6	64.9	1732
55	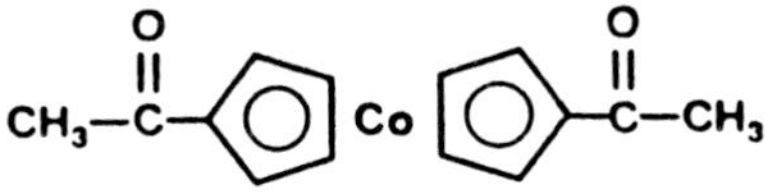 CH_3–C(=O)– Co –C(=O)–CH_3	10	40.5	204°	11.6	86.2	1810
25	 CH_3, CH_3, CH_3 Co	10	38.5	202°	13.7	83.6	1887
54	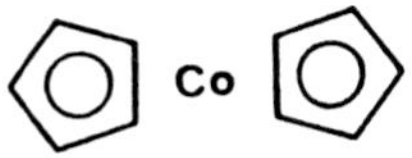 Co	15	46.0	185°	16.3	73.4	2024
32	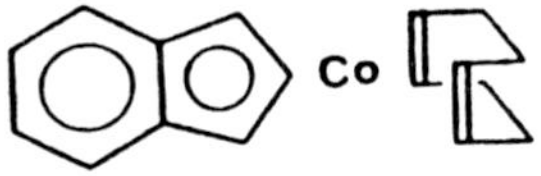 Co	18	36.5	136°	26.0	83.3	2431
23	 Co	10	28.0	171°	20.3	87.3	2908
34	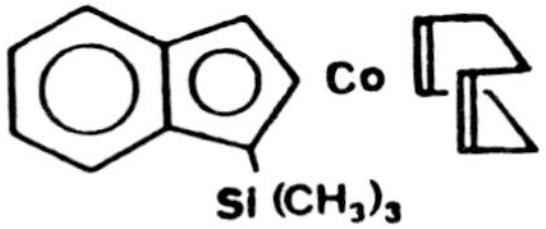 Co, $Si(CH_3)_3$	11	20.7	136°	19.4	94.6	3112
31	 $(CH_3)_3Si$– Co	15	43.0	201°	23.0	92.8	3499

increasing turn-over number, clearly indicates that the nature of the catalyst plays a decisive role in obtaining a maximum TON value.

Although the concentration of cobalt complex dissolved in the propionitrile and the amount of acetylene present at 25° were kept virtually constant, the highest TON achieved varies over a range of 50 to 3500. Certain [Co(η^4-diolefin) (η^5-Cp)] complexes were found to be particularly effective with respect to both the yield of 2-ethylpyridine and to the degree of conversion. Comparison of the results achieved with a series of different RCp—Co(I) systems indicates once again the key role of the substituent. The inhibitory effect of carbon monoxide and the disadvantages associated with cobaltocene, which acts solely as a precursor of the active Cp—Co moiety, are also apparent. The lack of chemoselectivity associated with the use of Otsuka's complex **21** mirrors itself in the extremely low yield of 2-ethylpyridine. In practice the efficiency of a catalyst system may be monitored by following the acetylene pressure during the reaction: active catalysts rapidly consume acetylene from the solution and as a result only moderate pressures are reached during the reaction whereas less active systems such as [Co(CO)$_2$(η^5-Cp)] leave much of the alkyne unreacted and as a result the acetylene pressure increases considerably. For this reason catalysts of unknown reactivity or whose reactivity cannot be estimated, can be conveniently tested by a reaction involving acetylene (safety-precautions discussed in Section 4.1.). It should be emphasized that the catalyst efficiencies summarized in Table 8 refer only to the co-cyclization of propionitrile with acetylene (Equation 30) and are not general for other alkyne-nitrile combinations. In the course of our work it became apparent that for each individual reaction a search must be made for the best catalyst since the nature of the organic substituents at the cobalt can cause quite dramatic effects upon *substrate specificity*. This is illustrated in Table 9 for the reaction of acetonitrile and acrylonitrile with acetylene.

Table 9

MAXIMUM TON IN THE REACTION OF $CH_3C{\equiv}N$ AND $CH_2{=}CH{-}C{\equiv}N$ WITH $HC{\equiv}CH$

No.	Catalyst	2-methylpyridine (N, CH_3)	2-vinylpyridine (N, $CH{=}CH_2$)
31	$(CH_3)_3Si$–Cp Co (cyclooctadiene)	3749	219
23	Cp Co (cyclooctadiene)	2166	440
32	indenyl Co (cyclooctadiene)	1871	568

As can be seen, the trimethylsilyl-modified Cp—Co catalyst **31** produces about 4000 moles of 2-methylpyridine per g-atom of cobalt whereas the same catalyst exhibits a turnover number of only 200 for the synthesis of 2-vinylpyridine. In contrast the indenyl-cobalt system **32** is preferred in the 2-vinylpyridine synthesis. This is an indication of the potential for optimizing a reaction by choice of a suitable organyl group.

3. THE APPLICATION OF THE COBALT-CATALYZED PYRIDINE SYNTHESIS

3.1. The prior state of the art

The cobalt-catalyzed reaction is a convenient and versatile route for the conversion of acetylene and cyano-compounds into 2-substituted pyridines. The industrial importance of pyridine and its derivatives has increased steadily since 1950. The U.S. production of pyridine and its alkylated homologues climbed from 3700 tons (40% of which being the parent compound) in 1958 to 14 000 tons in 1978. In 1980 the total worldwide figure exceeded 40 000 tons with a production capacity near 60 000 tons/a. An annual growth rate of about 4% has been predicted [58].

The traditional source of the pyridine bases in the early fifties was coal tar which contains 0.1%of a low boiling fraction consisting of 2-pyridine, bp 113.5–115.5°, as well as 2-methylpyridine (α-picoline), bp 128–129°, and a higher boiling cut, bp 143–146° from which 3-methylpyridine (β-picoline), 4-methylpyridine and 2,6-dimethylpyridine (2,6-lutidine) may be isolated. In addition up to 1 kg of pyridine bases may be recovered from the gas liberated during the production of 10 t of coke. Nowadays, coal tar and coal gas is the source of only ca. 2400 tons/a. (i.e., about 4% of the worldwide production) of pyridine and its derivatives while synthetic processes, account for some 57 700 tons [58].

The numerous synthetic pathways now available in the laboratory have been collected and comprehensively reviewed by R. A. Abramovitch [59]. The most important reactions for the industrial production of the pyridine bases rely on the reaction of aldehydes or ketones with ammonia in the vapour phase. This reaction is catalyzed by heterogeneous alumina or silica-alumina contacts which may be doped with a di- or trivalent metal oxide, halide, or phosphate [60], [61]. The condensation processes used commercially are not ideal and suffer from unsatisfactory selectivity. For example, the condensation of acetaldehyde with ammonia typically yields equal proportions of 2- and 4-methylpyridine. The addition of formaldehyde, which probably instigates an *in situ* formation of acrolein, results in the production of pyridine and 3-methylpyridine in a 3 : 2 ratio. (H_2-evolution is not observed) (Equation 32). The gas-phase co-condensation of acrolein with ammonia to give 3-methylpyridine (Equation 33) was developed by H. Beschke at Degussa (FRG) [62]. Under the reaction conditions water, formed as a by-product,

$$CH_3CHO + NH_3 \xrightarrow[\text{cat.}]{350-550^\circ} \text{(N, } CH_3\text{)} + \text{(}CH_3\text{, N)} + H_2O\ [+ H_2]$$

$$CH_3CHO/CH_2O + NH_3 \xrightarrow[\text{cat.}]{350-550^\circ} \text{(N)} + \text{(}CH_3\text{, N)} + H_2O\ [+ H_2] \qquad (32)$$

$$HC(=CH_2)-HC=O + CH(=CH_2)-CH=O + NH_3 \longrightarrow \text{(}CH_3\text{, N)} + 2\,H_2O \qquad (33)$$

causes the intermediate generation of formaldehyde (Equation 34) which reacts to give pyridine (Equation 35).

$$CH_2{=}CH{-}CH{=}O + H_2O \longrightarrow HCH{=}O + CH_3{-}CH{=}O \qquad (34)$$

$$HC(=CH_2)-HC=O + CH_3-CH=O + NH_3 \longrightarrow \text{(N)} + 2\,H_2O\ [+ H_2] \qquad (35)$$

The Rütgerswerke (FRG) have developed [63] a fairly selective (72%) pyridine synthesis by reacting crotonaldehyde, formaldehyde and ammonia in the presence of a carbon-covered alumosilicate catalyst (Equation 36).

$$HC(=CH-CH_3)-HC=O + CH_2=O + NH_3 \xrightarrow{72\,\%} \text{(N)} + 2\,H_2O\ [+ H_2] \qquad (36)$$

Nippon steel has developed an interesting liquid-phase process for the production of 2-methylpyridine and 2-methyl-5-ethyl-pyridine from ethylene and ammonia. The starting materials are reacted at 100–130° and 30–100 bar using a catalyst reminiscent of that used in the well known Wacker process viz Pd^{2+}/Cu^{2+}-redox system (Equation 37). 1.5 tons of ethylene and 0.6 tons of ammonia react to give

$$3\,C_2H_4 + 4\,Pd(NH_3)_4^{++} \longrightarrow \text{(N, } CH_3\text{)} + 4\,Pd + 8\,NH_4^+ + 7\,NH_3$$

$$(37)$$

$$4\,C_2H_4 + 4\,Pd(NH_3)_4^{++} \longrightarrow \text{(}C_2H_5\text{, N, } CH_3\text{)} + 4\,Pd + 8\,NH_4^+ + 7\,NH_3$$

1 ton of the product-mixture in which 2-methylpyridine predominates (4 : 1). Metallic palladium is precipitated in the first step and may be reoxidized by the copper(II) salt present in the solution (Equation 38).

$$4\,Pd + 8\,Cu(NH_3)_4^{++} \longrightarrow 4\,Pd(NH_3)_4^{++} + 8\,Cu(NH_3)_2^{+} \quad (38)$$

The copper(I) has to be reoxidized by air during which process the ammonia concentration is adjusted to allow for its consumption (Equation 37) and to complete the catalytic cycle (Equation 39).

$$8\,Cu(NH_3)_2^{+} + 8\,NH_4^{+} + 8\,NH_3 + 2\,O_2 \longrightarrow 8\,Cu(NH_3)_4^{++} + 4\,H_2O \quad (39)$$

A two-step process for the production of 2-methylpyridine has been commercialized by the DSM in which acrylonitrile is reacted with a large excess of acetone [65]. Initially a monocyanoethylation product is formed in the liquid phase in a process catalyzed by a primary amine and a weak acid and which occurs at 180° and 20.7 bar. The ring closure to 2-methylpyridine is catalyzed by a Pd-containing contact and is conducted in the vapour phase (Equation 40).

$$CH_3COCH_3 + CH_2{=}CHCN \longrightarrow CH_3COCH_2CH_2CH_2CN \longrightarrow \text{2-methylpyridine} \quad (40)$$

3.2. The industrial application of pyridine and some 2-substituted derivatives

Pyridine and alkyl pyridines are manufactured in the United States by, for example, Reilly Tar and Chemical Corp., Indianapolis, Nepera Chemical Co. Inc., Harriman N. Y., in Europe by Reilly at Tertre, Belgium, by Lonza at Visp, Switzerland, by DSM (Stamicarbon B. V.) at Geleen in the Netherlands and in Japan by Koei Chemical Co. Ltd. and Daicel Ltd. The production figures for *pyridine* were estimated in 1980 to lie between 5500 and 7500 tons for the U.S.A., at around 8000 tons for Western Europe and at ca. 3000 tons for Japan (60% for export) [58]. A total of 12 000–13 000 tons of pyridine is used worldwide for the production of the dipyridylium herbicides "diquat" (1,1′-ethylene-2,2′-dipyridylium bromide) (Equation 42) and "paraquat" (1,1′-dimethyl-4,4′-dipyridylium chloride) (Equation 41) using processes developed by ICI [66].

$$2\ \text{pyridine} \xrightarrow{Na} \text{4,4'-bipyridine} \longrightarrow CH_3\text{–}\overset{+}{N}\text{(4,4'-bipyridylium)}\overset{+}{N}\text{–}CH_3\ \ 2\,Cl^- \quad (41)$$

$$2\ \text{pyridine} \xrightarrow{Ni} \text{2,2'-bipyridine} \longrightarrow \text{1,1'-ethylene-2,2'-dipyridylium}\ \ 2\,Br^- \quad (42)$$

Reaction of pyridine with metallic sodium followed by air oxidation and hydrolysis yields 4,4′-dipyridyl which is transformed by addition of a methylating agent into the powerful contact herbicide and plant dessicant paraquat. 2,2′-Dipyridyl reacts with ethylene bromide to give diquat. These are used widely as herbicides: diquat being particularly effective for broad leaf plants and paraquat for grasses [67]. The production of one ton of the herbicide requires 1.2 tons of pyridine. 15% of the U.S. consumption is used as the starting material for the production of stearamide methylpyridinium chloride – an important waterproofing agent for textiles – which is prepared by heating stearamide, paraformaldehyde and pyridine hydrochloride in pyridine solution. Another outlet for pyridine is in acid-binding solvents. 1000 tons/a. of pyridine are hydrogenated to give piperidine which is used in the production of vulcanization accelerators and as a hardening agent for epoxy resins. A small amount of pyridine is used as the raw material for some pharmaceuticals (antihistamines and anti-infectives). The total 1980 consumption of *2-methylpyridine* (α-picoline) has been estimated at ca. 12 000 tons [58]. 50% is produced for the U.S. market whereas the demand in both Western Europe and Japan lies between 2000 and 2500 tons/a. A significant outlet for 2-methylpyridine is in the production of 2-chloro-6-(trichloromethyl) pyridine which is used as a nitrification inhibitor in agricultural chemistry and in the manufacture of 4-amino-3,5,6-trichloro-picolinic acid for the use as a defoliant. The major commercial outlet for 2-methylpyridine is its use as the starting material for the production of 2-vinylpyridine (Equation 43). The total yield of 2-vinylpyridine formed from

$$\text{2-}CH_3\text{-pyridine} \xrightarrow[-H_2O]{+O=CH_2} \text{2-}(CH=CH_2)\text{-pyridine} \qquad (43)$$

2-methylpyridine can be as high as 90%. 2-Vinylpyridine may also be obtained in almost quantitative yields by heating 2-alkylamino-pyridine derivatives (which are directly available by cobalt catalysis, see Table 10 below) with a supported (e.g. Al_2O_3) alkali metal hydroxide (Equation 44) [69].

$$\text{2-}(CH_2-CH_2-NRR^1)\text{-pyridine} \xrightarrow[100-500^\circ]{98.8\%} \text{2-}(CH=CH_2)\text{-pyridine} + HNRR^1 \qquad (44)$$

$R = R^1$ = alkyl, cycloalkyl, etc., RR^1N = heterocycle

The most important outlet for 2-vinylpyridine is in the manufacture of copolymers for use as tire-cord binders. The tire-cord is first treated with a resorcinol formaldehyde polymer and then with a terpolymer made from 15% 2-vinylpyridine, styrene and butadiene. This treatment affords the close bonding of tire-cord to rubber so necessary in the production of tires [68]. As a result the market for automobiles dictates the production. 2-vinylpyridine is also used as an additive in the drying of acrylic fibers: 1–5% of added co-polymerized 2-vinylpyridine serves as the reactive site for the dye.

The hydrochlorides and methiodides of a number of 2-alkylpyridine (Equation 45) have been examined for their effect on aqueous surface tension and for their

$$\text{2-R-pyridine} \xrightarrow{CH_3J} \text{N-}CH_3\text{-2-R-pyridinium } J^- \qquad R = C_nH_{2n+1} \tag{45}$$

antibacterial properties. Maximum activity in both cases is found with the salts of 2-pentadecyl pyridine [70]. 2-Alkylpyridines having C_{10}–C_{18}-alkyl chains have attracted some industrial interest as starting materials for further derivation. Pyridine may be converted into 2-aminopyridine using the so-called "Tchitchibabin reaction" in which it is reacted with sodium amide in dimethylaniline (Equation 46). The

$$\text{pyridine} + NaNH_2 \xrightarrow[\substack{8\text{-}10h \\ NaOH/H_2O}]{105\text{-}110^\circ} \text{2-}NH_2\text{-pyridine} + 1/2\ H_2 \tag{46}$$

product is obtained in an 85% yield by treating with aqueous NaOH followed by distillation [71]. 2-Aminopyridine is used in the manufacture of several chemotherapeutics, dyes for acrylic fibers, and as an additive for lubricants [72].

Direct chlorination of pyridine in the vapour phase at temperatures above 300° yields 2-chloropyridine [73] (Equation 47). 2-chloropyridine-*N*-oxide reacts

$$\text{pyridine} \xrightarrow[>300^\circ]{Cl_2} \text{2-Cl-pyridine} \tag{47}$$

with sodium hydrogensulphide (Equation 48) to give pyrithione which, in the form of its zinc salt, is added to hair cosmetics as a general antifungal agent [74].

$$\text{2-Cl-pyridine-}N^+\text{-}O^- \longrightarrow \text{2-SH-pyridine-}N^+\text{-}O^- \rightleftharpoons \text{N-OH-pyridine-2-thione} \tag{48}$$

3.3. Selected examples of the cobalt-catalyzed pyridine synthesis

The strength of the cobalt-catalyzed pyridine synthesis lies in the fact that it provides the only known one-step process for the selective preparation of the 2-substituted pyridines, i.e., those pyridine derivatives which are of greatest industrial significance. Moreover, the method is applicable to a broad variety of substituted alkynes and nitriles thereby giving access to a whole family of pyridine derivatives having 1, 2, 3 or 5 substituents in the ring. Selected examples are discussed below and details are to be found in Section 4.

3.3.1. PYRIDINE

The parent compound which was first prepared on a small scale by Sir William Ramsay way back in 1876 by passing a mixture of acetylene and gaseous HCN through a red hot iron tube [75], [76], may be prepared under mild conditions using the homogeneous η^6-borinato-cobalt catalyst **36** [77]. At present the turnover number for this reaction (Equation 49) is only about 100 cycles per cobalt but we

36

110°, 23 bar, 60 min
benzene

TON: 103

(49)

are hopeful that with further work it will be possible to develop a process having technical potential.

HCN is relatively cheap being a major by-product of the SOHIO-process for the ammonoxidation of propene to acrylonitrile (for each kg of propene, 0.15–0.20 kg of HCN are formed). The introduction of boron into the carbocyclic ligand attached to the cobalt enhances considerably the catalytic activity of the reaction. This is probably the result of the suppression of the protolytic 1,4-addition of HCN to the olefinic cobaltacycle; the cyano-substituted 1,3-dienes cannot be displaced by the acetylene and the catalytic activity is lost (Equation 50). Reactions of the

Co + HCN → Co (CN, H)

(50)

cobaltacycle with compounds having active hydrogen atoms e.g. Et_3SiH and pyrrole, have been reported by H. Yamazaki [78].

3.3.2. 2-ALKYLPYRIDINES

Our catalyst screening revealed that the trimethylsilyl-substituted derivative $[Co(\eta^4\text{-cod})(\eta^5\text{-}Me_3Si\text{-}Cp)]$ **31** [41] is the preferred system for the co-cyclization of acetylene and acetonitrile to give 2-methylpyridine (α-picoline) (Equation 51).

$$2\ HC{\equiv}CH + N{\equiv}C{-}CH_3 \xrightarrow[130-152^\circ,\ 20\ bar,\ 2\ h]{\mathbf{31}} \text{2-methylpyridine} \qquad (51)$$

$(CH_3)_3Si$ Co

TON: 4088

PROD: 6.445

This reaction is best performed in pure acetonitrile without additional solvents. The nitrile is saturated at 20–25° with acetylene at 17 bar. This allows the acetylene to be added in one batch at the start of the reaction (Method A, Table 10). At the

Table 10

PREPARATION OF 2-METHYLPYRIDINE

Method	A	B
Catalyst	$(CH_3)_3Si$–(cyclopentadienyl)Co(cyclooctadiene)	Co (indenyl)(cyclooctadiene)
Temperature	130–200°	130–152°
Acetylene pressure (25°)	17 bar	7 bar
Acetylene pressure (during the reaction)	55 bar	20 bar
Reaction time	240 minutes	120 minutes
Conversion (CH_3CN)	24.5%	25.2%
Yield based on CH_3CN-conversion	76.2%	76.8%
TON	3818	4088
Productivity	6.017 t /kg Co	6.445 t/kg Co

actual reaction temperature a maximum pressure of 60 bar may be reached which then slowly drops as the acetylene is consumed. We have never experienced any accident or explosion using high pressure acetylene – apparently the nitrile triple bond is able to stabilize the decomposition of acetylene in a manner similar to acetone. Nevertheless legal restrictions do exist, for example in N. America and Japan which limit the acetylene pressure to 20 bar. In such cases Method B (Table 10) can be used: at 25° the acetylene pressure is only 7 bar while at the reaction temperature a constant pressure of 20 bar of acetylene is maintained (resulting in a virtually constant concentration of acetylene in the solution) with the help of a compressor connected to the autoclave. Full experimental details for both methods are given in Section 4.2.2. 2-Undecylpyridine and other 2-alkyl pyridines can be prepared efficiently using **23** or **31** as the catalyst and the procedure described as Method B (Table 10). The product is formed in a yield as high as 94% and it can, moreover, be easily separated from the reaction mixture. This method should be compared with the conventional alkylation reactions [79], [80] where the yields lie in the range of 22% to 54% suggesting that the catalyzed co-cyclization (Equation 52) might be an attractive pathway for large-scale production (see Section 4.2.3.).

$$2\ HC{\equiv}CH + N{\equiv}C{-}(CH_2)_{10}CH_3 \xrightarrow[200^\circ,\ 25\,bar,\ 2h]{\mathbf{31}} \text{2-}(CH_2)_{10}CH_3\text{-pyridine} \qquad (52)$$

TON: 1194

PROD: 4.72

C. Botteghi [81] has applied the cobalt-catalyzed reaction in a unique synthesis of optically active 2-substituted pyridines (Equation 53) starting from optically pure

$$2\ HC{\equiv}CH + N{\equiv}C{-}C^{\bullet}H(CH_3){-}CH_2CH_3 \xrightarrow[140^\circ,\ >7\,bar,\ 2h]{\mathbf{23}} \text{2-}[C^{\bullet}H(CH_3)CH_2CH_3]\text{-pyridine} \qquad (53)$$

cyanides. The chiral centre is maintained during the cyclization with acetylene. This reaction has recently been extended to the synthesis of bipyridyl compounds having optically active substituents [82] and provides access to chiral ligands of potential interest in transition metal catalyzed asymmetric synthesis.

3.3.3. 2-VINYLPYRIDINE

Probably the most interesting application from the industrial point of view is the cobalt catalyzed one-step synthesis of 2-vinylpyridine (Equation 54). This

$$2\ HC{\equiv}CH + N{\equiv}C{-}CH{=}CH_2 \xrightarrow{\mathbf{23},\ \mathbf{54}} \text{2-vinylpyridine} \qquad (54)$$

valuable, fine chemical can be manufactured using equal amounts by weight of acetylene and acrylonitrile, both of which are comparatively inexpensive.

We first studied this reaction in 1974 using [Co(η^4-cod) (η^5-Cp)] **23** as the catalyst. At 100° and with a reaction time of 4 h and a nitrile concentration of 2.14 moles/l in toluene, a turn-over number of 63 was obtained [19b]. H. Yamazaki [21] in Japan and P. Hardt at Lonza AG in Switzerland [25] achieved total turn-over numbers between 38 and 48 using the less effective cobaltocene **54** under similar reaction conditions. Interestingly, a pseudo Diels-Alder reaction (Equation 55) involving the C=C double bond of the acrylonitrile is not observed.

$$\text{CpCo(diene)} + CH_2{=}CH{-}C{\equiv}N \not\longrightarrow \text{cyclohexene-}C{\equiv}N \qquad (55)$$

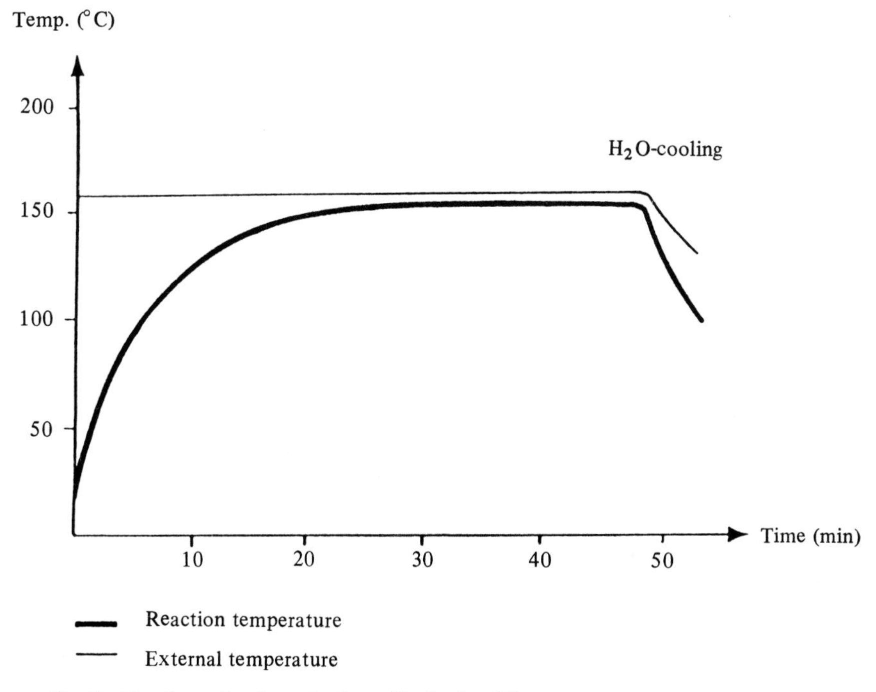

Fig. 7. The thermal polymerisation of 2-vinylpyridine.

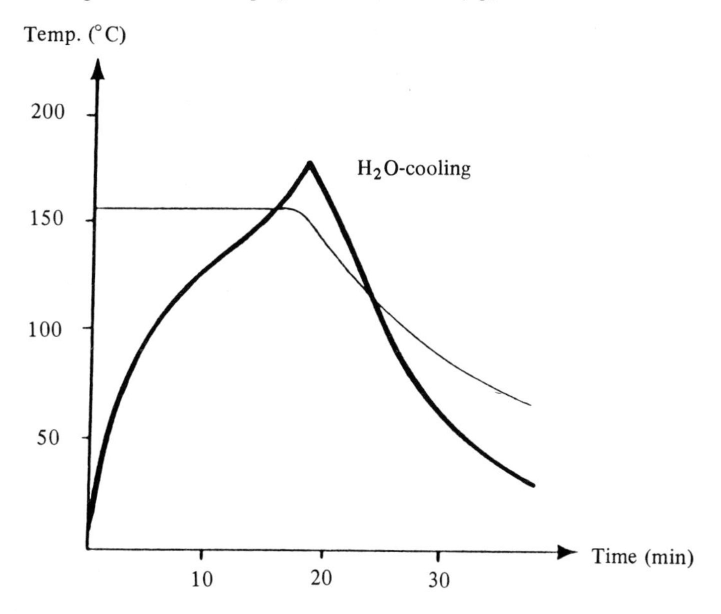

Fig. 8. The cobalt-catalyzed polymerisation of 2-vinylpyridine.

Since the product yield and the catalyst efficiency were not satisfactory, we concentrated our efforts on developing a technically feasible process using Cp—Co core catalysts. The Lonza AG., on the other hand, devoted their efforts to $[Co(\eta^5\text{-}Cp)_2]$ and its analogues [83]. Although cobaltocene is commercially available whereas a feasible access to the halfsandwich systems was not originally available (see, however, Section 2), we selected the latter type of complexes because they already contain the active Cp—Co species whereas cobaltocene has to be converted into the true catalyst by the transformation of one η^5-Cp ring into a η^4-diene group either by reaction with acetylene (by analogy to Equation 17) or with acrylonitrile (Equation 56).

$$2\ \text{Co} \xrightarrow{CH_2{=}CH{-}CN} \text{Co} \quad \text{Co} \quad (H,\ H;\ CH_2{-}CH{-}CN) \tag{56}$$

Moreover, cobaltocene is known to convert acrylonitrile into low molecular weight polyacrylonitrile [84]. All of the components present in the reaction mixture are sensitive to polymerization and the isolable yield of 2-vinylpyridine rapidly decreases on scaling up the procedure. In order to estimate the critical temperature range for thermal polymerization, we heated samples of pure 2-vinylpyridine in a magnetically stirred stainless steal autoclave under argon: in the absence of any stabilizing agents exothermal polymerization does not take place below 150° (Figure 7). However, the addition of 0.04 mol% $[Co(\eta^4\text{-cod})(\eta^5\text{-Cp})]$ **23** leads to a spontaneous exothermal reaction at about 140°, the bath temperature shoots up to above 155° and rapid cooling with cold water was necessary to prevent the system getting out of control (Figure 8). Thus 140° seems to be the upper limit for the Cp—Co-catalyzed conversion of pure acrylonitrile (Equation 54).

The addition of radical-traps such as *t*-butylcatechol, *N*-phenyl, *β*-naphthylamine or hydroxyquinone monoethylether does not prevent the polymerization of the acrylonitrile in the presence of the cobalt complex but can destroy the catalyst. Since the $[Co(\eta^4\text{-cod})(\eta^5\text{-Cp})]$ catalyst needs a reaction temperature higher than 140° in order to develop its full activity, this complex cannot be used for the conversion of pure acrylonitrile and acetylene into 2-vinylpyridine.

Selected data showing the effect of reaction temperature, reaction time, nitrile and catalyst concentration upon the yield of 2-vinylpyridine in the presence of $[Co(\eta^4\text{-cod})(\eta^5\text{-Cp})]$ have been brought together in Table 11.

By varying the parameters it is possible to increase the TON of the reaction from 63 to 208 at 150°. This is accompanied by a decrease in the reaction time to 30 min. The acrylonitrile is added as a 2 molar solution in toluene or benzene containing 0.2×10^{-2} moles of the catalyst **23**, the acetylene pressure is maintained at 17 bar during the reaction. If the acetylene concentration in the solution is further increased by raising the pressure to 30–40 bar then the turnover number increases to

Table 11

2-VINYLPYRIDINE SYNTHESIS WITH Co(η^4-cod) (η^5-Cp) **23** AT DIFFERENT REACTION PARAMETERS

Catalyst **23** (g)	Acrylo-nitrile (g)	Toluene (g)	Concentration **23** (10^{-2} moles/l)	Acrylo-nitrile (moles/l)	Reaction- temp.	time (min)	Acetylene uptake (g)
0.55	14.50	209.65	0.95	1.09	128°	60	11.30
0.55	15.15	208.55	0.95	1.14	128°	60	10.50
0.60	14.70	210.00	0.99	1.07	130°	60	12.60
0.65	15.05	208.35	1.08	1.09	128°	120	16.95
0.25	14.10	209.25	0.41	1.03	130°	60	7.75
0.60	16.70	206.20	0.99	1.21	148°	30	14.70
0.35	10.20	247.90	0.50	0.64	128°	60	7.80
0.15	13.10	210.05	0.25	0.952	135°	60	7.15
0.20	14.05	209.45	0.33	1.021	150°	30	9.45
0.10	14.95	207.50	0.17	1.09	150°	30	6.60
0.10	13.70	210.05	0.17	0.995	150°	30	4.60
0.1226	14.85	210.10	0.203	1.08	150°	30	6.33
0.2133	15.60	207.10	0.354	1.13	150°	30	9.69
0.1948	32.21	191.60	0.323	2.34	150°	30	11.54
0.1036	15.35	209.50	0.172	1.12	150°	30	6.0
0.2035	33.20	193.40 [b]	0.338	2.41	150°	30	9.42
0.1440	14.00	212.70 [b]	0.239	1.02	150°	30	6.0
0.166	15.50	208.20	0.275	1.12	145°	30	7.70
0.1055	14.10	208.55	0.175	1.024	150°	30	4.85
0.2025	15.15	206.35	0.336	1.101	150°	30	8.45
0.1463	14.45	212.00 [b]	0.243	1.059	150°	30	5.50
0.2025	14.50	208.40	0.336	1.053	150°	30	8.70
0.2015	15.40	210.90	0.334	1.119	140°	30	6.0
0.2010	13.90	208.35	0.334	1.001	150°	30	7.90
0.1999	13.45	207.50	0.3317	0.977	150°	30	8.45
0.2009	14.00	208.00	0.3334	1.017	150°	30	9.0

[a] Based on converted acrylonitrile.
[b] Toluene replaced by benzene.

Acrylo-nitrile		Residue	2-Vinylpyridine		2-VP/Benzene	2-Vinylpyridine	
recovered	conversion		Product	Yield	Selectivity	TON	Yield [a]
(g)	(%)	(g)	(g)	(%)	(molar ratio)	(moles/g-atom Co)	(%)
5.91	59	3.45	12.93	76	2.9/1	52	44.8
5.61	63	4.90	13.32	71	5/1	54	44.7
5.75	61	4.30	13.01	73	2.4/1	42	44.5
4.50	70	5.85	16.06	77	2.1/1	55	53.9
7.89	44	1.70	9.75	79	3/1	87	34.8
5.03	70	7.75	16.12	70	5/1	60	49.0
5.03	51	2.25	7.93	76	1.8/1	50	32.8
7.88	40	1.35	8.54	83	2.6/1	126	33.2
2.13	42	1.90	10.88	93	3.8/1	120	39.1
9.29	38	1.65	9.41	34	3.9/1	208	31.9
8.69	37	1.40	7.47	75	3.9/1	165	27.7
9.23	38	1.60	9.09	82	3.8/1	164	31.0
7.84	50	2.15	12.27	79	3.8/1	127	39.5
21.15	34	3.30	15.99	73	7.9/1	181	25.0
10.02	35	1.15	8.76	83	3.8/1	187	29.1
21.92	34	3.54	13.48	61	–	147	20.7
8.28	41	1.30	8.88	78	–	136	32.0
10.57	32	1.40	9.11	93	3.3/1	121	29.8
9.41	39	0.80	6.64	71.4	4.1/1	139	23.6
8.193	46	2.10	9.96	72.2	4/1	109	33.2
9.19	36	1.10	8.29	79.7	–	125	28.7
8.10	44	1.70	11.0	86.7	4/1	120	38.1
10.64	31	1.25	8.63	91.4	3.4/1	95	28.3
8.27	41	2.10	9.23	87.9	3.1/1	101	36.0
7.67	43	2.0	10.14	88.6	3.1/1	112	38.1
7.71	45	2.10	9.93	79.7	3.0/1	110	35.9

Table 12

THE EFFECT OF THE COBALT CATALYST UPON THE 2-VINYLPYRIDINE SYNTHESIS

No.	Catalyst	Temp.	Turnover AN (%)	Yield 2-VP Ref. to AN (%)	Ref. to AC (%)	TON	2-VP concentration (%)	2-VP: Benzene molar ratio
23	Co	160°	32.0	66.0	66.5	443	6.6 in benzene	3.5/1
32	Co	140°	25.0	88.6	80.3	480	7.0 in benzene	4.0/1
36[a]	B Co	120°	25.6	79.8	77.2	1326	35.2 in AN	6.5/1
36[a]	"	120°	14.2	61.6	85.1	2164	16.8 in AN	6.6/1
36[a]	"	120°	12.7	74.3	71.4	1896	17.7 in AN	5.2/1
36[a]	"	120°	12.6	75.3	73.8	1935	17.7 in AN	6.9/1

[a] In pure AN.

440. Under these conditions the indenyl-Co system **32** is even more effective and a TON of 480 can be obtained [41], [55], [85].

However, the low nitrile concentration limits the concentration of 2-vinylpyridine in the solvent (benzene or toluene) to ca. 7% and (even worse) 5 tons of [Co(η^4-cod)(η^5-Cp)] would be needed to produce 1000 tons of product. One important parameter which governs the efficiency of the reaction is the stationary concentration of acetylene in the solution. As can be seen in Table 13 (see Section 4.1 below), the pure nitriles, in contrast to benzene or toluene, are excellent solvents for acetylene. However, before we could use acrylonitrile both as reagent and as solvent it was necessary to find a catalyst active below the polymerization temperature of the acrylonitrile (ca. 140°). A major breakthrough was the discovery that the η^6-borinato group, which is electronically equivalent to the η^5-cyclopentadienyl ligand, is a highly effective organic ligand [77], [86]. For example, complex **36** catalyzes the reaction (Equation 57) even at 100°. This means that the acrylonitrile

36

120° . 51 bar . 2h

TON: 2164

PROD: 2.78

(57)

can be used in the absence of any other solvent with the result that the increased concentration of acetylene in the reaction mixture leads to a considerable enhancement in productivity.

In Table 12 we have compared the efficiency of some of our catalysts. Replacement of **23** by the more active η^5-indenyl system **32** leads to an improvement in yield mainly because side reactions are suppressed. The optimum temperature range using the η^6-borinato ligand **36** was found to be around 120°. The lower conversion of acrylonitrile in this case is a consequence of its use as the solvent. The yield with **36** is approximately the same as in the reaction catalyzed by **32** while the higher activity of the η^6-borinato-cobalt catalyst shows itself in the four-fold increase in turn-over number. As a result the concentration of 2-vinylpyridine in the acrylonitrile solvent increases by up to five times. A welcome bonus is the improved chemoselectivity and as a result far less benzene is formed.

The life of the catalyst is limited by the formation of side-products (produced by the vinylation of acrylonitrile (Equation 58) and to a lesser extent – of the vinylpyridine (Equation 59)) which act as poisons. These side products are formed

$$NC-\overset{H}{C}=CH_2 + 2\,HC\equiv CH \longrightarrow \text{CN} + \text{CN} \qquad (58)$$

$$+ \; HC\equiv CH \longrightarrow \qquad (59)$$

only to a very small extent but, being activated olefins, successfully compete with acetylene for the coordination sites at the cobalt: the addition of a few drops of the vinylation products shown in Equation 58 or Equation 59 to the reaction mixture causes complete suppression of the catalytic reaction even under drastic reaction conditions.

3.3.4. 2,2′-DIPYRIDYL

Our screening has shown that the trimethylsilyl substituted system **31** is the best catalyst for the preparation of 2,2′-dipyridyl from acetylene and 2-cyanopyridine (Equation 60). Since this reaction has to be carried out in benzene or toluene,

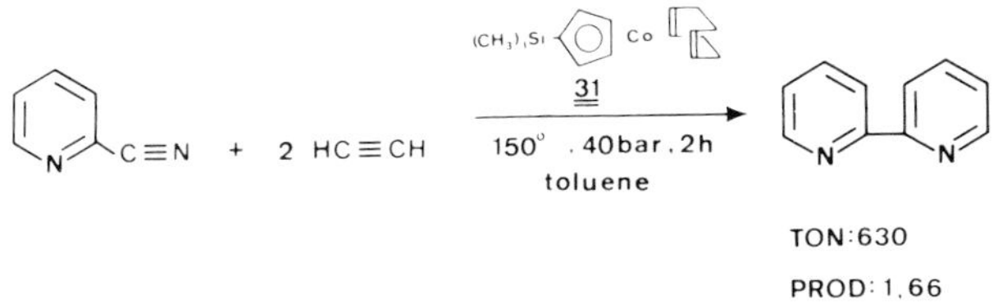

$$\text{2-NC-C}_5\text{H}_4\text{N} + 2\,HC\equiv CH \xrightarrow[150°,\ 40\,bar,\ 2h,\ \text{toluene}]{(CH_3)_3Si\text{-}C_5H_4Co(COD)\ \underline{\underline{31}}} \text{2,2'-dipyridyl} \qquad (60)$$

TON: 630

PROD: 1.66

high acetylene pressures have to be applied in order to maintain a high stationary acetylene concentration.

Bifunctional starting materials such as α,ω-dinitriles can also be reacted (Equation 61). The product is a mixture of bis-(2-pyridyl) compounds.

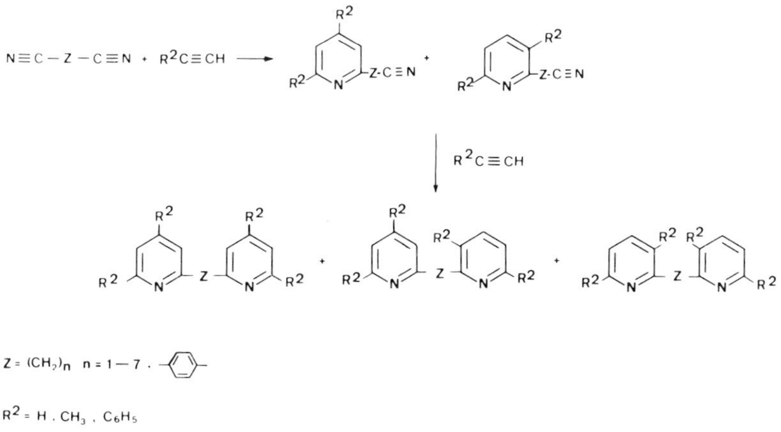

$$N\equiv C-Z-C\equiv N + R^2C\equiv CH \longrightarrow \text{4,6-}R^2_2\text{-2-(Z-C}\equiv N)\text{pyridine} + \text{3,6-}R^2_2\text{-2-(Z-C}\equiv N)\text{pyridine} \xrightarrow{R^2C\equiv CH} \text{bis-(2-pyridyl)-Z isomers} \qquad (61)$$

$Z = (CH_2)_n$ $n = 1-7$, $-C_6H_4-$

$R^2 = H, CH_3, C_6H_5$

3.3.5. 2-AMINO- AND 2-ALKYLTHIO-PYRIDINES

A wide variety of substituents at the cyano group are tolerated by the cobalt catalyst and, for example, monomeric cyanamide reacts with acetylene in the presence of the η^6-borinato cobalt catalyst **36** to give 2-aminopyridine (Equation 62). Alkylthiocyanates can also be used as the cyano component and react [87] to give

$$2\ HC{\equiv}CH + N{\equiv}C{-}NH_2 \xrightarrow[130^\circ,\ 40\,bar,\ 2\,h,\ DMF]{\mathbf{36}} \text{2-aminopyridine} \quad \text{TON: 245, PROD: 0.39} \tag{62}$$

alkylthiopyridines (Equation 63) which are otherwise accessible only by multistep synthetic pathways [88]. Although we have not attempted to optimize this example, the cobalt catalyzed reaction seems to offer an easy entry into the pyrithione systems.

$$2\ HC{\equiv}CH + N{\equiv}C{-}SCH_3 \xrightarrow[130^\circ,\ 48\,bar,\ 2\,h]{\mathbf{36}} \text{2-}SCH_3\text{-pyridine} \quad \text{TON: 58} \tag{63}$$

3.3.6. DIALKYL- AND TRIALKYL-PYRIDINES

The reaction of monosubstituted alkynes with nitriles gives a mixture of isomeric collidine derivatives (trialkylpyridines) (Equation 64).

$$R^1C{\equiv}N + 2\ R^2C{\equiv}CH \longrightarrow \text{2-}R^1\text{-4,6-}R^2_2\text{-pyridine} + \text{2-}R^1\text{-3,6-}R^2_2\text{-pyridine} \tag{64}$$

This reaction may also be carried out with two different alkynes. For example, the co-cyclization of acetylene and propyne with acetonitrile yields a mixture of dimethylpyridine (lutidines) in addition to 2-methylpyridine and the isomeric collidines. The co-cyclization (Equation 65) is not selective and appears to occur statistically.

$$CH_3C{\equiv}N + CH_3C{\equiv}CH + HC{\equiv}CH \longrightarrow \text{2,6-}(CH_3)_2\text{-pyridine} + \text{2,4-}(CH_3)_2\text{-pyridine} + \text{2,3-}(CH_3)_2\text{-pyridine} \tag{65}$$

K. P. C. Vollhardt [18b] has introduced an interesting variation by reacting bifunctional alkynes with nitriles. For example, 1,6-heptadiyne and 1,7-octadiyne undergo a two-step co-cyclization with nitriles to give annelated pyridine derivatives (Equation 66).

$$\text{1,7-octadiyne} \xrightarrow{CpCo(CO)_2\ \mathbf{50}} [\text{cobaltacyclopentadiene}\ Co\,Cp] \xrightarrow[-\,CpCo]{+\,R^1{-}C{\equiv}N} \text{3-}R^1\text{-5,6,7,8-tetrahydroisoquinoline} \tag{66}$$

3.3.7. UNSUCCESSFUL REACTIONS

In spite of the general insensitivity of the cobalt catalysts towards functional groups in the substrates, certain *limitations* do exist. Thus disubstituted alkynes, such as but-2-yne or tolane, undergo only a few catalytic cycles to give penta-substituted pyridines before the catalyst dies (Equation 67), while the reaction between cyanogen

$$R^1C{\equiv}N + 2\,R^2C{\equiv}CR^2 \xrightarrow{\mathbf{23}} \text{(penta-substituted pyridine: } R^1 \text{ at C-2}, R^2 \text{ at C-3,4,5,6)} \qquad (67)$$

and phenylacetylene to give phenyl-substituted bipyridyl derivatives occurs only in a stochiometric manner (Equation 68).

$$N{\equiv}C{-}C{\equiv}N + 4\,C_6H_5{-}C{\equiv}CH \xrightarrow{\mathbf{23}} \text{(tetraphenyl-2,2'-bipyridyl)} \qquad (68)$$

No reactions have been observed which involve either cyanogen chloride or nitriles having a carbonyl group bonded to the cyanide function (Equations 69 and 70).

$$Cl{-}C{\equiv}N + 2\,R^2C{\equiv}CH \not\rightarrow \text{(2-chloro-pyridine with } R^2 \text{ substituents)} \qquad (69)$$

$$R^1{-}\overset{O}{\overset{\|}{C}}{-}C{\equiv}N + 2\,R^2C{\equiv}CH \not\rightarrow \text{(2-acyl-pyridine, } C(=O)R^1 \text{, with } R^2 \text{ substituents)} \qquad (70)$$

The cobalt catalyzed reactions of alkynes and cyano compounds which we have studied are brought together in Table 17 (see below). Several examples are discussed in experimental detail in Section 4.2.

4. EXPERIMENTAL DETAILS

4.1. Apparatus

In order to achieve a sufficient concentration of acetylene in solution, we work with compressed acetylene at 10–60 bar. Special safety precautions must, however, be observed when working with acetylene under "high pressure" (i.e. 1.5 bar!), since acetylene (and propyne) can spontaneously decompose, even in the absence of air or oxygen, in a highly exothermic manner to give carbon and hydrogen. It is necessary that an explosion or detonation be confined to the autoclave and its nearest environment and that any recoil of the pressure wave to the compressor or the gas cylinders be prevented. The technical know-how for the handling of acetylene under pressure has been described in detail elsewhere [89]. The most important precautions consist in the separation of the compressor and the gas cylinders from the reaction vessel by a safety screen, in the installation of several recoil guards, and in the use of stainless steel wherever direct contact with acetylene occurs. For our experiments we use the set up shown in Figure 9. The gas is compressed in two steps to a maximum of 28 bar at about 20° and is then led through thin steel capillaries into the autoclave which stands in a room enclosed by 30 cm reinforced concrete walls. The pressure and the temperature during the reaction are controlled externally. The course of the reaction is monitored automatically using a temperature/pressure diagramme (Figure 9). In a typical example the autoclave is rapidly (10 min) heated to 125°. The start of the catalytic reaction is associated with a peak in the temperature and the pressure traces. After 30 min the autoclave is cooled down to 25° by passing cold water through a coil situated inside in the autoclave (Figure 10).

The advantages of using the experimental set up shown in Figures 9 and 10 are as follows:

- satisfactory heat transfer as a result of magnetic stirring in the autoclave and electric heating;
- precise temperature control (± 1°) enables exothermal reactions to be identified immediately;
- efficient water cooling prevents the catalytic reaction getting out of control;
- use of electronic balances allows the uptake of acetylene to be monitored during the reaction and enables a mass balance to be established.

As previously mentioned, an important parameter for the cobalt catalyzed reaction is a sufficiently high stationary acetylene concentration in the solution. Since no data on the solubility of compressed acetylene in organic solvents were availabe in the literature, we have determined values relevant to our work. A selection of our data is summarized in Table 13. As can be seen from the table, there is a more than proportional increase in the solubility of acetylene in the various solvents on raising the pressure. In the case of toluene, for example, the stationary concentration of acetylene increases eight fold on raising the pressure from 5 to 25 bar. By good fortune, nitriles turned out to be excellent solvents for compressed acetylene and solutions containing around 40% of acetylene can be obtained by using, for example, acetonitrile or acrylonitrile. For this reason we were able to carry out the catalysis in the absence of additional solvents and under high pressure acetylene. The solubility of compressed acetylene appears to be nitrile-dependent and has to be determined in each individual case (see Figure 11).

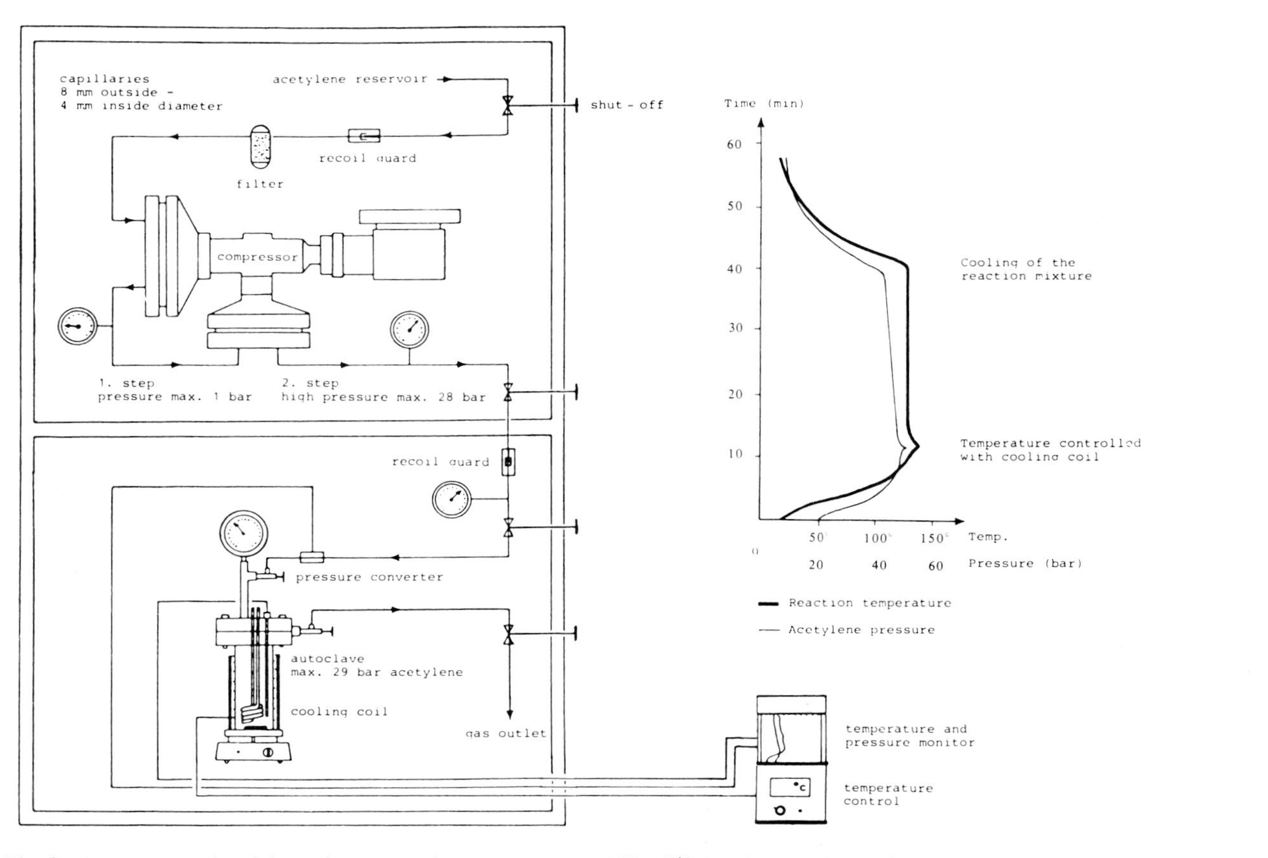

Fig. 9. An apparatus involving a 2-step membrane compressor MKZ 8/3 (Andreas Hofer, Mülheim a.d. Ruhr, Germany) for working with acetylene under high pressure.

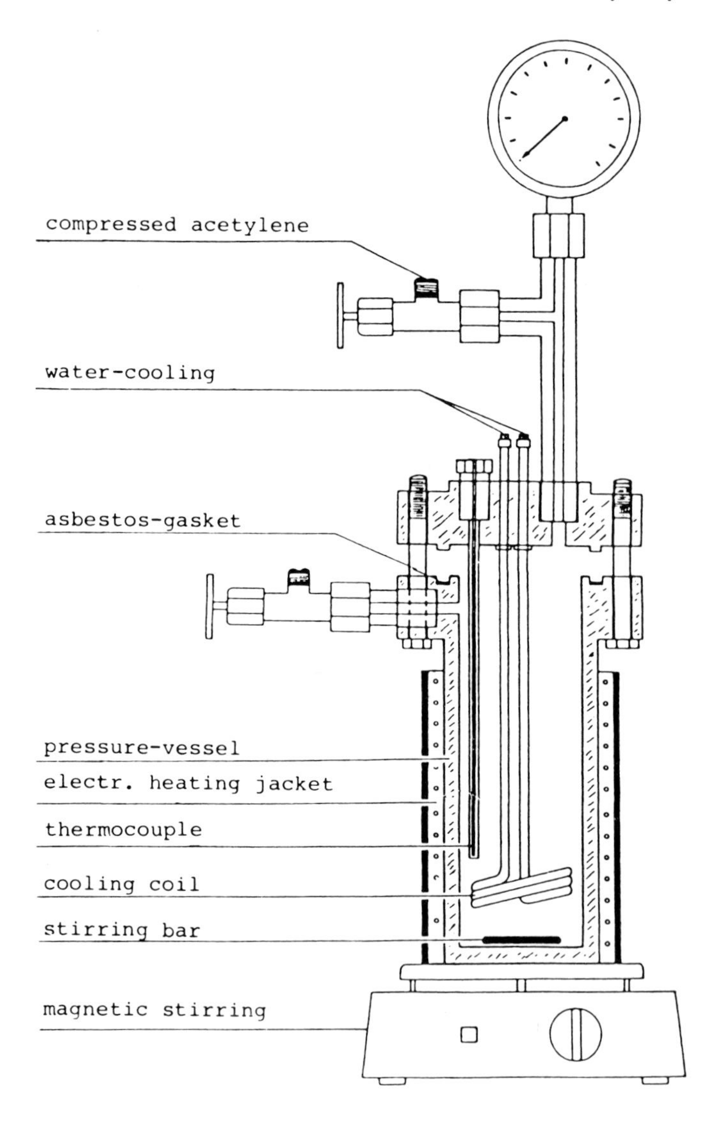

Fig. 10. An antimagnetic chromium-nickel steel autoclave with magnetic stirring.

4.2. Some synthetic procedures

4.2.1. PYRIDINE FROM HCN AND ACETYLENE

0.350 g (1.094 mmoles) of [Co(η^4-cod)(η^6-1-phenylborinato)] **36** is dissolved in 260 ml of benzene and treated with 30 ml (0.737 moles) of liquid HCN. The resulting solution is siphoned at room temperature into the autoclave (Figure 10)

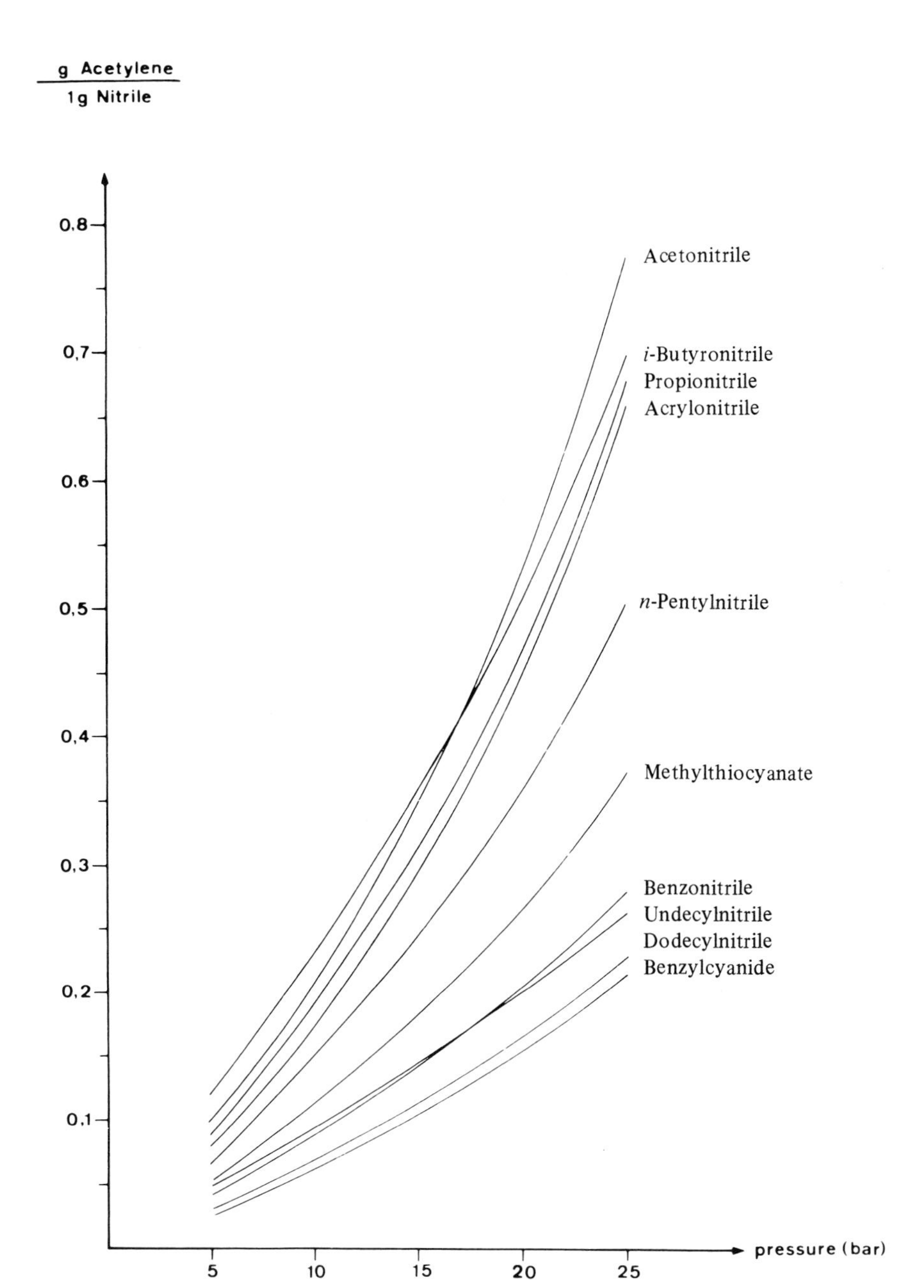

Fig. 11. The effect of pressure upon the solubility of acetylene in various nitriles at 25°.

Table 13

SOLUBILITY OF ACETYLENE AT 25° UNDER PRESSURE (G C_2H_2/KG SOLUTION)

	Pressure (bar)				
Solvents	5	10	15	20	25
Toluene	30	71	118	175	242
Benzene	59	96	138	193	265
Pyridine	56	112	169	224	281
N-Methylpyrrolidone	89	149	201	242	–
DMF	127	213	279	340	–
Acetone	119	216	301	379	–
Nitriles					
Benzylcyanide	32	64	96	131	174
Dodecylnitrile	35	71	105	140	174
Undecylnitrile	54	93	132	170	210
Benzonitrile	44	88	130	174	216
Methylthiocyanate	48	98	152	212	274
n-Pentylcyanide	67	134	200	266	333
Propionitrile	83	166	245	320	389
Acrylonitrile	78	157	236	315	394
i-Butyronitrile	116	202	278	346	415
Acetonitrile	74	175	268	355	431

and is saturated at 6.5 bar with approximately 15.2 g (0.58 moles) of acetylene. The solution is then heated over 18 min to a reaction temperature of 110° the pressure rising to 23 bar. Reaction is allowed to continue for 60 min, the autoclave is cooled to room temperature by passing water through the cooling coil and unreacted HCN and acetylene are vented at room temperature through an $FeSO_4$ solution.

239.7 g of crude product are formed. The volatile components are condensed at 100° and 0.2 Pa, leaving 0.4 g of residue. Gas chromatography (Instrument: Perkin-Elmer F22; conditions: column: 50 m CW 20 M; detector: FID; temperature: 200°–70°/240°; 6 p. min; carrier gas: He 1 bar)

The yield based on HCN is 15.2%. Catalyst efficiency is 103 moles of pyridine/gram atom of cobalt.

4.2.2. 2-METHYLPYRIDINE (α-PICOLINE) FROM ACETONITRILE AND ACETYLENE

Method A (acetylene pressure 55 bar, see Table 10): 0.0414 g (0.136 mmoles) of [Co(η^4-cod) (η^5-trimethylsilyl-Cp)] **31** dissolved in 114.1 g (2.783 moles) of acetonitrile is siphoned into the autoclave (Figure 10) at 25° and the solution is then saturated with acetylene compressed to 17 bar. The solution is heated over 45 min to 130°, the pressure increasing to 55 bar. An exothermic reaction occurs, the inside temperature rises to 150° and the pressure drops to 48.5 bar. The temperature is gradually increased to 200° over 120 min; the pressure increasing to 51 bar. The final acetylene pressure after 120 min at 200° is 44 bar. Reaction time: 240 min. Reaction temperature: 130°–200°.

The autoclave is cooled to 20° (the pressure dropping to 5 bar) and unreacted acetylene is vented. The resulting crude product (147.85 g) is withdrawn from the autoclave and the volatile components (146.5 g) condensed at 0.1 Pa and 100° (fraction 1) leaving 0.55 g of residue.

Gas chromatography (for instrument and conditions, see Section 4.2.1.) shows the following compositions:

Fraction 1	%	g	mmoles
Benzene	5.83	8.54	109.5
Acetonitrile	58.8	86.14	2101.0
2-Methylpyridine	33.0	48.35	519.89

yield based on
nitrile conversion: 48.35 g = 519.84 mmoles = 76.2%
converted acetonitrile: 27.96 g = 682.00 mmoles = 24.2%
TON: 3818
Selectivity
(pyridine/benzene): 4.7/1

Method B (acetylene pressure 20 bar, see Table 10): 0.040 g (0.1316 mmoles) of complex **31**, dissolved in 113.8 g (2.7756 moles) of acetonitrile are siphoned into the autoclave (Figure 10) and saturated at 25° with acetylene at 7 bar. The mixture is heated over 32 min to 130°, the pressure increasing to a maximum of 23 bar. After the exothermic reaction has started, the temperature is increased to 150° and then held at 150–152°. During the reaction (110 min) the acetylene pressure is kept constant at 20 bar by means of the compressor which supplies acetylene at the same rate as it is consumed.

Crude reaction mixture: 139.4 g
fraction 1: (bp to 100° at 0.1 Pa): 139.0 g
residue: 0.1 g

GC results (for instrument and conditions see Section 4.2.1.):

Fraction 1	%	g	mmoles
Benzene	1.45	2.02	25.9
Acetonitrile	61.2	85.07	2074.8
2-Methylpyridine	36.0	50.04	538.1

yield based on	
nitrile conversion:	50.04 g = 538.1 mmoles = 76.8%
converted acetonitrile:	28.73 g = 700.8 mmoles = 25.2%
TON:	4088
Selectivity	
(pyridine/benzene):	20.8/1

Work-up procedure for 2-methylpyridine: From a reaction of 77 g (1.878 moles) of acetonitrile with acetylene at 20 bar (Method B), a 133.7 g fraction was obtained with the composition (GC) shown below:

	%	g	mmoles
Benzene	4.83	6.46	82.8
Acetonitrile	28.8	38.51	999.0
2-Methylpyridine	64.4	86.10	926.0

yield based on	
nitrile conversion:	86.10 g = 926 mmoles = 98.6%
converted acetonitrile:	38.49 g = 939 mmoles = 50.0%
TON:	2505
Selectivity	
(pyridine/benzene):	11.2/1

131.0 g of this sample were separated using a 0.5 m packed column and the following fractions obtained:

fraction 1:	boiling up to 126°/1 bar:	43.4 g
fraction 2:	br 126–130°/1 bar:	7.0 g
fraction 3:	bp 130°/1 bar:	76.4 g
residue:		2.7 g

A GC analysis shows fraction 1 to consist to 81% of acetonitrile and to 15.4% of benzene, fraction 2 contains 96.8% 2-methylpyridine and 2.9% acetonitrile, fraction 3 is pure (99.9%) 2-methylpyridine and of the residue 88.6% is 2-methylpyridine.

4.2.3. 2-UNDECYLPYRIDINE FROM DODECYLNITRILE AND ACETYLENE

0.0972 g (0.4185 mmoles) of complex **23** are dissolved in 123.0 g (0.680 moles) of dodecylnitrile ($C_{11}H_{23}CN$) and siphoned into the reaction vessel (Figure 10). The mixture is saturated with acetylene at 22 bar and 24°. The reaction mixture is heated to 110° over 55 min, the pressure rising to 36.5 bar. The start of the catalytic reaction is accompanied by an increase in the temperature to 140° and a pressure drop to 22 bar. The pressure is then maintained at 22–25.5 bar and the temperature of the reaction mixture is slowly (90 min) raised to 167°. Reaction time: 150 min. Reaction temperature: 100–167°. After venting the autoclave, 152.1 g of crude product were isolated.

Work-up procedure: 4 runs were combined to give a 551 g sample having a bp of 155°/0.15 Pa. Separation using a 0.5 m packed column gave the following fractions:

fraction 1:	bp up to 82°/1 mbar:	3.7 g
fraction 2:	bp up to 96°/1 mbar:	127.1 g
fraction 3:	bp up to 125°/1 mbar:	18.2 g
fraction 4:	bp up to 127°/1 mbar:	363.2 g
residue		13.0 g

GC analysis of fraction 4 showed it to consist of pure (99.9%) 2-undecylpyridine (2005 mmoles). (Instrument: Perkin-Elmer F22; conditions: column: 65 m CW 20 M/H 1613 A; detector: FID; temperature: 200°–270°–240°; carrier gas: He 1 bar.)

4.2.4. 2-VINYLPYRIDINE FROM ACRYLONITRILE AND ACETYLENE

0.0321 g (0.0975 mmoles) of complex **36** are dissolved in 128.2 g (2.419 moles) of acrylonitrile and siphoned into the reaction vessel (Figure 10). The mixture is saturated with acetylene at 17 bar and 24°. The autoclave is heated to 85°, the pressure rising to 41 bar. After the catalytic reaction has started, the temperature is increased to 120° over 25 min, the pressure rising to 51 bar. After a further 100 min at 120° the pressure drops to 48 bar.

Reaction time:	125 min
Reaction temperature:	85°–120°
Crude product:	139.3 g
Condensate (100°/0.015 Pa):	128.7 g volatile material (fraction 1)
Residue:	0.69 g

GC analysis (Instrument: Varian 3700; conditions: column: 49 m CW 20 M/H 1613 B; detector: FID; temperature: 200°–70°/240°; 6p. min – 220°; carrier gas: N_2 0.7 bar):

	%	g	mmoles
Benzene	1.80	2.50	32
Acrylonitrile	79.3	110.0	2075
2-Vinylpyridine	16.0	22.19	211

yield based on	
nitrile conversion:	16.0 g = 211 mmoles = 61.5%
converted acrylonitrile:	18.2 g = 343 mmoles = 14.2%
TON:	2164
Selectivity:	6.1/1

Work-up procedure: 99.0 g of a sample containing 64.8% of acrylonitrile, 29.4% of 2-vinylpyridine and 3.6% of benzene are separated using a 0.5 m packed column.

fraction 1:	bp up to 60°/80 mbar:	67.7 g
fraction 2:	bp 60–64°/67 mbar:	1.7 g
fraction 3:	bp 64–65°/67 mbar:	24.1 g
residue		2.0 g

GC analysis of fraction 3 shows it to consist of pure (99.8%) 2-vinylpyridine (229 mmoles, yield 81.9%).

4.2.5. 2,2′-DIPYRIDYL FROM 2-CYANOPYRIDINE AND ACETYLENE

0.036 g (0.1184 mmoles) of complex **31** is dissolved in 87.9 g toluene and 52.4 g (0.5038 moles) 2-cyanopyridine added. The mixture is siphoned into the reaction vessel (Figure 10) and saturated with acetylene at 16 bar and 25°. The autoclave is heated up to 150° over 50 min, the pressure rising to 42 bar. The temperature is then raised to 200° over 80 min and the reaction mixture kept at this temperature for a further 40 min during which time the pressure slowly drops to 40.5 bar.

Reaction time:	120 min
Reaction temperature:	150°–200°
Crude product:	144.5 g
Condensate (100°/0.015 Pa):	143.3 g
Residue:	1.0 g

GC analysis of the condensate (Instrument: Perkin-Elmer F22; conditions: column: 50 m CW 20 M/H 1484; detector: FID; temperature: 200°–70°/240°; 6p. min – 250°; carrier gas: He 1 bar):

	%	g	mmoles
Benzene	1.50	2.15	27.6
Toluene	60.36	86.5	
2-Cyanopyridine	29.03	41.6	400
2,2′-Dipyridyl	8.12	11.64	74.6

yield based on
converted 2-cyanopyridine: 11.64 g = 74.6 mmoles = 71.9%
converted 2-cyanopyridine: 10.80 g = 103.8 mmoles = 20.6%
TON: 630
Selectivity: 2.7/1

Work-up procedure: 143.9 g of crude product were fractionated under vacuum.
fraction A: bp up to 108°: 1.4 g
fraction B: bp 108°–115°: 74.7 g
fraction C: higher boiling product: 64.2 g

63 g of fraction C were fractionated on a 0.5 m column packed with glass beads.
fraction 1: bp up to 40°/6 mbar: 7.1 g
fraction 2: bp 40–74°/6 mbar: 43.1 g 2-cyanopyridine
fraction 3: bp 74–79°/3 mbar: 5.6 g 2-cyanopyridine
residue: 7.1 g

The column is washed out with ether and the solution added to the residue which is then concentrated to give after recrystallization 7 g of pure 2,2′-dipyridyl.

4.2.6. 2-AMINOPYRIDINE FROM CYANAMIDE AND ACETYLENE

14.6 g (0.34 moles) of monomeric cyanamide are dissolved in 128.7 g of dimethylformamide. 0.0515 g (0.1609 mmoles) of complex **36** is added and the mixture siphoned into the reaction vessel (Figure 10). The solution is saturated with acetylene at 12 bar and 25°. The autoclave is heated to 80° over 13 min, the pressure increasing to 28 bar. After the catalysis has started, the temperature is continuously increased over 2 h to 132°. The pressure reaching 41 bar.

Reaction time: 120 min
Reaction temperature: 80°–132°
Crude product: 146.6 g

Results of a fractional distillation:
fraction 1: bp up to 32°/6 mbar: 127.40 g
fraction 2: bp up to 68°/6 mbar: 5.90 g
residue: 12.25 g

GC analysis (Instrument: Perkin-Elmer F22; conditions: column: 50 m CW 20 M/H 1484 C; detector: FID; temperature: 240–70/240; 8p. min – 260; carrier gas: He 1 bar):

	%	g	mmoles
fraction 1:			
Benzene	1.07	1.36	17.5
DMF	97.9	124.7	
2-Aminopyridine	0.40	0.51	5.4
fraction 2:			
Benzene	1.87	0.11	1.4
DMF	43.7	2.6	
2-Aminopyridine	54.0	3.19	33.9

yield based on
nitrile conversion: 3.70 g = 39.4 mmoles = 80.7%
converted cyanamide: 2.05 g = 48.8 mmoles = 14.3%
TON: 245
Selectivity: 2.1/1

4.2.7. 2-METHYLTHIOPYRIDINE FROM METHYLTHIOCYANATE AND ACETYLENE (NOT OPTIMIZED)

0.0485 g (0.1516 mmoles) of complex **36** are dissolved in 158.9 g (2.177 moles) methylthiocyanate and siphoned into the reaction vessel (Figure 10). The reaction mixture is saturated with acetylene at 20 bar and 26°. The autoclave is heated to 80° over 87 min, the pressure increasing to 37 bar. After the start of the reaction the temperature is continuously increased to 130° over 120 min. The pressure slowly rising to 48 bar.

Reaction time: 120 min
Reaction temperature: 80°–130°
Crude product: 158.0 g
Condensate (100°/0.15 Pa): 157.1 g

GC analysis (Instrument: Siemens/1; conditions: column: 20 m SE–30/H 1514; detector: FID; temperature: 230°–60°/300°; 6p. min – 300°; carrier gas: He 0.6 bar):

	%	g	mmoles
Benzene	0.062	0.1	1.25
Methylthiocyanate	96.3	151.3	2072
Methylthiopyridine	0.7	1.1	8.8

yield based on the	
reacted cyano compound:	1.1 g = 8.8 mmoles = 8.5%
converted cyano compound:	7.6 g = 104.1 mmoles = 4.8%
TON:	58
Selectivity:	7.0/1

5. MECHANISTIC CONSIDERATIONS

H. Yamazaki and Y. Wakatsuki [90] have contributed quite considerably to our understanding of the course of the cobalt-catalyzed pyridine synthesis by isolating a number of phosphane stabilized Cp—Co complexes which may be regarded as plausible intermediates in the catalytic cycle – step-wise addition of two alkyne molecules to the central metal atom, ring closure to give a cobaltacyclopentadiene moiety, insertion of the C≡N triple bond followed by elimination of the product (Figure 12) [24].

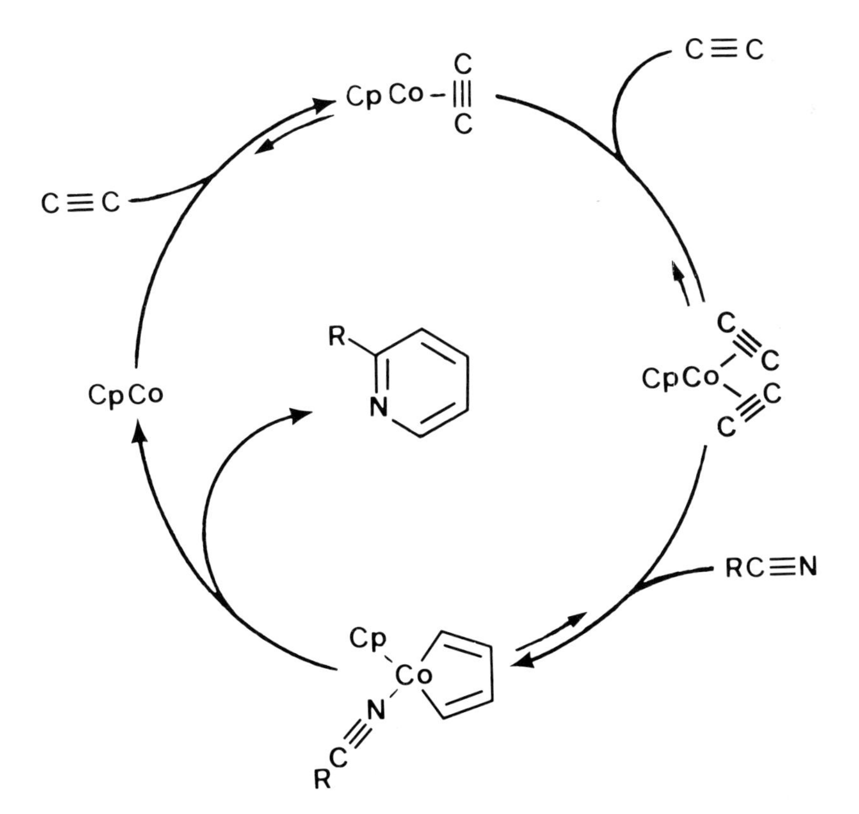

Fig. 12. Cobalt(I)-catalyzed reaction of alkyne and nitrile. The catalytic cycle.

The reaction of the triphenylphosphine stabilized Cp—Co core with diphenylacetylene is an elegant model for this sequence (Equation 71). Displacement of the

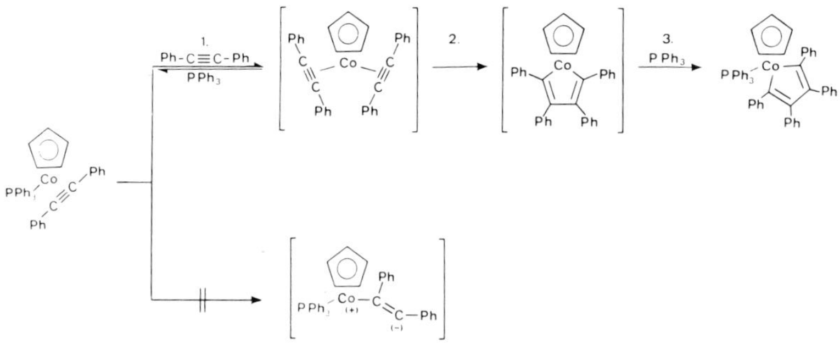

(71)

phosphine from the metal occurs prior to coupling of the alkyne molecules (step 1) in a manner similar to the olefin dissociation shown in Equation 23. A consequence of this equilibrium is that the addition of excess phosphine (or other donor ligands [24]) reduces the rate of reaction. This observation, together with the fact that polar solvents have no influence upon the rate, suggest that a polar intermediate is not involved and that the key step is the formation of a cobaltacyclopentadiene intermediate (step 2). This can be isolated as a stable phosphine complex (step 3) which furthermore reacts with the nitrile to give the expected pyridine derivative [91].

We have studied the reaction order with respect to the nitrile in order to exclude an intermediate of the type $[\overline{CoCH{=}CHCR{=}N}(\eta^5\text{-Cp})]$ in which a nitrile has inserted into the Co-alkyne fragment. Solutions of **23** in different amounts of propionitrile were siphoned into an autoclave and connected to a propyne cylinder. A constant propyne concentration was maintained and nitrile consumption and pyridine formation were monitored at 87° and 109° by GC analysis (Figure 13).

As can be seen from Figure 13, the rate of dimethyl, ethyl-pyridine formation is independent of the nitrile concentration, i.e., the reaction is zero order with respect to the nitrile. The effect of the concentration of the alkyne in solution upon the rate has also been studied. The rate of formation of dimethyl, ethyl-pyridine and trimethylbenzene using complex **23** as the catalyst (compare Equation 14) was monitored by GC in a batch reactor and the results are shown in Figure 14.

The reaction was carried out in the presence of a large excess of the nitrile, i.e., the nitrile concentration in solution is essentially constant. The experimentally determined reaction order with respect to the alkyne is 1.7 with an Arrhenius energy of activation of 96 KJ/mole of pyridine or benzene formed in the catalytic cycle which leads to a frequency factor of 5×10^8 1/(mole sec) [92]. The analogous study of trimethylbenzene formation gave a reaction order of 1.8 with respect to the propyne [93]. The results of these kinetic studies can be summarized as follows:

- The cobalt-catalyzed pyridine formation and alkyne cyclotrimerization depend upon the square of the concentration of the alkyne in solution.

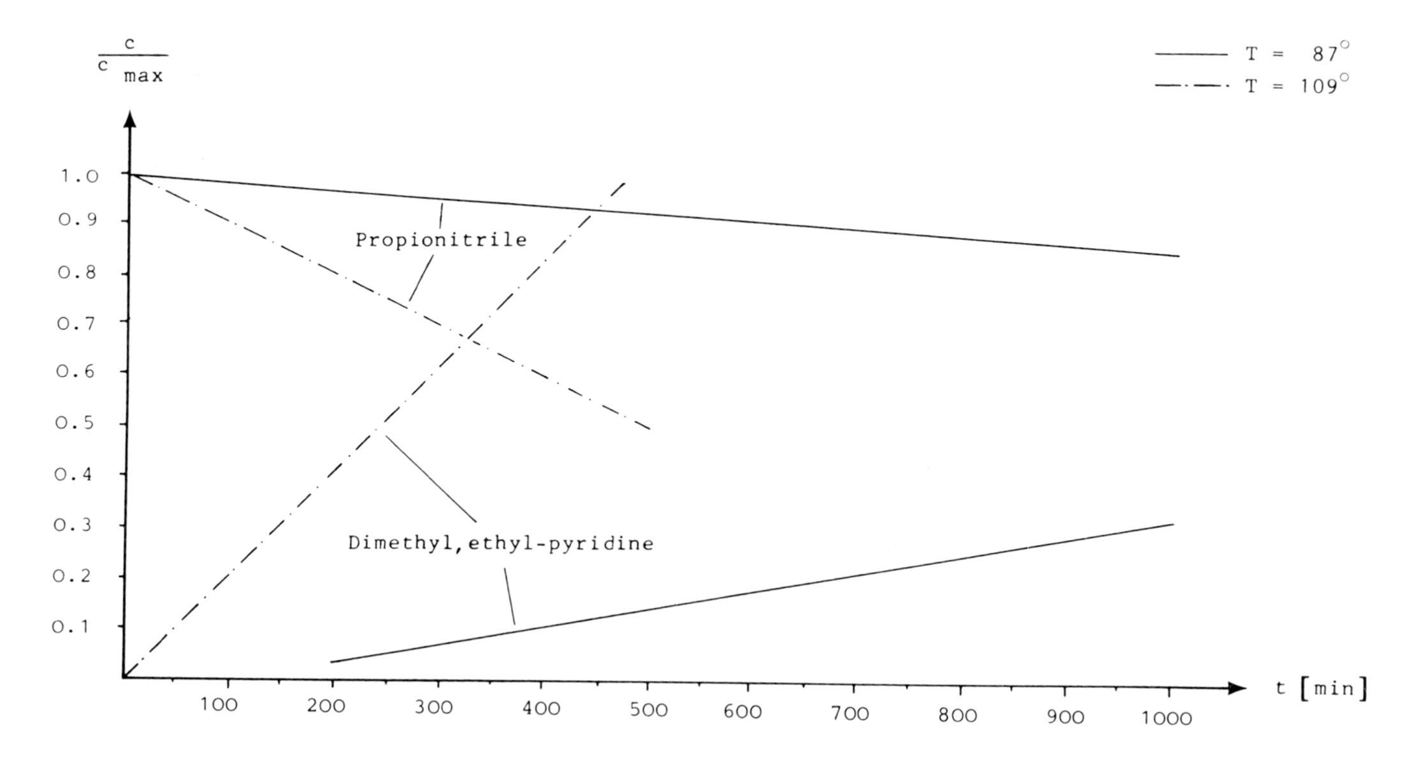

Fig. 13. The rate of formation of dimethyl,ethyl-pyridine at 87° and 109° at constant propyne concentration (c = concentration of propionitrile or dimethyl, ethyl-pyridine).

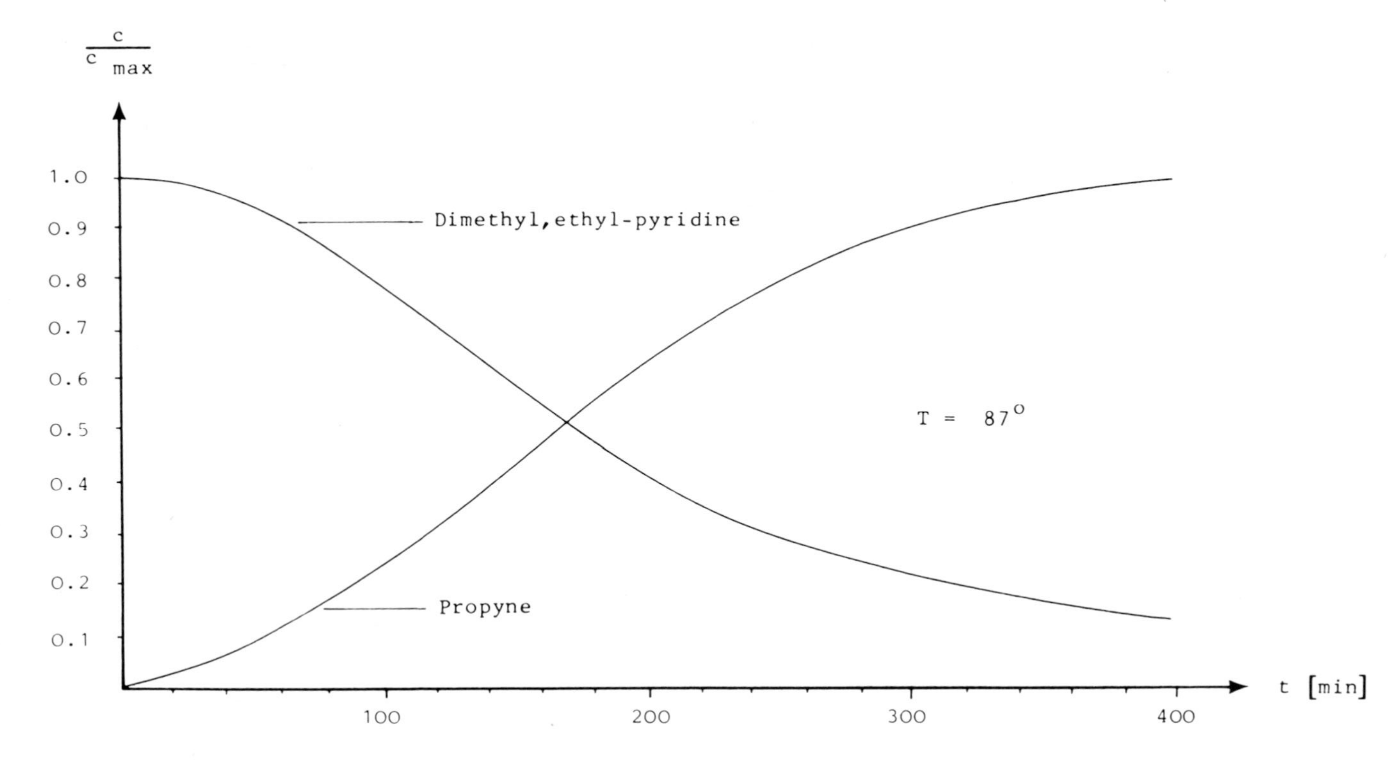

Fig. 14. The rate of formation of dimethyl, ethyl-pyridine and the rate of consumption of propyne at 87° and at constant propionitrile concentration (c = concentration of propyne or dimethyl, ethyl-pyridine).

– The rate of pyridine formation is independent of the nitrile concentration. Or in other words, it is unlikely that a nitrogen-containing intermediate is involved in the rate determining step.
– A common cobaltacyclopentadiene intermediate (cf. Equation 71) is responsible for both the pyridine and the benzene formation and may be regarded as a junction between the hetero- and carbocyclic pathways.

According to K. P. C. Vollhardt [18i] the cobaltacyclopentadiene intermediate has a greater affinity for nitriles than for alkynes and as a result the reaction proceeds preferentially to give pyridine rather then benzene [122].

A puzzling (though reproducible) observation is found in the initiation phase of the catalyzed reaction: instead of the line shape normally expected for second-order reactions (Figure 15a), the curve showing the pyridine concentration has an inflexion (Figure 15b). This effect could be due to association phenomena leading to the "active species" and suggest that more than one cobalt atom is involved in the catalyst formation. However, we have tested this using binuclear complexes such as **57** and **58** and the results do not account for the kinetic effect.

The nitrile triple bond can react in two ways with the mononuclear cobaltacycle:
– A Diels-Alder type of addition (Figure 16a) through a cobaltanorbornadiene intermediate followed by reductive elimination to yield the product and regenerate the Cp—Co core [94].
– Complexation of the nitrile to the cobalt atom (end-on or side-on) followed by insertion into the cobalt—carbon bond to give a seven-membered intermediate (Figure 16b).

H. Yamazaki *et al.* originally favoured the Diels-Alder addition (Figure 16a) [95] (without, however, presenting any experimental evidence) but later [91] they adopted the pathway involving the 7-membered cobaltacycle (Figure 16b). For the mechanistically-related cyclotrimerization of alkynes, R. G. Bergman *et al.* [94] have obtained kinetic evidence indicating that both pathways are possible; however, the Diels-Alder reaction seems to occur only with alkynes having strongly dienophilic character. Since alkylcyanides are known to be poor dienophiles, the pathway shown as Figure 16a appears to be unlikely. The observation that the thermal Diels-Alder addition of $CH_2{=}CHC{\equiv}N$ to, for example, butadiene occurs almost exclusively to the C=C double bond to give a cyclohexene derivative, whereas in the cobalt-catalyzed formation of 2-vinylpyridine via cobaltacyclopentadiene, reaction occurs exclusively through the C≡N triple bond (compare Equations 54 and 55), also points to a different mechanism. On the other hand, model-complexes supporting the intermediate shown in Figure 16b are not known. A cobaltacarbocycle, however, has been assumed to be formed by H. Yamazaki [96] during the reaction of alkynes with the cobaltacyclopentene complexes (Equation 72).

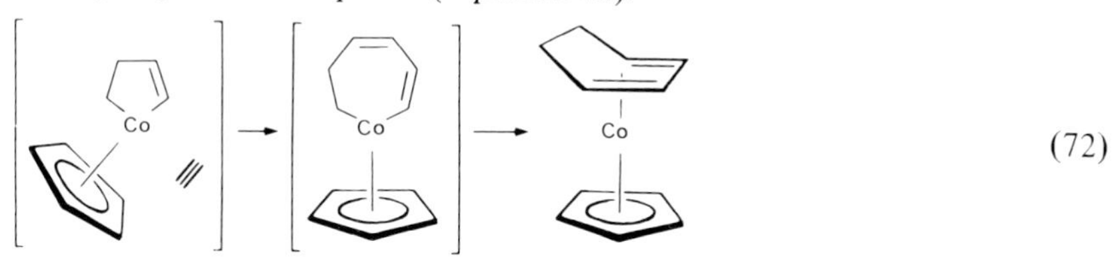

(72)

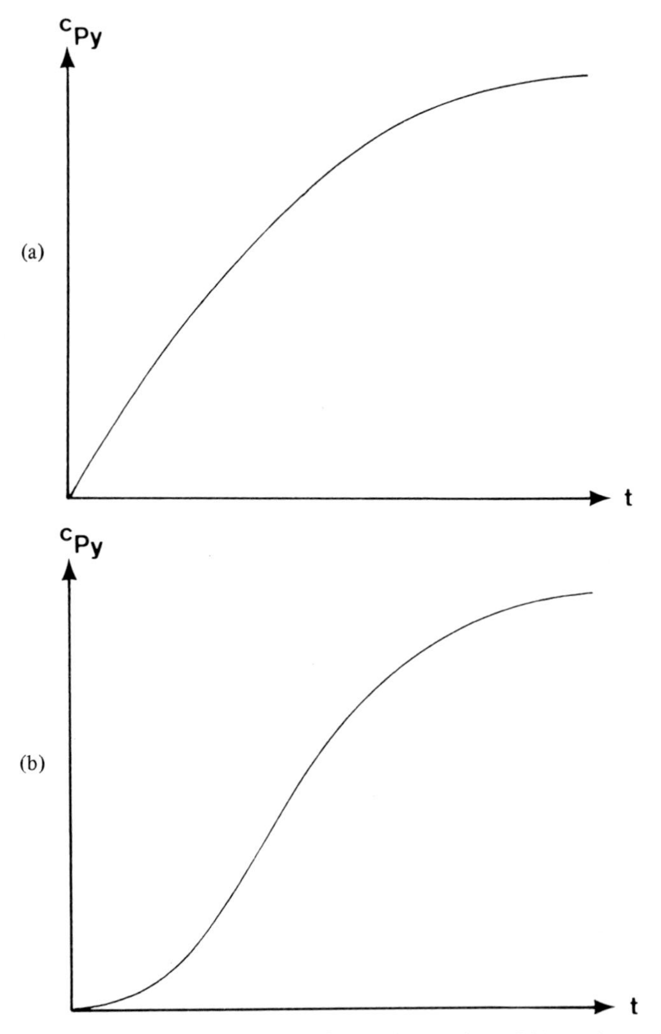

Fig. 15. (a) Typical line shape for 2nd order reactions. (b) Experimentally found line shape.

Interestingly, the reversal of this reaction has also been observed [97]: reaction of a substituted benzene coordinated to platinum is accompanied by ring opening (Equation 73).

(73)

$L = $ e.g. $P(CH_3)_3$

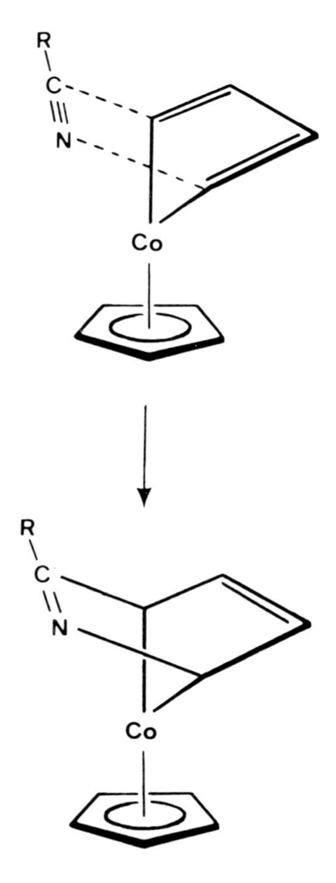

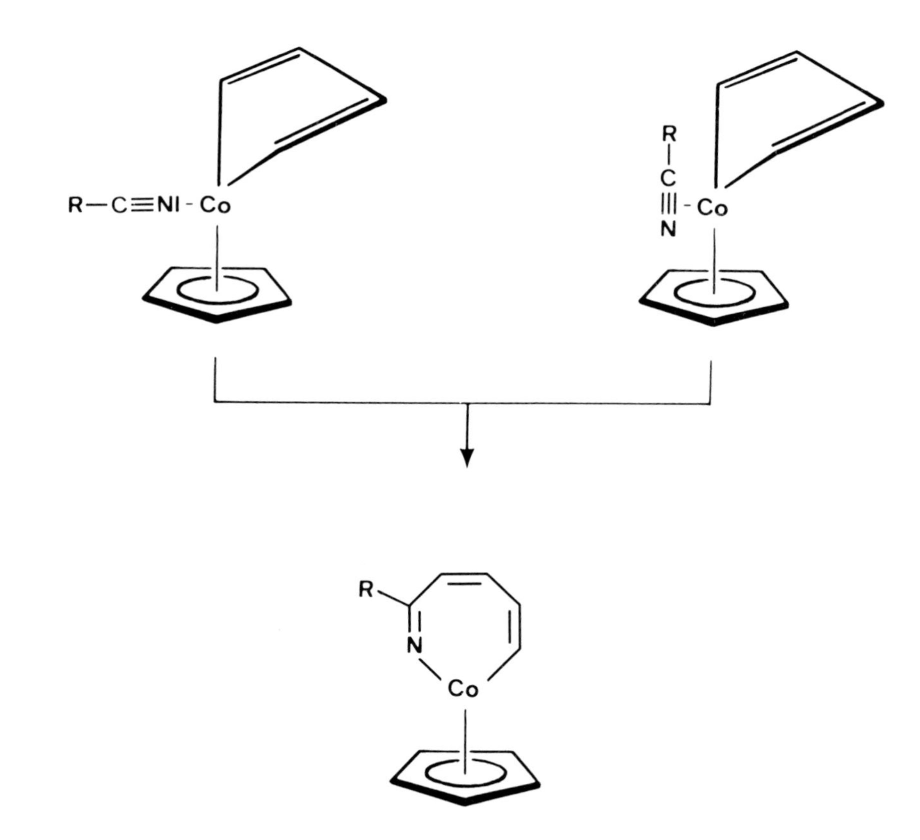

Fig. 16. The reaction of a nitrile with the cobaltacyclopentadiene intermediate: (a) Diels-Alder addition; (b) Insertion into the Co—C bond.

Analysis of the products formed by reacting monosubstituted alkynes with nitriles (see Equation 18) suggests that in the postulated seven-membered intermediate (Figure 16b) bonding to the Co atom occurs exclusively through the nitrogen atom and indeed R. Hoffmann and A. Stockis have carried out calculations which suggest that in such metallocycles heteroatoms more electronegative than carbon will preferentially adopt sites α to the metal [98].

Although no direct evidence is available as to the mechanism of the final reductive elimination step which leads to product formation, it has been suggested [99] that "mononuclear metallacycloheptatrienes are not likely to be very stable, since *cis*-reductive elimination of benzene from d^6-complexes can be a symmetry allowed and (probably) thermodynamically-favourable reaction" (Equation 74).

L_nM (metallacycloheptatriene) → L_nM + (benzene) (74)

Finally, it should be mentioned that rearrangement of the Cp-cobaltacyclopentadiene intermediate to the thermodynamically more stable [Co(η^4-butadiene)(η^5-Cp)] complex (Equation 75) (which is catalytically inactive) is a thermal forbidden process.

(Cp)Co(cobaltacyclopentadiene) ⇸ (Cp)Co(cyclobutadiene) (75)

In addition to contributing to our understanding of the course of the reaction, the results of our kinetic studies have the following practical consequences:

- Since the reaction rate is controlled by the square of alkyne concentration, increasing the concentration e.g. raising the pressure, variation of the solvent, will increase the TON.
- A high nitrile-alkyne ratio in the reaction mixture, i.e., using the nitrile as solvent, will enhance the probability of C≡N-insertion and increase the pyridine yield.

6. EXPERIMENTAL OPTIMIZATION OF THE CATALYTIC TURNOVER NUMBER

6.1. Introduction [92]

In order to improve any new catalytic process it is necessary to optimize systematically the reaction parameters. Since the cobalt catalyzed pyridine synthesis follows a rather complex time-dependence curve (apparent from the sigmoidal form of the concentration-curve, Figure 15b), any mathematical treatment of the kinetic data designed to calculate an optimal set of parameters for the catalysis is bound to fail. This case, which is not uncommon in molecular catalysis, is best handled by the

Simplex strategy which uses an essentially experimental approach to optimize the reaction parameters. The general direction of parameter variation has, initially, to be selected using a 2^n factorial design [100], [101].

Systematic optimization strategies [102] have, over the last 20 years been increasingly applied to chemical problems both in the laboratory and in industry. The mathematical background was, however, derived about 200 years ago by I. Newton [103a], L. Euler [103b], J.-L. Lagrange, C. F. Gauss and W. R. Hamilton. In spite of a number of theoretical contributions [106], [107], only a few practical examples of the solution of optimization problems by applying statistical methods have been reported in the chemical literature and these involve synthetic organic chemistry [104], [105]. The application of a factorial design for selecting the conditions in an experimental series is not, however, limited to problems involving chemical engineering and we have found it to be a very effective and time-saving approach for planning experiments in the laboratory since the number of experimental runs can be kept to a minimum. Significance-tests have, moreover, confirmed the reliability of the results [108].

In contrast to the scale up of an industrial process, where economics are of primary importance, the final goal of optimizing laboratory results is to reach the "technological optimum", i.e., maximum product yield, minimum formation of unwanted side products or, in the case of catalysis, a maximum in catalyst productivity. The evaluation of the productivity of the various cobalt complexes appeared to be the most important goal in our field since the synthesis of some of them require sophisticated and time consuming methods. In contrast, the nitrile and alkyne substrates are generally commercially available. We decided to optimize the turnover number (TON) of these processes which have potential industrial interest. As an example we will discuss here the optimization of the TON of complex **23** in the reaction of propyne and propionitrile to give dimethyl, ethyl-pyridine (Equation 31).

The apparatus we have developed is shown in Figure 17 and consists of 4 units: (1) a continuous-flow reactor linked to a process chromatography equipment consisting of (2) the analytical instruments (3) a control panel and (4) the data processor.

Solutions of the educts and catalyst **23** are pumped through the system which is controlled by electronic balances. The actual reaction is performed in an 87 ml continuous-flow reactor from which samples are taken automatically and analyzed by GC using glass capillary columns [109]. The analytical data are then reduced and processed automatically.

6.2. Description of the 2^n factorial design [110]

The following parameters could influence the TON of the catalyst:

(1) reaction temperature (T)

(2) mean reaction time (τ)

(3) concentration of catalyst in solution (c_{cat})

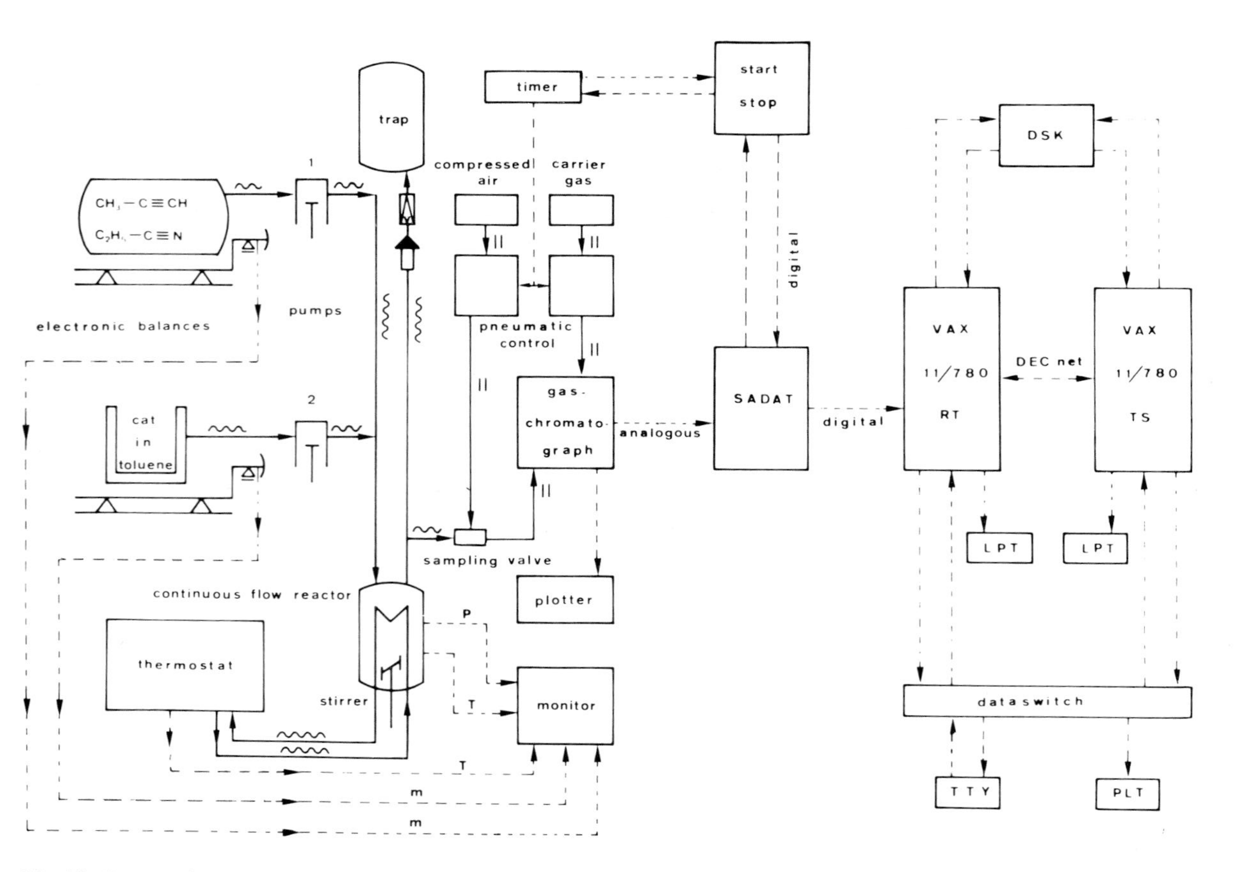

Fig. 17. Process chromatographic investigation of homogeneous catalyzed organic reactions.

(4) concentration of alkyne ($c_{C \equiv C}$)
(5) concentration of nitrile ($c_{C \equiv N}$)

Parameter 5 can be ignored since the kinetic measurements show the reaction rate to be independent of the nitrile concentration. In order to test the significance of the variables 1–4 upon the reaction (Equation 31), a 2^4 factorial experiment has to be carried out. The initial values for the different parameter combinations were derived from our kinetic data. The results were interpreted using the Yates-scheme [111], and are summarized in Table 14 and have a 95% significance level (Fisher-test). It became apparent that the mean reaction time (parameter 2) in the range 87–435 min could also be neglected since it had no effect on the TON of the catalyst. Parameters 1, 3 and 4 do, however, have an influence and an increase in the temperature and the alkyne-concentration combined with low concentrations of catalyst result in an increase in the TON.

It can be seen from Table 14 that a rather strong interaction exists between the catalyst concentration and the reaction temperature and this accounts for the non-linear response of the reaction to the variation of these parameters. The dependence of the response value Z_R (= TON) on the significant parameters and the interaction may be described by the following regression equation (Equation 76).

$$
\begin{aligned}
Z_R = {} & 8.03 - 3.94 \cdot \frac{c_{cat} - 62.4}{41.6} + 3.68 \cdot \frac{c_{C \equiv C} - 160}{80} + 5.65 \cdot \frac{T - 102.5}{12.5} \\
& - 1.70 \cdot \frac{c_{cat} - 62.4}{41.6} \cdot \frac{c_{C \equiv C} - 160}{80} \\
& - 3.99 \cdot \frac{c_{cat} - 62.4}{41.6} \cdot \frac{T - 102.5}{12.5} \\
& + 2.14 \cdot \frac{c_{C \equiv C} - 160}{80} \cdot \frac{T - 102.5}{12.5}
\end{aligned}
\tag{76}
$$

The concentrations have been measured as grams of substrate per liter of solvent while the temperature is in °C. The model shown in Equation 76 it satisfies the "lack-of-fit" test [112] at a safety level of 95%.

6.3. Optimization using the simplex method

We have chosen the *S*implex *S*elfdirecting *D*esign (SSD) for our experimental optimization [113]. This method is superior to the Gauss-Seidel strategy in terms of rapidity while the gradient strategy (which is generally thought to be the quickest method) is more complicated and less reliable.

Table 14

2^4 FACTORIAL DESIGN

Run No.	Code [a]	Data	1	2	3	4	Factor or interaction	Effect	MSQ mean square = $2^4 \cdot (\text{effect})^2$	Significance MSQ > 35.33
1	(1) [b]	0.08	0.53	3.00	55.11	128.54	M	8.03	1031.69	yes
13	a [c]	0.45	2.47	52.11	73.43	−63.00	CAT	−3.94	248.38	"
7	b	1.12	14.17	16.05	−29.63	58.80	C≡C	3.68	216.68	"
9	ab	1.35	37.94	57.38	−33.37	−27.14	CAT/C≡C	−1.70	46.24	"
2	c	11.25	2.85	0.60	25.71	90.44	T	5.65	510.76	"
14	ac	2.92	13.20	−30.23	33.09	−63.90	CAT/T	−3.99	254.72	"
8	bc	29.92	17.32	−0.15	−13.71	34.22	C≡C/T	2.14	73.27	"
10	abc	8.02	40.06	−33.22	−13.43	−23.56	CAT/C≡C/T	−1.47	34.57	no
3	d	1.05	0.37	1.94	49.11	18.32	τ	1.14	20.79	"
16	ad	1.80	0.23	23.77	41.33	−3.74	CAT/τ	−0.23	0.85	"
5	bd	7.05	−8.33	10.35	−30.83	7.38	C≡C/τ	0.46	3.38	"
12	abd	6.15	−21.90	22.74	−33.07	0.28	CAT/C≡C/τ	0.02	0.01	"
4	cd	14.02	0.75	−0.14	21.83	−7.78	T/τ	−0.49	3.84	"
15	acd	3.30	−0.90	−13.57	12.39	−2.24	CAT/T/τ	−0.14	0.31	"
6	bcd	31.28	−10.72	−1.65	−13.43	−9.44	C≡C/T/τ	−0.59	5.57	"
11	abcd	8.78	−22.50	−11.78	−10.13	3.30	CAT/C≡C/T/τ	0.21	0.70	"

[a] = For abbreviations and reading [108].
[b] (1) = All factors run on low level.
[c] a, b, c, d = Parameters run on high level.

The principle of the Simplex optimization strategy may easily be demonstrated with a 2 dimensional example (Figure 18). The starting point is three values (1–2–3)

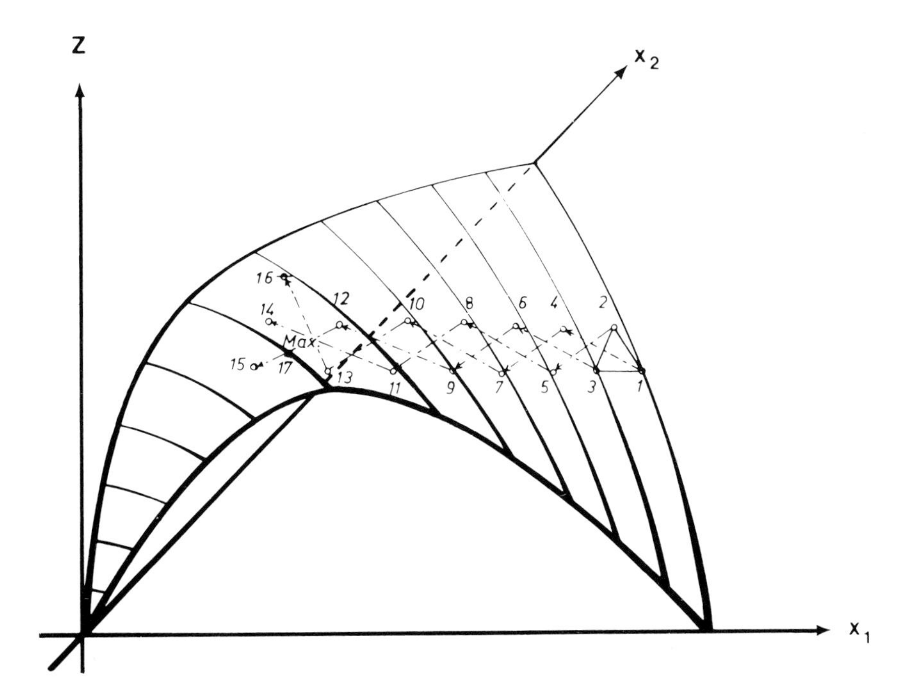

Fig. 18. Optimization by the 2-dimensional Simplex Method.

which form a triangle; point 1, having the coordinates x_{11} and x_{21}, obviously yields the lowest value and the results should improve if parameters are selected which fall on the dotted line leading to point 4.

Point 4 is selected simply by doubling the distance between point 1 and the centre of gravity of the line 2–3. The parameters for the next experiments are determined from triangle 2–3–4 by an analogous procedure. As indicated in Figure 18, the coordinates of the optimum are thus approached in a step-wise manner. In the case where the new parameter values yield a lower result (point 16 in Figure 18), the coordinates from the preceding experiment which gives the next lowest result serve as the starting point. Near the maximum the Simplex starts to rotate, i.e., any further parameter variation leads to a lower result. The optimum is finally encircled by decreasing the magnitude (f) of the consecutive steps in the parameter variation. Equation 77 leads to the coordinates of the new parameter combination x_{N+1+n}, x_j being the required value.

$$\underline{x}_{N+1+n} = \frac{r}{N} \cdot \sum_{i=n}^{N+n} \underline{x}_i - \left(1 + \frac{r}{N}\right) \cdot \underline{x}_j \;;\; n = 1, 2, 3, \cdots \;;\; r = f + 1 \qquad (77)$$

In the case of 3 factors a tetrahedron results ("3-Simplex"), and an N-Simplex would have $N+1$ edges, i.e., $N+1$ experiments have to be carried out prior to the experimental optimization of a reaction having N factors [102a]. After the 2^4 factorial experiment had revealed that the mean reaction time was not relevant to the optimization of the TON for catalyst **23** in our reaction (Equation 31), a start-Simplex consisting of 4 points had to be selected from the factorial experimental data. Each parameter has to occur once with a low value while the other two parameters had high values. The fourth point of our start-Simplex consists of a combination in which all parameters had high values (Table 14, run No. 9, 14, 8 and

Table 15

SIMPLEX SELFDIRECTING DESIGN (LIST OF THE NUMERICAL VALUES)

Run No.	c_{cat} (g/l)	Propyne (g)	T (°C)	Conversion (%)	TON
1	104	240	90	19	1.4
2	104	80	115	85	2.9
3	104	240	115	87	8.0
4	20.8	240	115	74	29.9
5	48.5	133	140	93	10.5
6	11.6	324	132	68	50.8
7	7.6	242	134	54	42.8
8	4.6	320	124	17	21.2
9	22.1	234	130	76	21.9
10	9.0	288	126	46	36.2
11	6.6	282	135	49	52.7
12	8.3	283	141	72	60.4
13	10.0	374	138	65	55.6
14	5.0	314	144	58	78.7
15	8.9	368	147	69	59.4
16	4.8	271	150	68	82.7
17	3.2	210	143	46	76.1
18	2.4	266	148	38	89.9
19	4.9	364	152	61	92.7
20	3.1	318	156	70	203.2
21	2.1	361	154	39	199.7
22	4.3	437	160	81	253.3
23	1.4	404	161	46	261.0
24	2.1	476	161	54	263.5
25	3.1	518	167	62	282.9
26	1.2	590	165	41	320.5
27	2.9	660	168	69	294.0
28	2.6	660	170	76	351.4
29	1.4	763	170	44	355.4
30	0.6	695	170	13	124.7
31	1.1	695	170	28	305.3
32	3.4	798	170	73	277.9
33	3.0	798	170	74	312.1
34	2.0	798	170	64	398.5

10). The stepmagnitude (*f*) had to be decreased several times during the optimization process and the Simplex was adapted to the experimental limitations of the parameter variation by means of contraction and reflection procedures. Table 15 lists the numerical values derived from our optimization experiments.

The results of the TON-optimization using the catalyst **23** in the dimethyl, ethyl-pyridine synthesis may be summarized as follows (Figure 19):

(1) A TON of 400 moles of pyridine per g-atom of cobalt can be achieved under steady state conditions by using the best set of reaction conditions (see Table 15). Thus corresponds to a thirteen-fold improvement on the start-Simplex.

(2) The optimum concentration of the catalyst in the feed stream was found to lie between 1.5 and 2.5 g/l. The maximum TON was found at 2.0 g/l which corresponds to 0.25 mmoles of catalyst **23** present in the reaction volume.

(3) The optimum concentration of propyne in the feed stream is 360 g/l corresponding to 800 g propyne in 1 l propionitrile.

(4) The propyne conversion can be varied between 45% and 75% depending upon the reaction temperature.

To summarize, the production of dimethyl, ethyl-pyridines using catalyst **23** is in practice best performed using a molar nitrile: cobalt ratio of 500 and a molar alkyne: cobalt ratio of 1500. The preferred reaction temperature in our case was 170° but was limited by experimental factors; higher temperatures could conceivably lead to higher TON values.

7. SYNTHESIS OF THE CATALYSTS

7.1. Preparation procedures

7.1.1. η^3-ALLYLCOBALT OLEFIN COMPLEXES

7.1.1.1. *[Co(η^4-butadiene) ($\eta^{2,3}$-5-methylheptadienyl)]* **20** [26]

Anhydrous $CoCl_2$ (65 g, 0.5 moles) is suspended at –30° under argon in 1.5 l liquid butadiene in a 4 l three-necked flask. 500 ml dry ethanol are added from a dropping funnel over 15 min. The resulting dark blue suspension is stirred rapidly at –30° and a freshly prepared solution of $NaBH_4$ (38 g, 1 mole) in 1.5 l dry ethanol is added over 4 h. During the reaction gas is evolved and the colour changes first to pink, then to greenish-grey and finally becomes brown. Stirring is continued for 2–3 h and the mixture is then allowed to stand for 15 h at –30°. The supernatant liquid is carefully separated from the precipitate (NaCl), concentrated under high vacuum at –80° to –30° during which part of the red-brown product usually begins to precipitate. Crystallization is completed at –90° to –100° and the crude product is collected and washed with dry pentane (4 × 100 ml). The resulting red-brown crystals of **20** are dried at –30° under high vacuum.

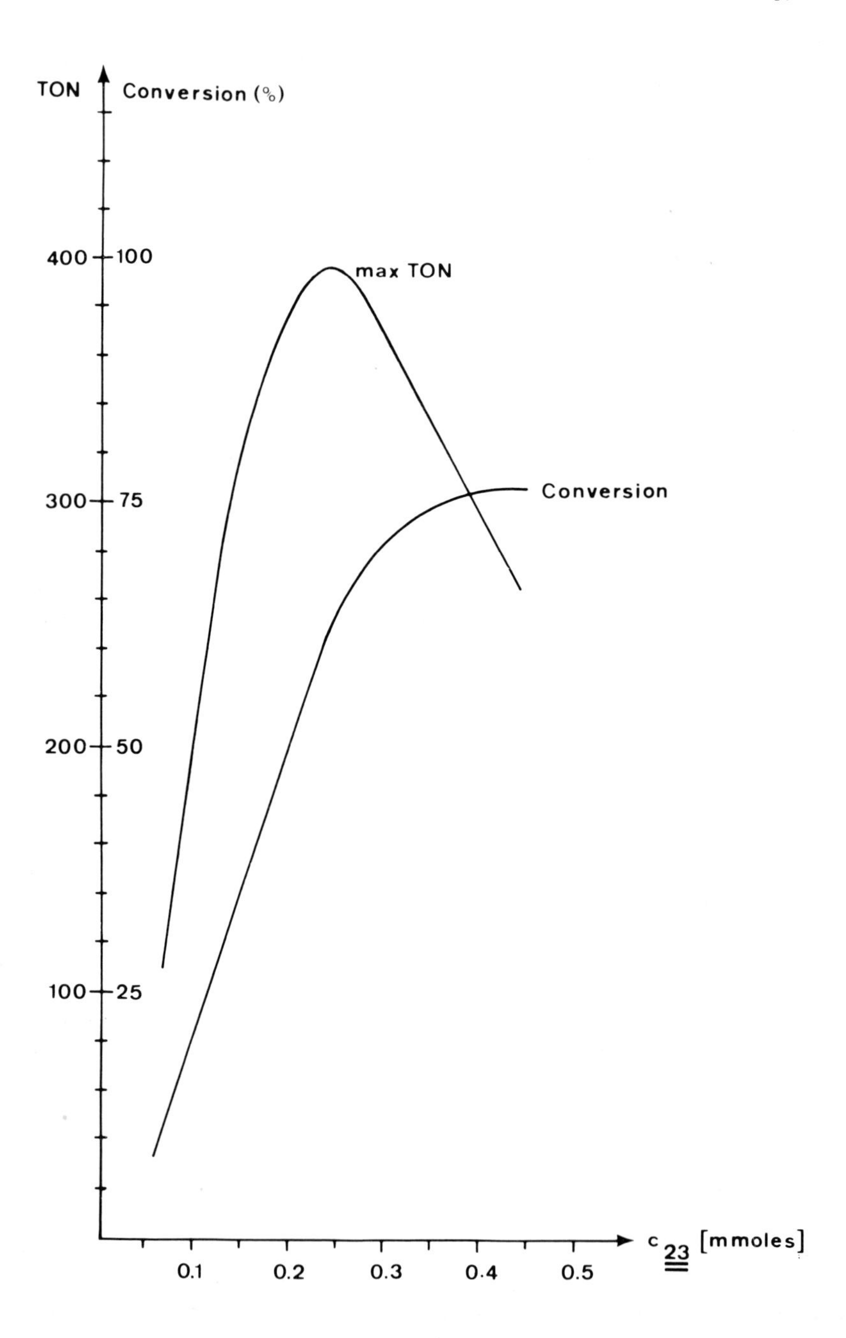

Fig. 19. Optimization (graphical) of the synthesis of dimethyl, ethyl-pyridine catalyzed by **23**.

Yield: 77.7 g of **20** (350 mmoles; 70% theory).
The compound is air sensitive, decomposes at about $-30°$ and should be stored at $-80°$.

7.1.1.2. *[Co(η^4-cod)(η^3-cyclooctenyl)]* **21**

This complex is prepared by the one-pot reduction of $Coacac_3$, in the presence of cod, with triethylaluminium (Variant C1 of Equation 25):
Cobalt(III)acetylacetonate (112g, 0.31 moles), cod (123 g, 1.1 moles) (dried over $NaAlEt_4$) and 1 l of pentane are vigorously stirred in a 2 l three-necked flask under argon. The resulting green suspension is cooled at 0° to $-5°$ and triethylaluminium (107 g, 0.94 moles) is added from a dropping funnel over 45 min. After 10% of the reducing agent has been added, the colour changes to brown and vigorous gas evolution is observed. At the end of the addition a dark brown solution is obtained which is stirred for a further 2.5 h at +20°. $Alacac_3$, unreacted $Coacac_3$ and a brown residue are removed by filtration through a glass-frit. The clear filtrate is held at $-30°$ for 14–16 h which causes small, colourless crystals of $Alacac_3$ to precipitate and these are filtered off at $-30°$. Cooling the solution over 2 h to $-60°$ to $-65°$ causes the product to crystallize (if necessary this can be initiated by scratching or by seeding with a dry crystal). Temperatures below $-65°$ cause the co-precipitation of Et_2Alacac and should be avoided.

The supernatant liquid is filtered off and the product washed at $-65°$ to $-70°$ (3 × 150 ml pentane). The brown-black crystals of **21** obtained after drying for 3.5 h at +20° in high vacuum are analytically pure.
Yield: 59.8 g of **21** (216.7 mmoles, 69.9% theory).

7.1.2. CATALYSTS CONTAINING η^5-CYCLOPENTADIENYL- AND η^6-BORINATO LIGANDS

7.1.2.1. *[Co(η^4-cod)(η^5-Cp)]* **23**

This versatile catalyst may be obtained in an overall yield of 86%, based on the cobalt carbonyl used, by applying the two-step Method A, discussed in Section 2.3 (Equations 21 and 22). A convenient alternative is one-pot reduction of the cobalt salt in the presence of the olefin (Method C, Section 2.3, Equation 25):

Method C1, using triethylaluminium:
A mixture of cyclopentadiene (6.8 g, 103 mmoles), cod (27 g, 250 mmoles) and triethylaluminium (42.4 g, 370 mmoles) in 200 ml pentane is vigorously stirred under argon at 0° to $-10°$. Cobalt(III)acetylacetonate (35.6 g, 100 mmoles) is added in small portions. A vigorous gas evolution occurs. After the addition of the cobalt salt, the mixture is allowed to warm up to room temperature over 2 h and is then filtered through a glass-frit and the filtrate cooled to $-50°$. The orange-brown

product crystallizes out in about 16 h. Then the supernatant liquid is removed, and the crystals washed with pentane (2 X 50 ml) at –50° and dried under vacuum.
Yield: 17.8 g of **23** (76.7 mmoles, 76.7% theory).

Method C2, using anthracene-activated magnesium [47c], [116]:
The reactivity of magnesium metal is dependent, among others, upon the extent and purity of the active surface [114]. Although many attempts have been made to increase the reactivity by the addition of activators and accelerators [115], until now no completely satisfactory preparative procedure has been described for the production of transition metal complexes using magnesium as the reducing agent. Together with B. Bogdanović, we have demonstrated that the addition of a catalytic amount of anthracene to the reaction mixture in THF provides a highly active magnesium which can serve as a powerful reducing agent [116] and have used this reagent to prepare [Co(η^4-diolefin) (η^5-Cp)] complexes in a one step procedure (Equation 25). The application of magnesium seems to be more attractive for this purpose than any other alkali or alkaline-earth metals in terms of economy, toxicity, safety and availability.

Experimental procedure:
1.1 g (6.2 mmoles) of anthracene, 300 ml of THF and 0.1 ml of methyliodide are added with stirring in an inert gas atmosphere at 20° to 7.2 g (300 mmoles) of magnesium powder (particle size 0.15 mm). After about 2 h, the reaction mixture is activated by suspending the reaction vessel for 3 h in an ultrasonic bath (continuous peak HF output 240 Watt, 35 kHz). The mixture of orange magnesium-anthracene is then heated with stirring to 60°. After the addition of 27.0 g (250 mmoles) of cod and 7.3 g (111 mmoles) of monomeric cyclopentadiene, the source of heat is removed and 35.6 g (100 mmoles) of solid cobalt(III) acetylacetonate added over 30 min. An exothermic reaction takes place and the temperature increases to 73°, the mixture becoming deep red-brown. After cooling to 23°, the reaction mixture is filtered through a G-3 glass-frit and evaporated to dryness (max. bath temperature 30°). The residue is taken up in 500 ml of pentane, filtered and washed with a total of 500 ml pentane in several portions. The clear red filtrate is evaporated to dryness (max. bath temperature 30°) under vacuum and the residue sublimed at 0.15 Pa (sublimation begins at 30° and was continued up to 60°). The sublimate is dissolved in ca. 100 ml of pentane and the complex deposited by cooling to –80° for 16 h. The supernatant liquid is removed and the crystals of **23** are washed with pentane (3 X 50 ml) at –80° and dried in high vacuum.
Yield: 18.4 g of **23** (79.1 mmoles, 79.1% theory).

7.1.2.2. *[Co(η^4-cod) (η^5-MeCp)]* **24**

Method C2:

Starting materials:	Magnesium	7.2 g	=	300.0 mmoles
	Anthracene	1.1 g	=	6.2 mmoles

THF	300	ml	
Methyliodide	0.1	ml	
Methyl-CpH	8.8	g	= 110.0 mmoles
Cod	27.0	g	= 250.0 mmoles
$Coacac_3$	35.6	g	= 100.0 mmoles

Yield: 17.5 g of **24** (71.1 mmoles, 71.1% of theory).

Characterization of **24**:
Yellow-brown crystals (mp –15°) from pentane by recrystallization at –80°.

^{1}H-NMR ($CDCl_3$, 80 MHz):
δH_1: 4.58 (m)
δH_2: 4.27 (m)
δH_3: 3.22 (m)
δH_4: 2.34 (m)
δH_5: 1.57 (m)
δH_6: 1.55 (m)

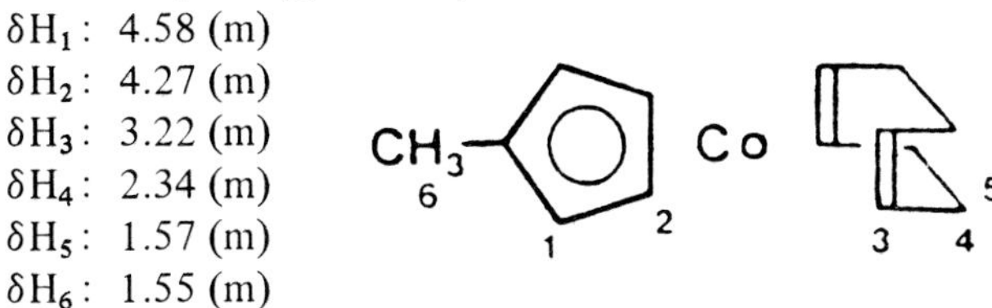

^{13}C-NMR (d_8-toluene):
δC_1: 64.7 (d, $J_{C,H}$ = 153 Hz)
δC_2: 32.4 (t, $J_{C,H}$ = 125 Hz)
δC_3: 85.2 (d, $J_{C,H}$ = 176 Hz)
δC_4: 83.2 (d, $J_{C,H}$ = 173 Hz)
δC_5: 95.6 (s)
δC_6: 12.3 (q, $J_{C,H}$ = 126 Hz)

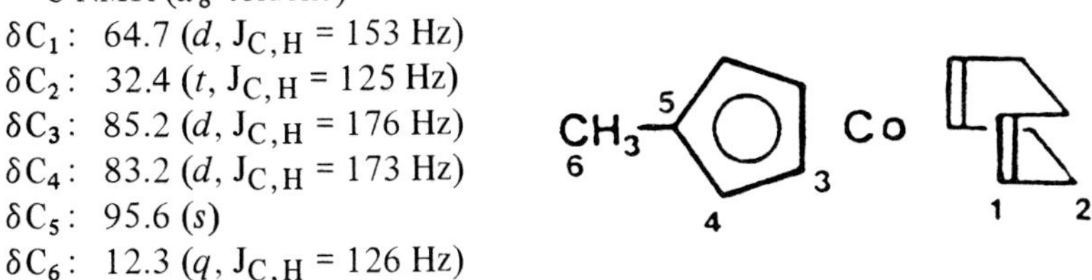

Mass spectrum:
m/z: 246 (M^+, 78%); 216 (90%); 138 (100%); 59 (53%).

7.1.2.3. *[Co(η^4-cod)(η^5-t-BuCp)]* **25**

Method C2:

Starting materials:				
Starting materials:	Magnesium	3.6	g	= 150.0 mmoles
	Anthracene	0.55	g	= 3.1 mmoles
	THF	150	ml	
	Methyliodide	0.05	ml	
	t-Butyl-CpH	7.0	g	= 57.4 mmoles
	Cod	13.5	g	= 125.0 mmoles
	$Coacac_3$	17.8	g	= 50.0 mmoles

Yield: 6.0 g of **25** (20.83 mmoles, 41.7% of theory).

Characterization of **25**:
Dark brown crystals (mp 45°) by sublimation at a bath temperature of 80° to 150°/0.15 Pa, and recrystallization from pentane at –80°. Solutions of **25** are very air sensitive.

Elemental analysis:
Observed: C = 70.93%; H = 8.62%; Co = 20.52%
Calculated: C = 70.82%; H = 8.74%; Co = 20.44%

^{1}H-NMR (d_8-toluene, 80 MHz):
δH_1: 4.60 (t, J = 2.2 Hz)
δH_2: 3.39 (t, J = 2.2 Hz)
δH_3: 3.41 (m)
δH_4: 2.43 (m)
δH_5: 1.65 (m)
δH_6: 1.39 (s)

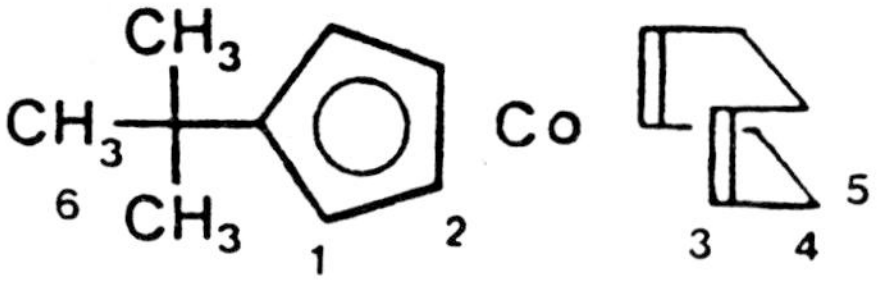

CH3 CH3 6 CH3 1 2 Co 3 4 5

^{13}C-NMR (d_8-toluene):
δC_1: 63.5
δC_2: 32.4
δC_3: –
δC_4: 83.6
δC_5: 80.4
δC_6: 31.8
δC_7: 32.3

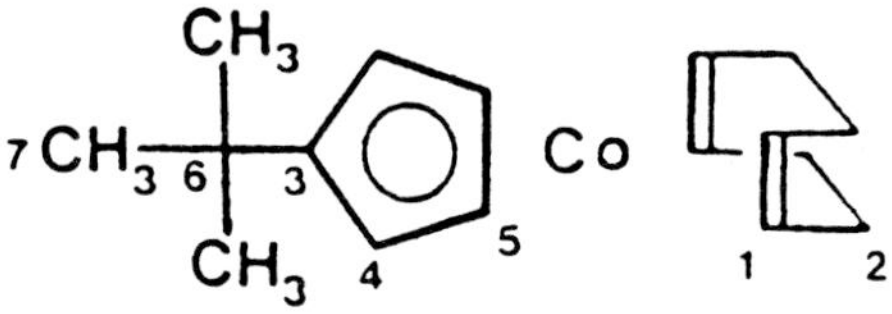

CH3 7CH3 6 3 CH3 4 5 Co 1 2

Mass spectrum:
m/z: 288 (M^+, 61%); 231 (93%); 229 (100%); 164 (54%); 137 (28%); 125 (31%); 59 (47%).

7.1.2.4. *[Co(η^4-cod)(η^5-Me$_5$Cp)]* **27**

Method A:
Step 1 (see Equation 21):
[$Co_2(CO)_8$] (25.9 g, 75.7 mmoles) is dissolved in 300 ml cyclohexane and acetyl-pentamethyl cyclopentadiene [117] (10.4 g, 58.4 mmoles) added. The solution is kept for 3 d at 80°, the solvent evaporated and the residue extracted with pentane (3 X 150 ml). The mixture is filtered and concentrated to 60 ml. The organocobalt carbonyl is isolated by chromatography through an Al_2O_3 column (neutral, activity II) using pentane as the eluent. Evaporation gives red-brown crystals mp 54°.
Yield: 4.0 g (16 mmoles, 28% theory).

Step 2:
[$Co(CO)_2(\eta^5$-$Me_5Cp)$] (1.49 g, 5.6 mmoles) are dissolved in 150 ml cod and irradiated (Hg UV-lamp). 236.8 ml (10.57 mmoles) CO are evolved over 3 h and the colour of the solution changes to bright orange-brown. Cod is removed *in vacuo*, the yellow-brown crystalline residue extracted with a 1 : 1 ether-pentane mixture and the solution cooled to –80°. Complex **27** precipitates as orange crystals (mp 166°) which were dried *in vacuo*.
Yield: 1.3 g of **27** (4.3 mmoles, 76.7% theory).

Characterization of **27**:
Elemental analysis:

Observed:	C = 71.67%;	H = 8.83%;	Co = 19.46%
Calculated:	C = 71.52%;	H = 8.94%;	Co = 19.53%

^{13}C-NMR (d_8-toluene):
δC_1: 70.5
δC_2: 32.6
δC_3: 91.6
δC_4: 9.2

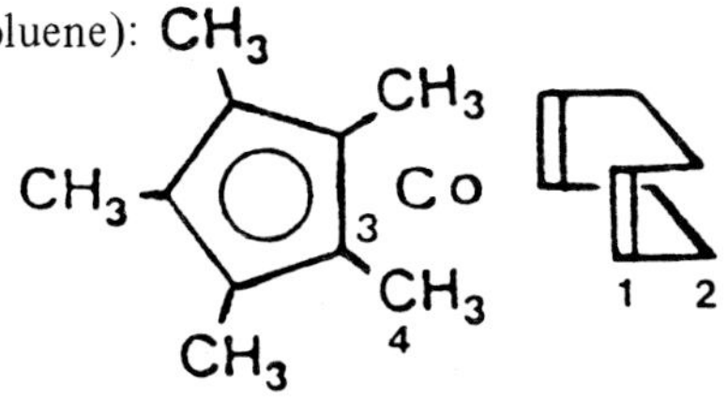

CH3 CH3 CH3 Co 3 CH3 4 CH3 1 2

Mass spectrum:
m/z: 302 (M^+, 100%); 285 (94%); 272 (84%); 194 (57%); 192 (72%); 133 (78%); 59 (34%).

7.1.2.5. *[Co(η⁴-cod) (η⁵-PhCp)]* **28**

Method C2:

Starting materials:			
	Magnesium	7.2 g	= 300.0 mmoles
	Anthracene	1.1 g	= 6.2 mmoles
	THF	300 ml	
	Methyliodide	0.1 ml	
	Phenyl-CpH	15.0 g	= 105.0 mmoles
	Cod	28.5 g	= 264.0 mmoles
	$Coacac_3$	35.5 g	= 99.7 mmoles

Yield: 0.3 g of **28** (0.98 mmoles, 1% theory).
Isolated as copper-red flakes (mp 61–62°) by chromatography using a 70 cm SiO_2 column (eluent: pentane) followed by recrystallization from pentane at –50°.

Characterization of **28**:
Elemental analysis:

Observed:	C = 74.80%;	H = 7.00%;	Co = 18.74%
Calculated:	C = 74.03%;	H = 6.82%;	Co = 19.16%

^{13}C-NMR (d_8-toluene)
δC_1: 66.1 (d, $J_{C,H}$ = 154 Hz)
δC_2: 32.1 (t, $J_{C,H}$ = 125 Hz)
δC_3: 81.5 (d, $J_{C,H}$ = 173 Hz)
δC_4: 84.5 (d, $J_{C,H}$ = 174 Hz)
δC_5: 99.1 (s)
δC_6: 135.7 (s)
δC_7: 128.7 (d)
δC_8: 125.8 (d)
δC_9: 126.4 (d)

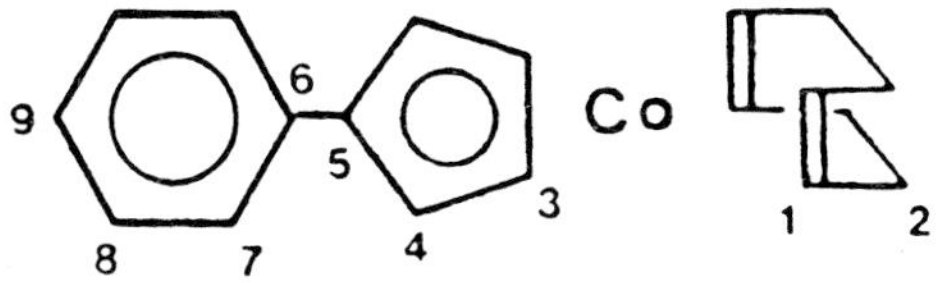

9 6 5 Co 3 4 8 7 1 2

Mass spectrum:
m/z: 308 (M^+, 98%); 278 (68%); 200 (100%); 141 (64%); 59 (42%).

7.1.2.6. *[Co(η^4-cod) (η^5-Me_3SiCp)]* **31**

Method C2:

Starting materials:	Magnesium	7.2 g	= 300.0 mmoles
	Anthracene	1.1 g	= 6.2 mmoles
	THF	300 ml	
	Methyliodide	0.1 ml	
	Me_3Si-CpH	15.2 g	= 110.0 mmoles
	Cod	27.0 g	= 250.0 mmoles
	$Coacac_3$	35.6 g	= 100.0 mmoles

Yield: 21.0 g of **31** (70.0 mmoles, 70% theory) mp 64°.
The catalyst has been described in [41].

7.1.2.7. *[Co(η^4-cod) (η^5-indenyl)]* **32**

Method B (see Equation 24):
3.8 g (17.1 mmoles) of Natta's complex **20**, dissolved in 100 ml pentane, is mixed with 1.98 g (17.1 mmoles) indene at –30°. The reaction mixture is allowed to reach room temperature and stirred for a further 2 h. About 5 ml cod is then added. The clear red solution is heated under reflux for ca. 24 h. After cooling, the precipitate, if any, is filtered off and the solution is concentrated to about 30 ml and slowly cooled to –50°. Brown-red crystals are deposited. The supernatant mother liquor is then removed and the crystals (red-brown needles) dried for 1 h at room temperature under high vacuum (0.01 Pa).
Yield: 2.4 g of **32** (8.5 mmoles, 50% theory) mp 98°.
The catalyst has been described in [41].

Method C2:

Starting materials:	Magnesium	2.6 g	= 108.3 mmoles
	Anthracene	1.1 g	= 6.2 mmoles
	THF	300 ml	
	Methyliodide	0.1 ml	
	Indene	29.0 g	= 250.0 mmoles
	Cod	27.0 g	= 250.0 mmoles
	$CoCl_2$	13.0 g	= 100.0 mmoles

Yield: 20.8 g of **32** (73.8 mmoles, 73.8% theory) mp 98°.

Method C3 (using lithium metal as reducing agent):
Finely divided lithium (1.4 g, 200 mmoles) is suspended under argon in 300 ml THF and heated to 65°. Cod (27.0 g, 250 mmoles) and indene (29.0 g, 250 mmoles) are added. The source of heat is removed and $CoCl_2$ (13.0 g, 100 mmoles) added over 10 min. An exothermic reaction occurs and the colour changes through red to dark red-brown. The mixture is allowed to cool to room temperature, filtered through a glass-frit and the solvent removed (40°/0.1 Pa). The residue is extracted with pentane

(3 X 100 ml) and the red-brown filtrate cooled to –80°. The product is deposited overnight.
Yield 20.7 g of **32** (73.4 mmoles, 73.4% theory).

7.1.2.8. *[Co(η^4-cod) (η^5-1-Me-indenyl)]* **33**

Method C2:

Starting materials:			
	Magnesium	7.2 g	= 300.0 mmoles
	Anthracene	1.1 g	= 6.2 mmoles
	THF	300 ml	
	Methyliodide	0.1 ml	
	1-Methylindene	13.8 g	= 107.0 mmoles
	Cod	27.3 g	= 253.0 mmoles
	Coacac$_3$	35.6 g	= 100.0 mmoles

Yield: 14.3 g of **33** (48.3 mmoles, 48.3% of theoretical), mp 72°–73°.

Characterization of **33**:
Elemental analysis:

Observed:	C = 72.97%;	H = 7.09%;	Co = 19.93%
Calculated:	C = 72.92%;	H = 7.14%;	Co = 19.98%

IR-analysis:
2820–3000; 1468; 1452; 1424; 1370; 1335; 1318; 1205; 845; 810 cm^{-1}.

^{1}H-NMR (d_8-toluene)
δH_1: 5.59 (*d*, J = 2.8 Hz)
δH_2: 3.82 (*d*, J = 2.8 Hz)
δH_3: 7.20 (*m*)
δH_4: 1.30 (*s*)
δH_5: 3.21 (*m*)
δH_6: 2.12 (*m*)
δH_7: 1.50 (*m*)

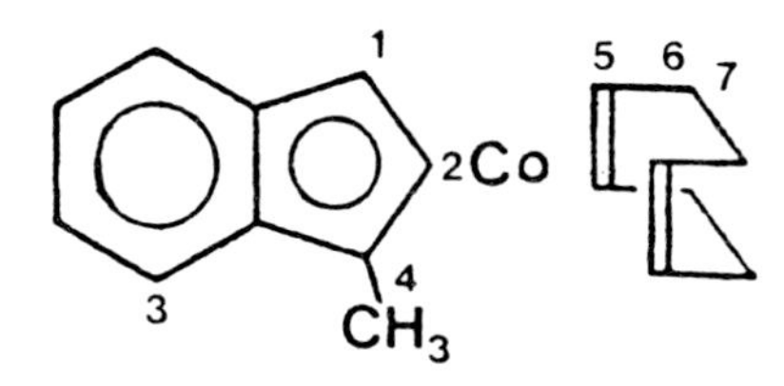

^{13}C-NMR (d_8-toluene)
$\delta C_{1/4}$: 69.7
δC_2: 31.9
δC_3: 30.8
$\delta C_{4/1}$: 67.7
δC_5: 85.0
δC_6: 89.8
δC_7: 73.9
δC_8: 105.6
δC_9: 105.4
$\delta C_{10/13}$: 121.1
$\delta C_{11/12}$: 124.0
$\delta C_{13/10}$: 123.4
δC_{14}: 9.8

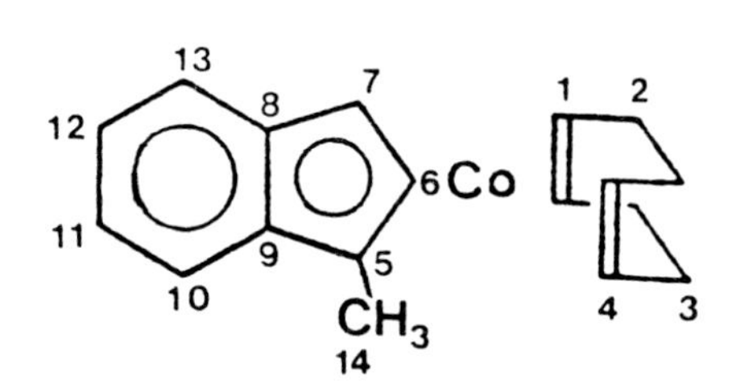

Mass spectrum:
m/z: 296 (100%); 266 (44%); 188 (86%); 129 (42%); 128 (60%); 113 (38%); 59 (39%).

7.1.2.9. *[Co(η^4-cod) (η^5-1-Me$_3$Si-indenyl)]* **34**

Method C2:

Starting materials:	Magnesium	7.2 g	= 300.0 mmoles
	Anthracene	1.1 g	= 6.2 mmoles
	THF	300 ml	
	Methyliodide	0.1 ml	
	1-Me_3Si-indene	21.6 g	= 115.0 mmoles
	Cod	27.0 g	= 250.0 mmoles
	Coacac$_3$	35.6 g	= 100.0 mmoles

Yield: 17.9 g of **34** (50.6 mmoles, 50.6% of theory) as a dark red viscous oil (bp 150–190°, 0.1 Pa).

Characterization of **34**:
Elemental analysis:

Observed:	C = 67.84%;	H = 7.66%;	Co = 16.55%;	Si = 7.86%
Calculated:	C = 67.77%;	H = 7.68%;	Co = 16.63%;	Si = 7.92%

^{1}H-NMR (d_8-toluene)
δH_1: 5.29 (*d*, J = 2.7 Hz)
δH_2: 3.97 (*d*, J = 2.7 Hz)
δH_3: 7.50 (*m*)
δH_4: 7.08 (*m*)
δH_5: 3.50 (*m*)
δH_6: 3.03 (*m*)
δH_7: 2.08 (*m*)
δH_8: 1.35 (*m*)
δH_9: 0.34 (*s*)

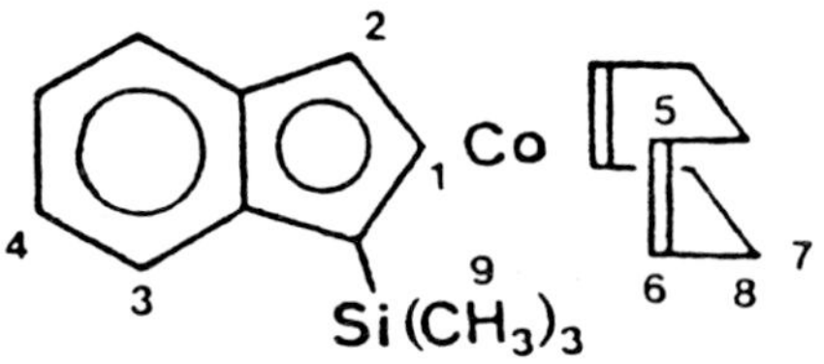

^{13}C-NMR (d_8-toluene)
$\delta C_{1/2}$: 68.4
$\delta C_{2/1}$: 66.1
$\delta C_{3/4}$: 31.6
$\delta C_{4/3}$: 31.3
δC_5: 78.8
δC_6: 94.5
δC_7: 78.9
$\delta C_{8/9}$: 108.8
$\delta C_{9/8}$: 108.5
$\delta C_{10/13}$: 125.0
$\delta C_{13/10}$: 124.7
$\delta C_{11/12}$: 123.6
$\delta C_{12/11}$: 123.5
δC_{14}: –0.1

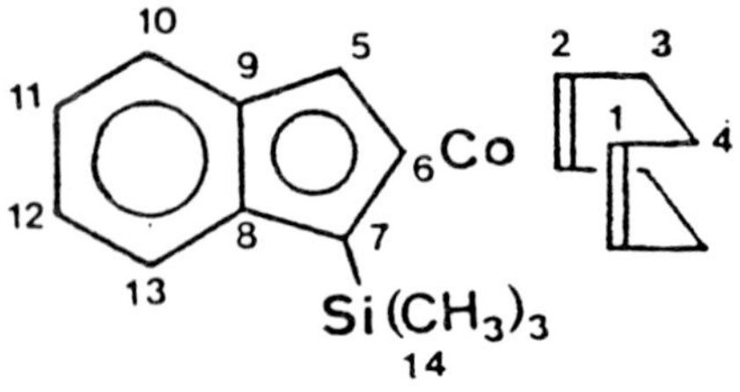

Mass spectrum:
m/z: 354 (M^+, 100%); 279 (63%); 246 (25%); 59 (17%).

7.1.2.10. *[Co(η^4-cod) (η^6-1-phenylborinato)]* **36**

Method C2:

Starting materials:			
	Magnesium	146 mg	= 6 mmoles
	Anthracene	24 mg	= 0.13 mmoles
	THF	6 ml	
	Methyliodide	1 drop	
	1-Phenyl-1,4-dihydroborabenzene *	340 mg	= 2.27 mmoles
	Cod	550 mg	= 5.09 mmoles
	$Coacac_3$	710 mg	= 1.99 mmoles

Yield: 270 mg of **36** (0.84 mmoles, 42.2% theory) mp 169°.

7.1.2.11. *[Co(η^4-cyclopentadiene) (η^5-Cp)]* **40**

Method C2:

Starting materials:			
	Magnesium	14.4 g	= 600 mmoles
	Anthracene	1.1 g	= 6.2 mmoles
	THF	500 ml	
	Methyliodide	0.1 ml	
	Cyclopentadiene	105.6 g	= 1600 mmoles
	$Coacac_3$	71.0	= 199 mmoles

Yield: 23.7 g of **40** (124.5 mmoles, 62% theory) mp 99° [57].

7.1.2.12. *[Co(η^4-norbornadiene) (η^5-indenyl)]* **53**

Method C:

Starting materials:			
	Magnesium	7.2 g	= 300.0 mmoles
	Anthracene	1.1 g	= 6.2 mmoles
	THF	300 ml	
	Methyliodide	0.1 ml	
	Indene	29.0 g	= 250.0 mmoles
	Norbornadiene	23.0 g	= 250.0 mmoles
	$Coacac_3$	35.6 g	= 100.0 mmoles

Yield: 8.1 g of **53** (30.45 mmoles, 30.45% theory) as red-brown needles (mp 58°) from pentane at −80°.

Characterization of **53**:
Elemental analysis:
Observed: C = 72.32%; H = 5.52%; Co = 22.12%
Calculated: C = 72.19%; H = 5.64%; Co = 22.15%

* We were unable to repeat the literature preparation [118a]. The reaction described here was carried out using a sample kindly supplied by G. E. Herberich and E. Raabe [118b].

^{1}H-NMR (d_8-toluene, 80 MHz)
δH_1: 7.14 (*s*)
δH_2: 5.75 (*d*, J = 2.0 Hz)
δH_3: 4.03 (*d*, J = 2.0 Hz)
δH_4: 2.81 (*m*)
δH_5: 2.64 (*m*)
δH_6: 0.64 (*m*)

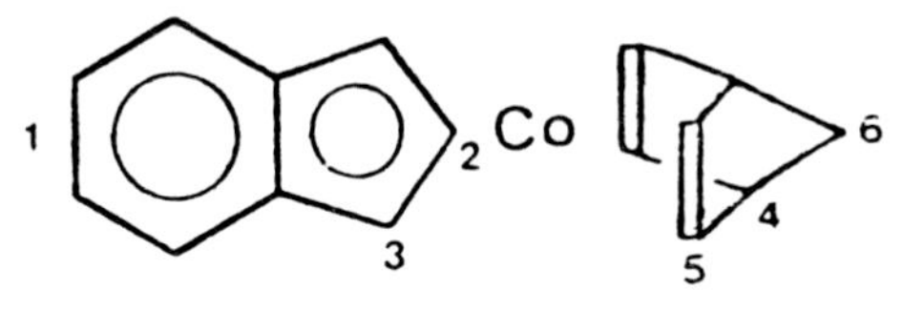

Mass spectrum:
m/z: 266 (M$^+$, 95%); 239 (24%); 174 (50%); 150 (97%); 115 (100%); 59 (40%).

7.2. Comparison of the methods for preparing the organocobalt catalysts

None of the methods discussed above and in the Experimental Section is equally suited to the synthesis of all organocobalt complexes. A comparative evaluation is shown in Table 16.

Table 16

COMPARISON OF THE METHODS FOR PREPARING THE ORGANOCOBALT CATALYSTS

No.	Complex	Method (Yield in %)					
		A			B	C1	C2
		(Eq. 21)	(Eq. 22)	Total	(Eq. 24)	(Eq. 25)	(Eq. 25)
20	Co	u		u	u		
21	Co	u		u	not det.	70	
23	Co	93 [48]	92 [50b]	86	not det.	76	79
24	CH_3 Co						71
25	CH_3 CH_3 CH_3 Co						42

Table 16 (continued)

No.	Complex	Method (Yield in %)					
		A			B	C1	C2
		(Eq. 21)	(Eq. 22)	Total	(Eq. 24)	(Eq. 25)	(Eq. 25)
27	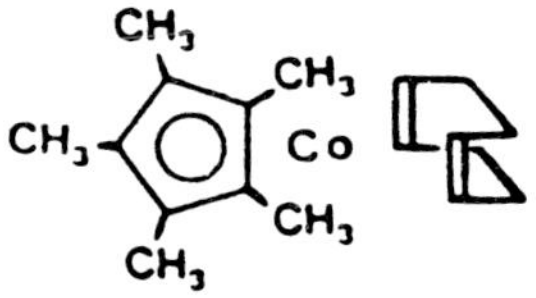 CH_3 CH_3 CH_3 CH_3 CH_3 Co	28 [52]	77	22			
28	 Co	62 [48]					1
31	 $(CH_3)_3Si$ Co	85 [117]	80 [41]	68	trace	43	70
32	 Co	u		u	50 [41]	81	85
33	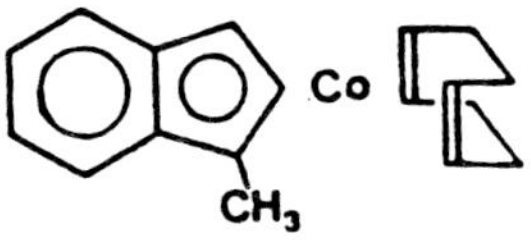 Co CH_3						48
34	Co $Si(CH_3)_3$						51
36	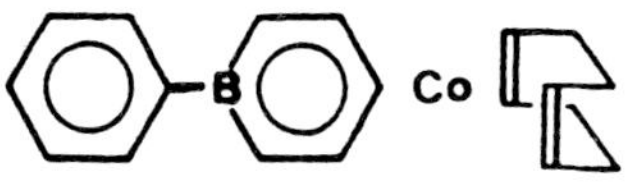 B Co	u		u			42
40	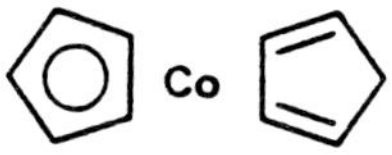 Co				not det.		63
53	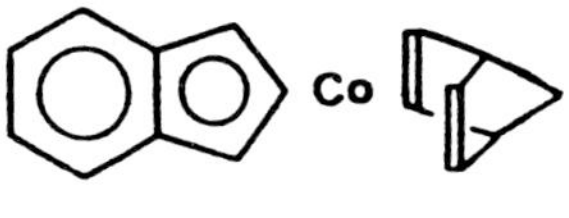 Co	u		u			31

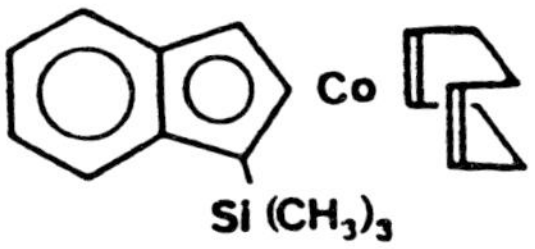

u = Unsuitable method.

Method A, which involves $[Co_2(CO)_8]$, is a two-step procedure (Equations 21–23). The initial preparation of $[Co(CO)_2(\eta^5\text{-Cp})]$ is a versatile reaction which normally gives good to satisfactory (40–90%) yields of the product. It cannot, however, be used to prepare indenyl-cobalt complexes and in many cases the product is an oil which has to be purified by careful vacuum distillation. The further reaction with cod (Equation 22) is an effective general method for preparing $[Co(\eta^4\text{-cod})(\eta^5\text{-RCp})]$ complexes in high (70–90%) yield. The reaction is, however, rather slow, even in high boiling solvents such as xylene, but can be accelerated by photolysis.

Method B, which involves Natta's complex **20**, is a useful route for explorating experiments but is only suitable for preparing complexes of substituted cyclopentadiene derivatives having pronounced C—H acidity.

Method C1, the one-pot reduction of cobalt salts with aluminium triethyl, has the disadvantage that the reaction must be conducted in an inert atmosphere and problems can arise in separating non-crystalline products (e.g. **24**) from the aluminium component. Nevertheless, it represents a convenient method for the large-scale, laboratory synthesis of **21**, **23** and **32**.

Method C2, a one-pot procedure using anthracene-activated magnesium, is probably the most versatile route available both in the laboratory and for large-scale production. The yield can be increased considerably by ultrasonic treatment of the magnesium but the method does require the availability of dry, pure THF.

8. RELATED REACTIONS

The cobalt catalyzed co-cyclization of alkynes with heterofunctional substrates is not limited to nitriles. H. Yamazaki and coworkers have recently published a comprehensive review [119b] of the organic reactions which can be brought about by reacting phosphine stabilized five-membered cobaltacycles (prepared from alkynes) with a variety of substrates. The products include 2-pyridones, 2-thiopyridones and the corresponding thia-analogue, as well as five-membered heterocyclic rings containing N, S, Se and P (Equation 78).

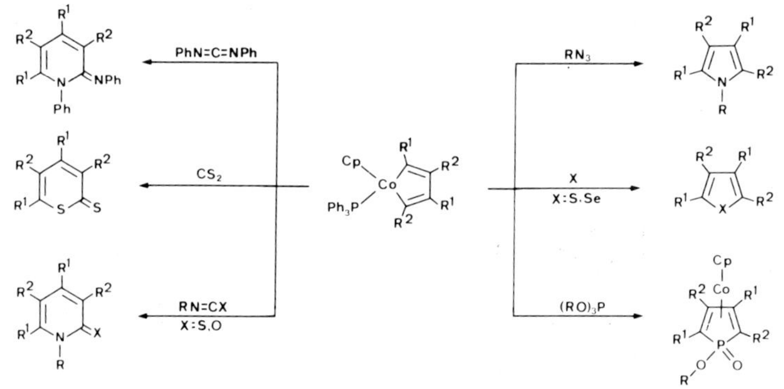

(78)

We have recently discovered that elemental sulphur can be effectively incorporated in the alkyne co-cyclizations [120] and a series of Cp—Co-dithiolatoethene complexes has been obtained using a simple one-pot procedure (Equation 79). In a

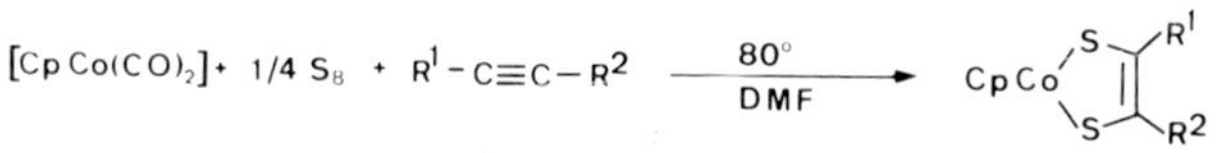

$$[CpCo(CO)_2] + 1/4\ S_8 + R^1-C{\equiv}C-R^2 \xrightarrow[DMF]{80^\circ} CpCo\langle S_2C_2R^1R^2\rangle \tag{79}$$

	Yield		Yield
$R^1 = H$, $R^2 = C_2H_5$	(15%)	$R^1 = R^2 = H$	(13%)
C_4H_9	(10%)	C_2H_5	(13%)
C_6H_{13}	(9%)	C_4H_9	(9%)
C_6H_5	(10%)	C_6H_5	(9%)
$C(=O)-CH_3$	(37%)	$COOCH_3$	(73%)
$COOH$	(20%)	$COOC_2H_5$	(71%)
$COOCH_3$	(22%)		
$COOC_2H_5$	(44%)		
$(CH_2)_4C{\equiv}CH$	(13%)		

typical run elemental sulphur (7 g, 218 mmoles) suspended in DMF and acetylene dicarboxylic acid dimethylester (14.2 g, 100 mmoles) is treated with $[Co(CO)_2(\eta^5\text{-}Cp)]$ (18 g, 100 mmoles), added over 3 min. A vigorous gas evolution is observed and the colour changes to dark violet. The mixture is heated at 80° for 6 h and the volatile components removed under vacuum at 100°. The residue is deposited on a 50 cm silica gel column and a violet fraction eluted with methylene chloride. Recrystallization from 200 ml of toluene gives the dithiolato-ethene complex as dark violet needles (24.2 g, 73.3 mmoles, 73.3% theory, mp 119°).

An X-ray diffraction study has confirmed the structure of the derivative in which the dithiolato ring is substituted by an ethyl ester (see Equation 80) [121a]. The C=C double bond was found to have the unusually short value of 1.37 Å. The

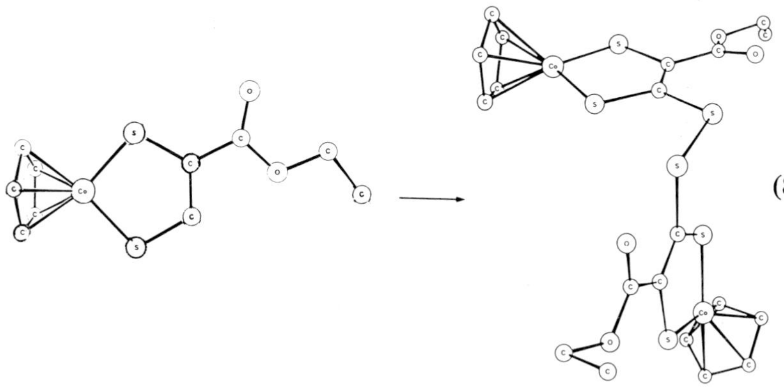

(80)

hydrogen atom present in the ring can undergo electrophilic substitution to give a complex in which two units of the parent complex are bridged by an S_2 moiety [121b] (Equation 80).

The reaction with acetylene dicarboxylic acid proceeds in a step-wise manner: decarboxylation occurs and instead of the disubstituted product, the monocarboxylic acid derivative is isolated. Here again the structure has been confirmed by an X-ray structural analysis [121a]. Traces of the unsubstituted dithiolato complex are also formed in the reaction mixture. This compound is, however, more readily prepared by direction reaction of acetylene and elemental sulphur with $[Co(CO)_2(\eta^5\text{-Cp})]$ (Equation 81).

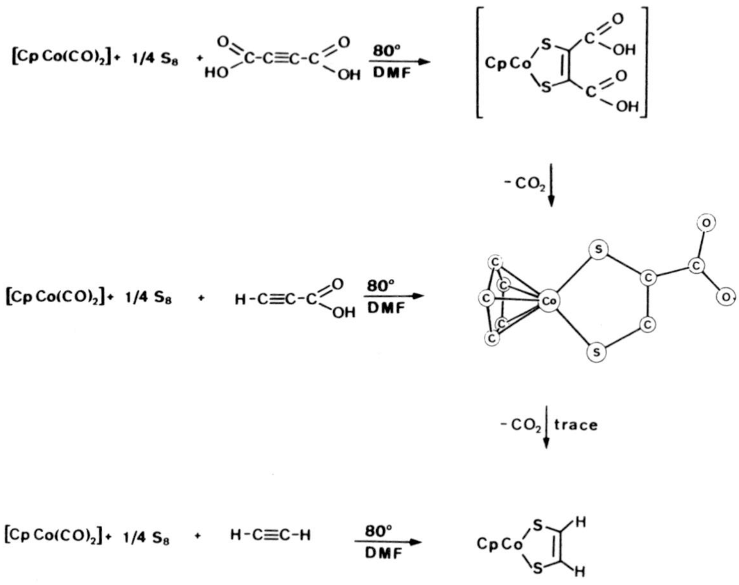

(81)

In the presence of excess alkynes and sulphur the Cp—Co core catalyses the formation of thiophenes (Equation 82). The discovery that elemental sulphour can

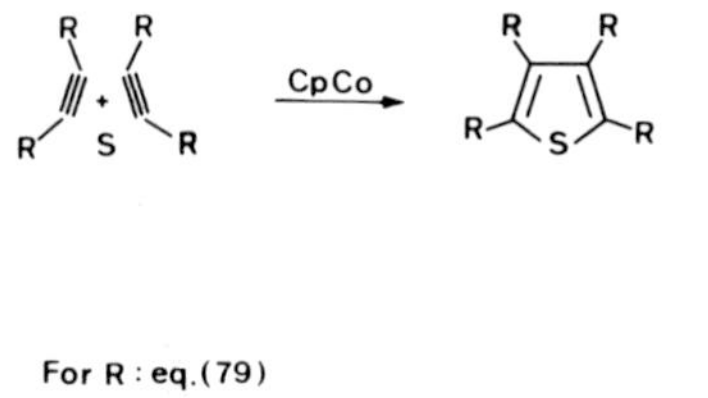

(82)

be incorporated into the catalytic cycle indicates that the inherent insensitivity of the Cp—Co core to polar donor atoms should make it possible to synthesize a broad range of heterocycles.

Table 17

THE COBALT-CATALYZED SYNTHESIS OF PYRIDINE AND ITS DERIVATIVES

Product	Alkyne	Nitrile	Solvent	Catalyst [a]
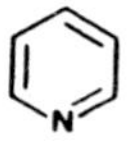 * pyridine	Acetylene	HCN	Benzene	18, 23, 31, <u>32</u>, <u>36</u>, 56
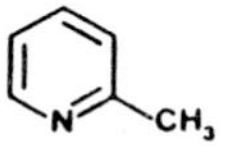 * 2-CH_3-pyridine	Acetylene	Acetonitrile		1, 5, 18, 20, 21, <u>23</u>, 24, 25, 26, 28, 30, <u>31</u>, 32, 34, 36, 37, 50, 51, 54 [21a], [25a], [25c], 55, 57
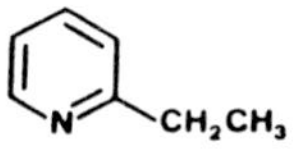 2-CH_2CH_3-pyridine	Acetylene	Propionitrile		20, 21, <u>23</u>, 24, 25, 26, 27, 28, 30, <u>31</u>, 32, 34, 36, 40, 42, 43, 44, 45, 46, 47, 48, 49, 50, 54 [21a], 55
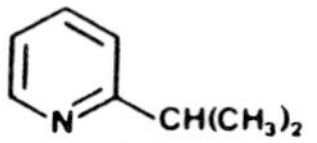 2-$CH(CH_3)_2$-pyridine	Acetylene	*i*-Butyronitrile		23, 54 [21a], [25a], [25c]
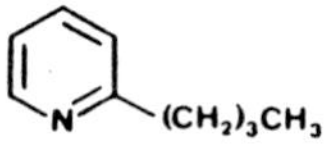 2-$(CH_2)_3CH_3$-pyridine	Acetylene	*n*-Butylcyanide		23
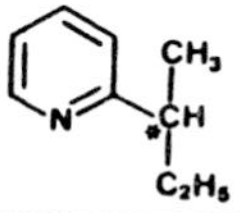 2-$\overset{*}{C}H(CH_3)(C_2H_5)$-pyridine	Acetylene	(S)-(+)-sec.-Butyl-cyanide		23 [81]

Table 17 (continued)

Product	Alkyne	Nitrile	Solvent	Catalyst [a]
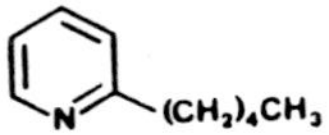 N, $(CH_2)_4CH_3$	Acetylene	*n*-Pentylcyanide		23
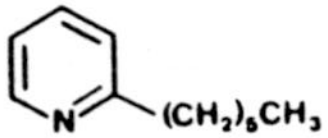 N, $(CH_2)_5CH_3$	Acetylene	*n*-Hexylcyanide		23
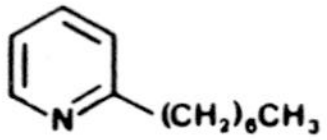 N, $(CH_2)_6CH_3$	Acetylene	*n*-Heptylcyanide		23
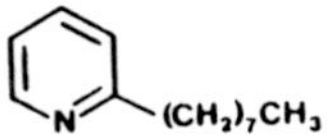 N, $(CH_2)_7CH_3$	Acetylene	*n*-Octylcyanide		23
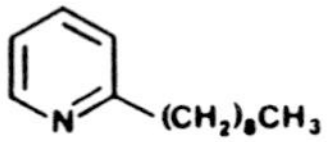 N, $(CH_2)_8CH_3$	Acetylene	*n*-Nonylcyanide		23
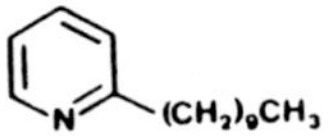 N, $(CH_2)_9CH_3$	Acetylene	*n*-Decylcyanide		23
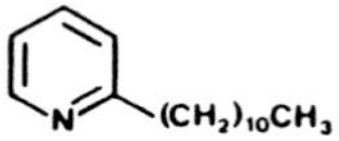 * N, $(CH_2)_{10}CH_3$	Acetylene	*n*-Undecylcyanide		23, 31, 32, 43, 54
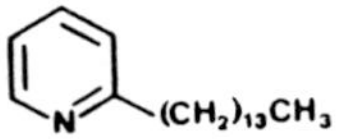 N, $(CH_2)_{13}CH_3$	Acetylene	*n*-Tetradecylcyanide		23, 31, 32, 43, 54
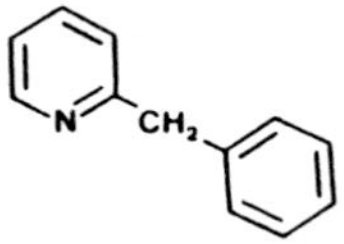 N, CH_2	Acetylene	Benzylcyanide		23, 54 [21a], [25a], [25c]

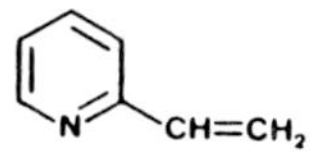 * N, $CH{=}CH_2$	Acetylene	Acrylonitrile	Toluene Benzene NMP	**10, 20, 23, 24, 26, 31, 32, 34, 36, 37, 40, 44, 47, 50, 52, 54** [21a], [25a], [25c], **56**
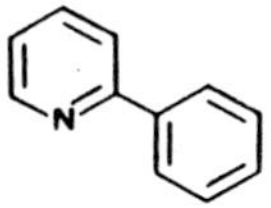 N	Acetylene	Benzonitrile		**23, 36, 54** [21a], [25a], [25c]
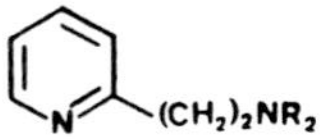 N, $(CH_2)_2NR_2$	Acetylene	$R_2N{-}(CH_2)_2{-}CN$ R = Alkyl, Aryl		**54** [25b]
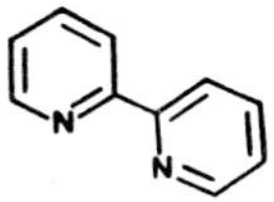 * N, N	Acetylene	2-Cyanopyridine	Toluene DMF	**23, 30, 31, 32, 36, 40, 54** [25a], [25c]
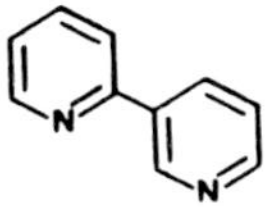 N, N	Acetylene	3-Cyanopyridine	Toluene	**23, 31**
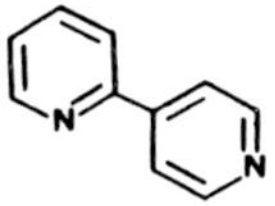 N, N	Acetylene	4-Cyanopyridine	Toluene	**23, 31**
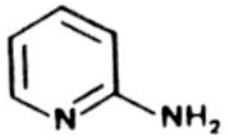 * N, NH_2	Acetylene	Cyanamide	DMF NMP	**32, 36**
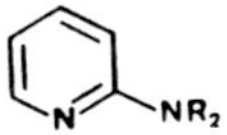 N, NR_2	Acetylene	$R_2N{-}CN$ R = Alkyl, Aryl		**54** [25a], [25c], [25d]

Table 17 (continued)

Product	Alkyne	Nitrile	Solvent	Catalyst [a]
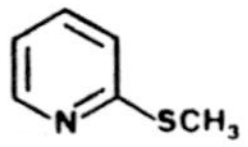 * SCH_3	Acetylene	Methylthiocyanate		**23, 36**
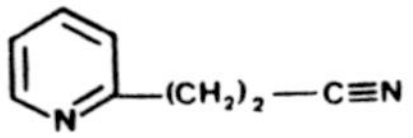 $(CH_2)_2—C≡N$	Acetylene	1,2-Dicyanoethane		**23**
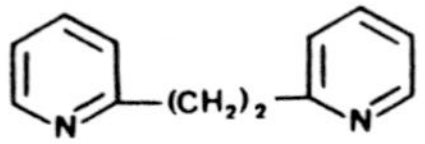 $(CH_2)_2$				
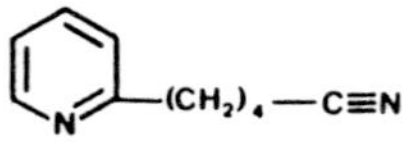 $(CH_2)_4—C≡N$	Acetylene	1,4-Dicyanobutane	Toluene	**23, 54** [25a], [25c]
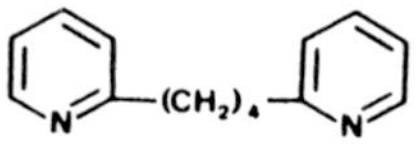 $(CH_2)_4$				
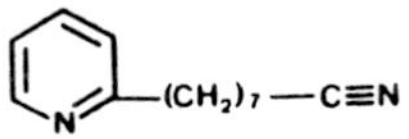 $(CH_2)_7—C≡N$	Acetylene	1,7-Dicyanoheptane	Toluene	**20, 23**
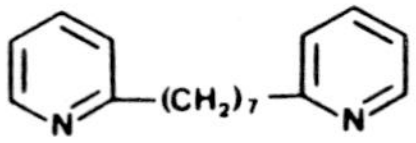 $(CH_2)_7$				

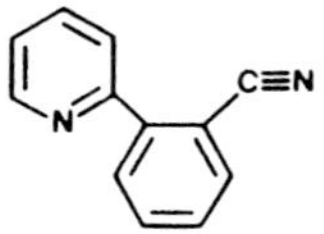 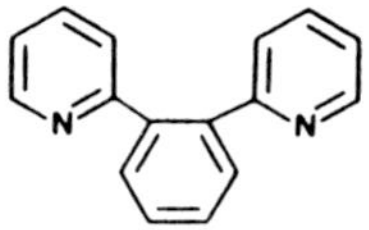	Acetylene	1,2-Dicyanobenzene	Toluene	**23**
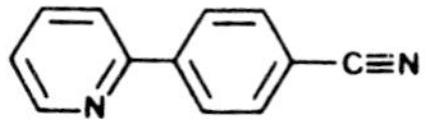 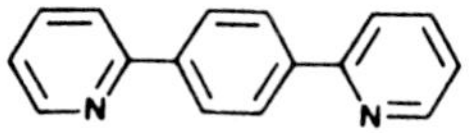	Acetylene	1,4-Dicyanobenzene	Toluene	**23, 54** [25a], [25c]
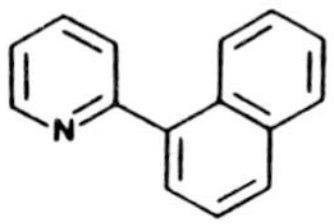	Acetylene	1-Cyanonaphthalene	Toluene	**31**
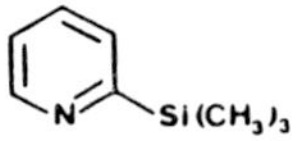	Acetylene	Trimethylsilylcyanide	Toluene	**23**

Table 17 (continued)

Product	Alkyne	Nitrile	Solvent	Catalyst [a]
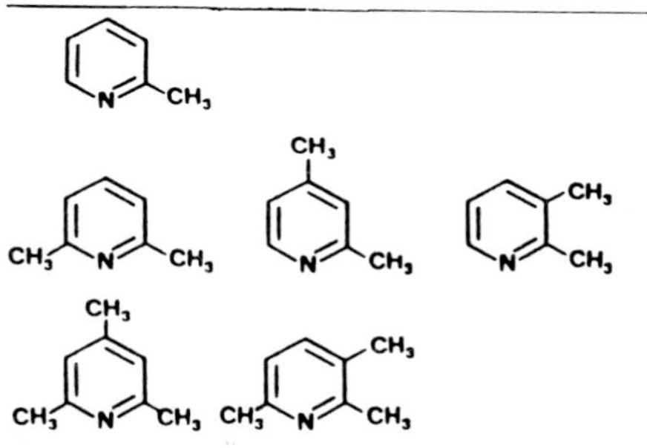	Acetylene/ Propyne	Acetonitrile		20
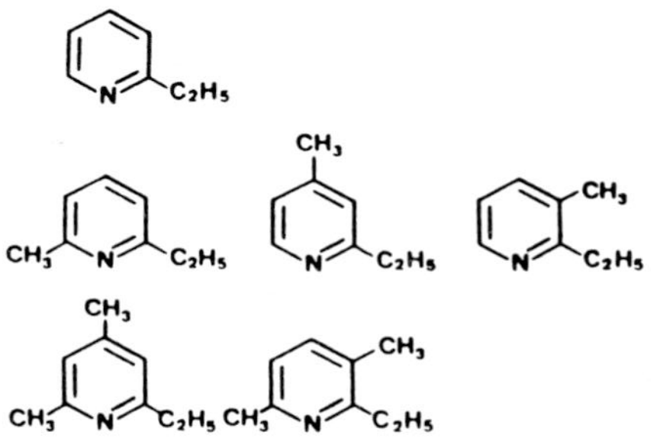	Acetylene/ Propyne	Propionitrile		20
	Propyne	Acetonitrile		9, 10, 20, 23, 37, 50, 54 [21a]

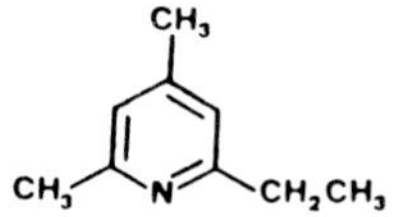	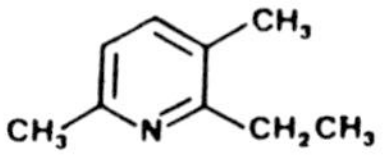	Propyne	Propionitrile		18, 20, 21, 23, 24, 25, 26, 28, <u>30</u>, 31, 32, 33, <u>34</u>, 35, 36, 37, 38, 40, 41
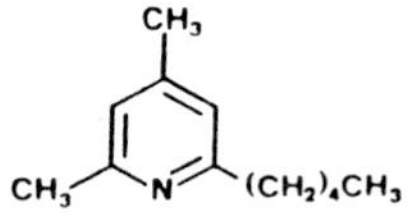	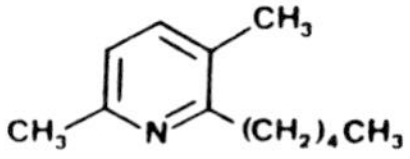	Propyne	*n*-Pentylcyanide		23
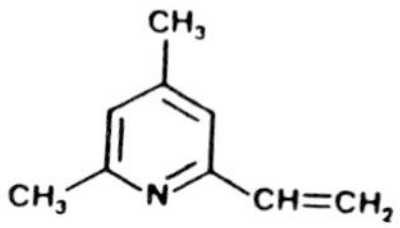	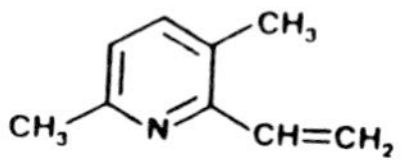	Propyne	Acrylonitrile	Toluene	20, 23
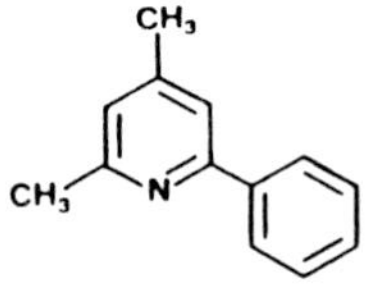	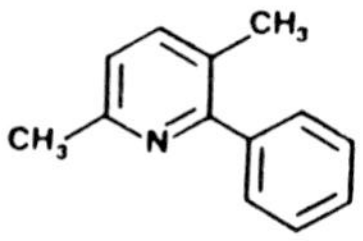	Propyne	Benzonitrile		20
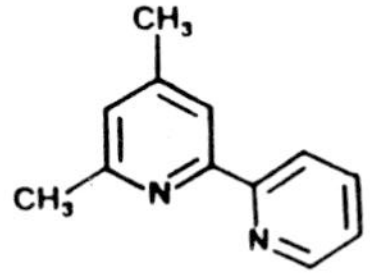	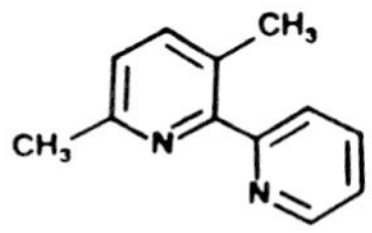	Propyne	2-Cyanopyridine	Toluene	20, 23
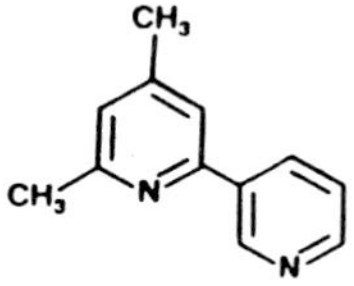	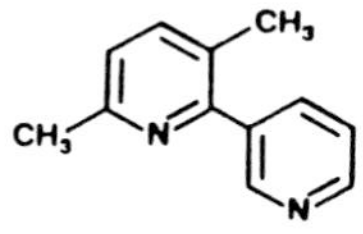	Propyne	3-Cyanopyridine	Toluene	20

Table 17 (continued)

Product	Alkyne	Nitrile	Solvent	Catalyst [a]
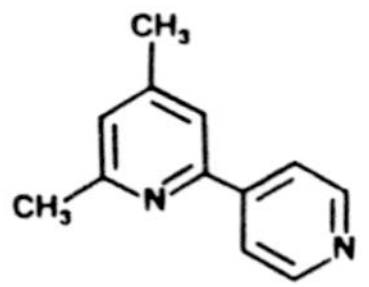 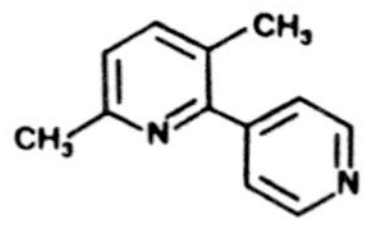 CH_3, CH_3, N, N ; CH_3, CH_3, N, N	Propyne	4-Cyanopyridine	Toluene	20
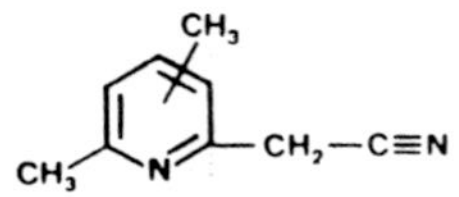 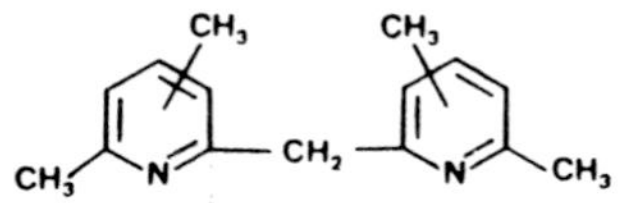 CH_3, CH_3, N, CH_2—C≡N ; CH_3, CH_3, N, CH_2, CH_3, CH_3, N	Propyne	Dicyanomethane	Toluene	20
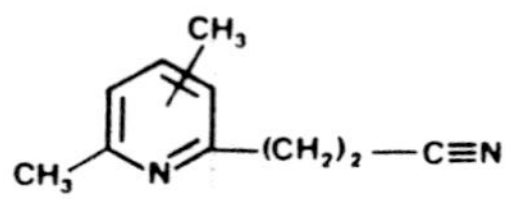 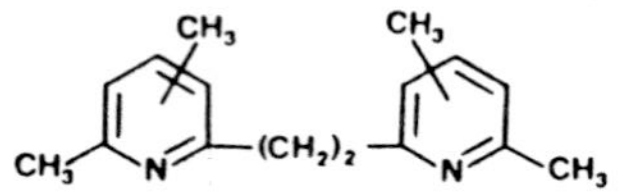 CH_3, CH_3, N, $(CH_2)_2$—C≡N ; CH_3, CH_3, N, $(CH_2)_2$, CH_3, CH_3, N	Propyne	1,2-Dicyanoethane	Toluene	20

$(CH_3)_2C_5H_2N-(CH_2)_4-C\equiv N$ $(CH_3)_2C_5H_2N-(CH_2)_4-C_5H_2N(CH_3)_2$	Propyne	1,4-Dicyanobutane	Toluene	20
$(CH_3)_2C_5H_2N-(CH_2)_6-C\equiv N$ $(CH_3)_2C_5H_2N-(CH_2)_6-C_5H_2N(CH_3)_2$	Propyne	1,6-Dicyanohexane	Toluene	20
$(CH_3)_2C_5H_2N-(CH_2)_7-C\equiv N$ $(CH_3)_2C_5H_2N-(CH_2)_7-C_5H_2N(CH_3)_2$	Propyne	1,7-Dicyanoheptane	Toluene	23

Table 17 (continued)

Product	Alkyne	Nitrile	Solvent	Catalyst [a]
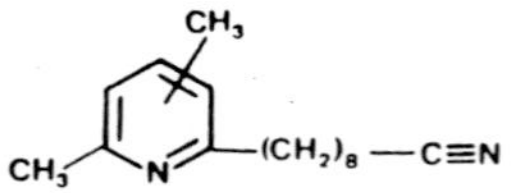 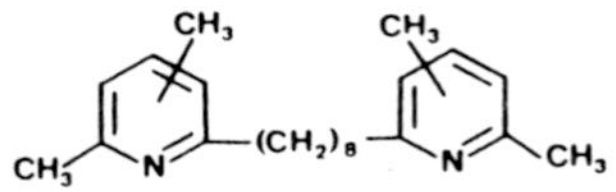 CH_3, CH_3, N, $(CH_2)_8$—C≡N	Propyne	1,8-Dicyanooctane	Toluene	20
CH_3, CH_3, N, $(CH_2)_8$, CH_3, N, CH_3				
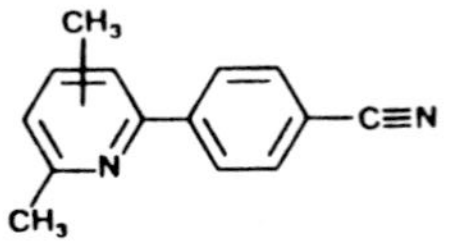 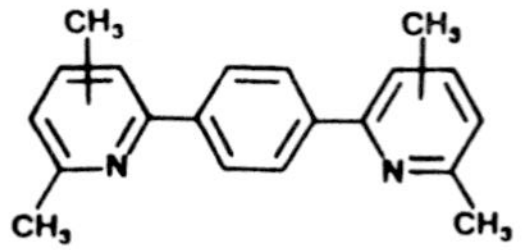 CH_3, N, CH_3, C≡N	Propyne	1,4-Dicyanobenzene	Toluene	20, 23
CH_3, N, CH_3, CH_3, N, CH_3				
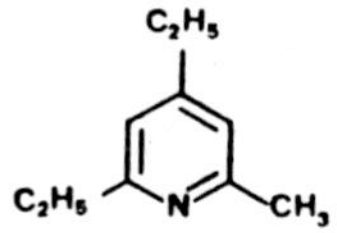 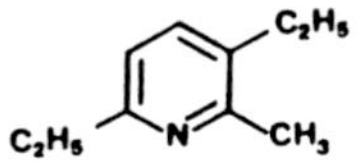 C_2H_5, C_2H_5, N, CH_3; C_2H_5, C_2H_5, N, CH_3	1-Butyne	Acetonitrile		20, 47 [21b], 54 [21a]
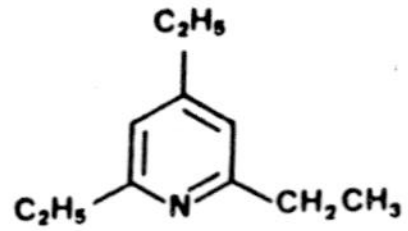 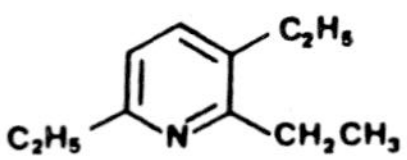 C_2H_5, C_2H_5, N, CH_2CH_3; C_2H_5, C_2H_5, N, CH_2CH_3	1-Butyne	Propionitrile		23, 24, 26, 32

C_2H_5; C_2H_5; N; $(CH_2)_2CH_3$	C_2H_5; C_2H_5; N; $(CH_2)_2CH_3$	1-Butyne	*n*-Butyronitrile	Toluene	54 [21a]
C_3H_7; C_3H_7; N; CH_3	C_3H_7; C_3H_7; N; CH_3	1-Pentyne	Acetonitrile	Toluene	45 [21b], 54 [21a]
C_4H_9; C_4H_9; N; CH_3	C_4H_9; C_4H_9; N; CH_3	1-Hexyne	Acetonitrile		52, 54 [21a]
C_4H_9; C_4H_9; N; SCH_3	C_4H_9; C_4H_9; N; SCH_3	1-Hexyne	Methylthiocyanate	Toluene	23
C_4H_9; C_4H_9; N; SC_2H_5	C_4H_9; C_4H_9; N; SC_2H_5	1-Hexyne	Ethylthiocyanate	Toluene	23
C_4H_9; C_4H_9; N; $SCH(CH_3)CH_3$	C_4H_9; C_4H_9; N; $SCH(CH_3)CH_3$	1-Hexyne	*i*-Propylthiocyanate	Toluene	23

Table 17 (continued)

Product	Alkyne	Nitrile	Solvent	Catalyst [a]
	1-Heptyne	Acetonitrile		9
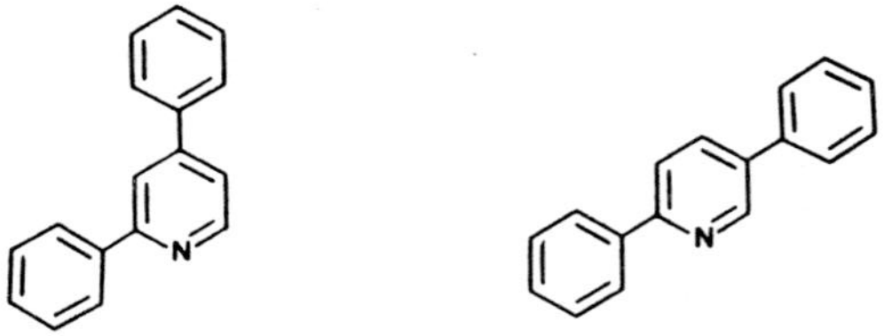	Phenyl-acetylene	HCN	Xylene	23
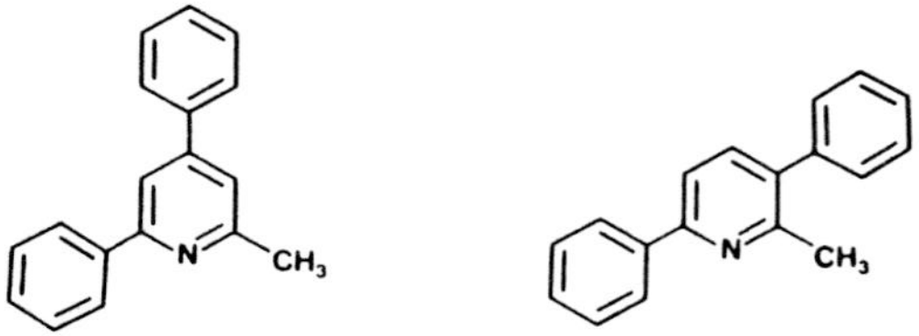	Phenyl-acetylene	Acetonitrile		1, 2, 3, 4, 5, 6, 9, 10, 11, 12, 14, 16, 17, 18, 19, 20, 22, 23, 47 [21b], 53, 58
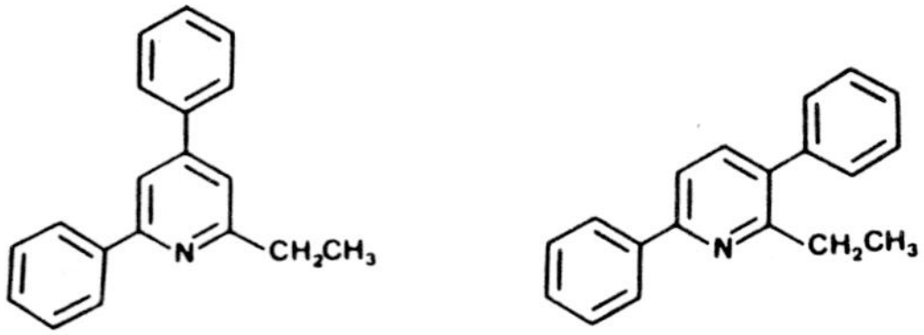	Phenyl-acetylene	Propionitrile		7, 8, 13, 15, 19, 20, 21, 23, 28, 29, 39

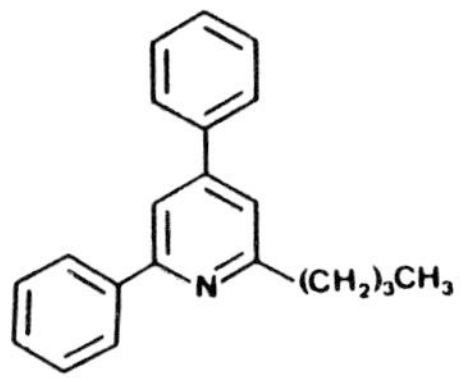 N, $(CH_2)_3CH_3$	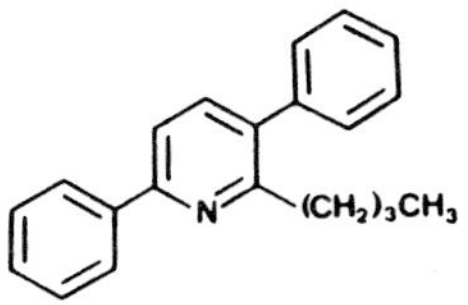 N, $(CH_2)_3CH_3$	Phenyl-acetylene	*n*-Butylcyanide	23
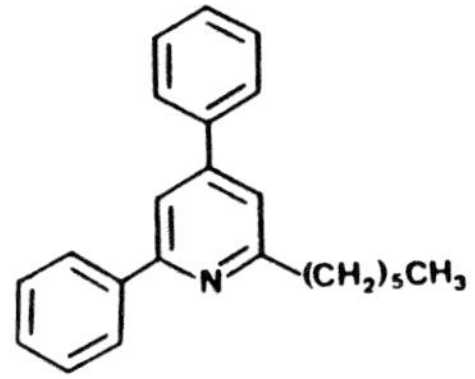 N, $(CH_2)_5CH_3$	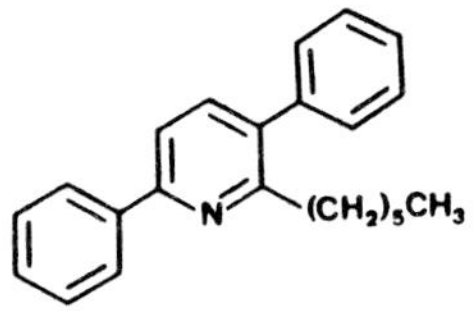 N, $(CH_2)_5CH_3$	Phenyl-acetylene	*n*-Hexylcyanide	23
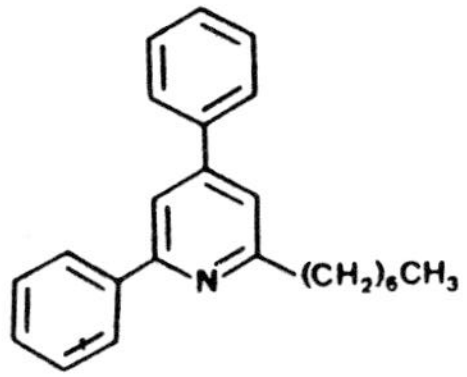 N, $(CH_2)_6CH_3$	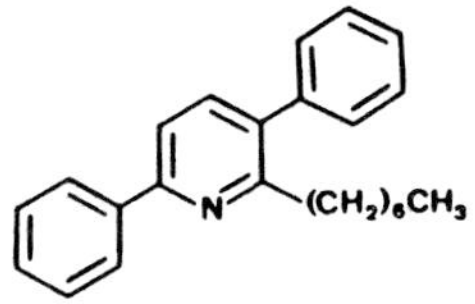 N, $(CH_2)_6CH_3$	Phenyl-acetylene	*n*-Heptylcyanide	20
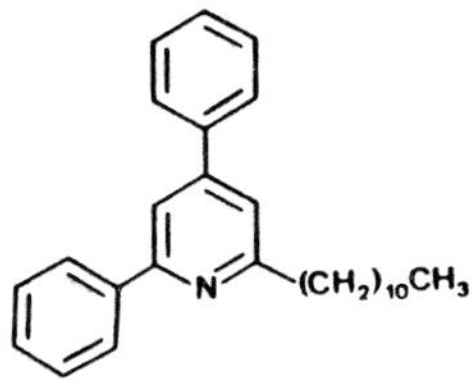 N, $(CH_2)_{10}CH_3$	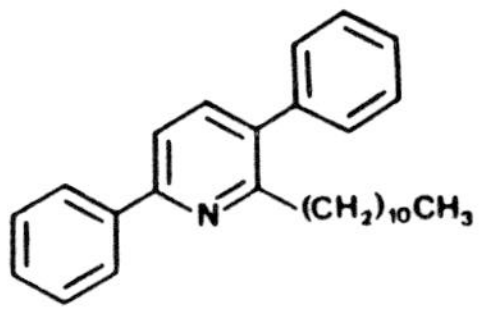 N, $(CH_2)_{10}CH_3$	Phenyl-acetylene	*n*-Undecylcyanide	23

Table 17 (continued)

Product		Alkyne	Nitrile	Solvent	Catalyst [a]
N, CH_2	N, CH_2	Phenyl-acetylene	Benzylcyanide		9
N, $CH{=}CH_2$	N, $CH{=}CH_2$	Phenyl-acetylene	Acrylonitrile	Toluene	20
N	N	Phenyl-acetylene	Benzonitrile	Toluene	10

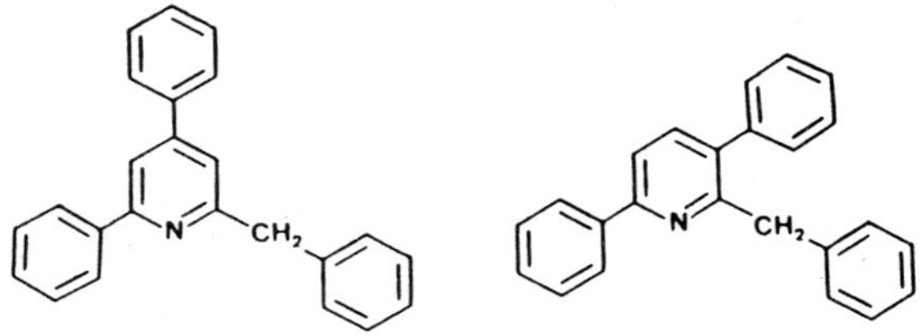

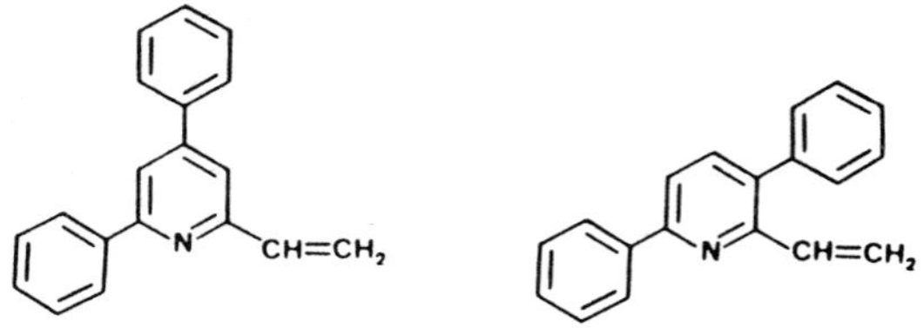

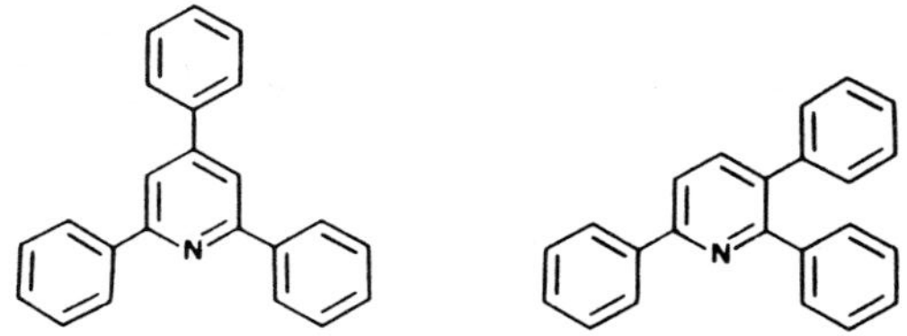

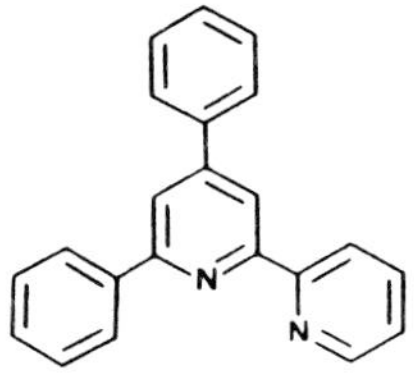	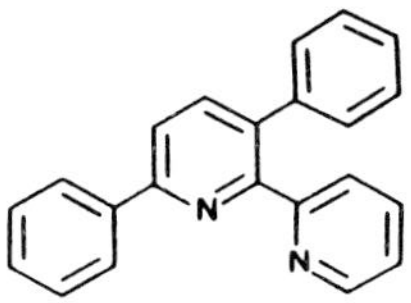	Phenyl-acetylene	2-Cyanopyridine	Toluene	**20**
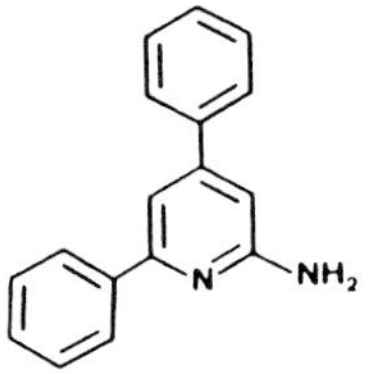	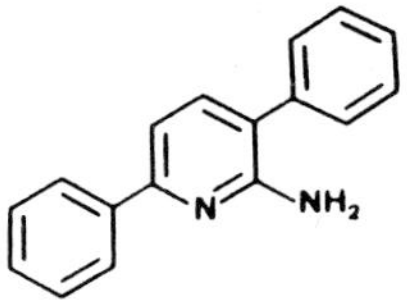	Phenyl-acetylene	Cyanamide	Toluene	**23, 54**
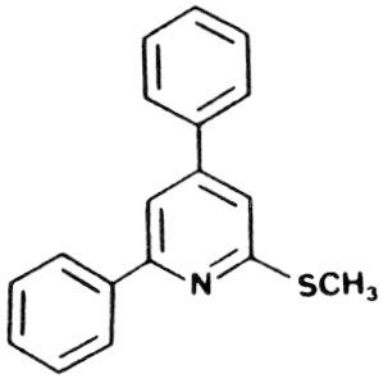	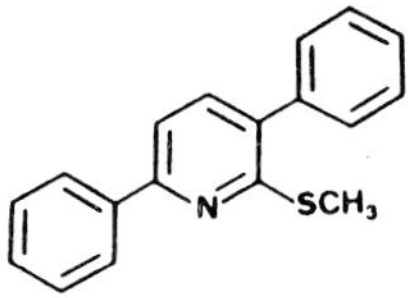	Phenyl-acetylene	Methylthiocyanate	Toluene	**20, 23**
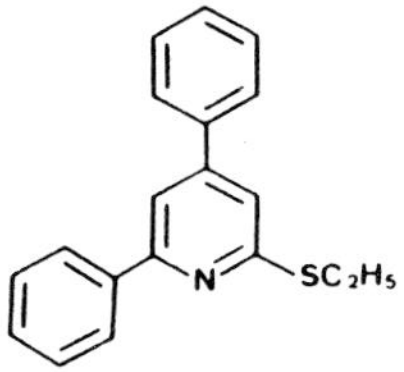	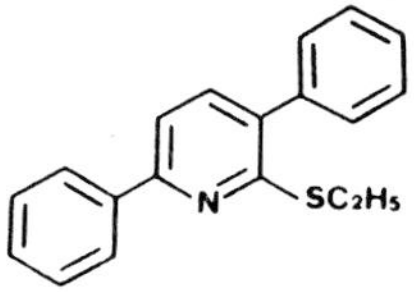	Phenyl-acetylene	Ethylthiocyanate	Toluene	**23**

Table 17 (continued)

Product	Alkyne	Nitrile	Solvent	Catalyst [a]
	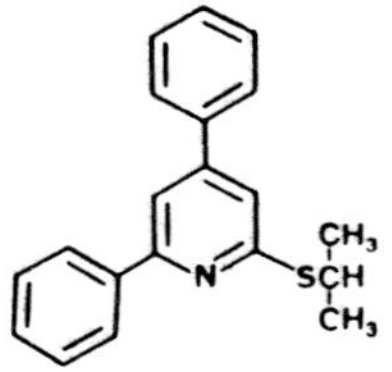 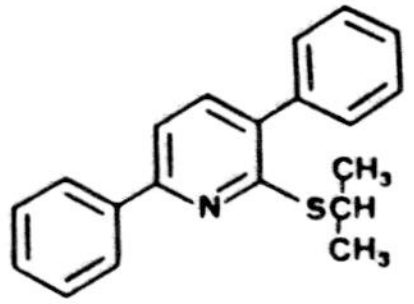 Phenyl-acetylene	*i*-Propylthiocyanate	Toluene	**23**
	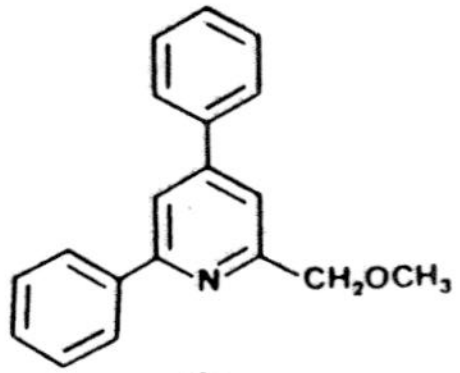 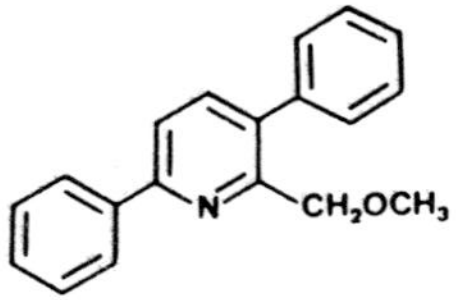 Phenyl-acetylene	Methoxyacetonitrile	Toluene	**23**
	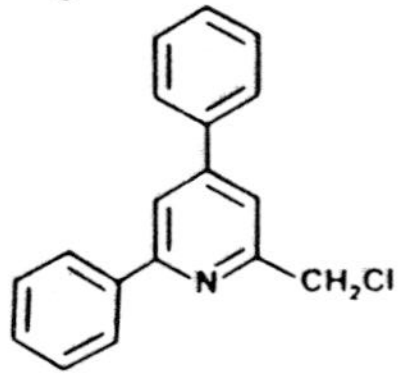 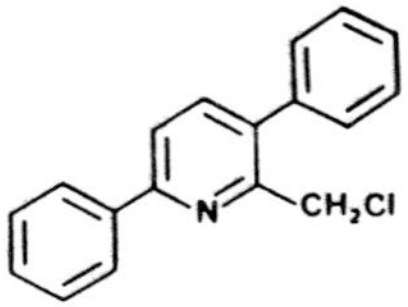 Phenyl-acetylene	Chloroacetonitrile	Toluene	**23**
	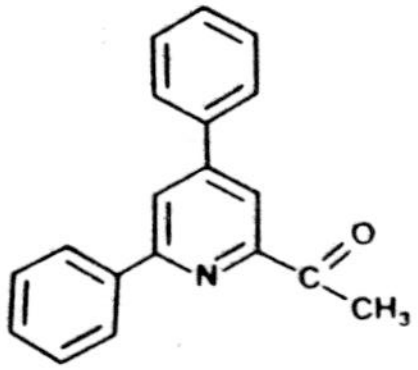 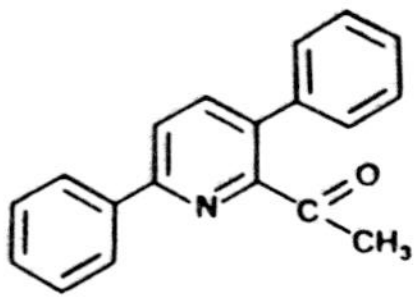 Phenyl-acetylene	Acetylcyanide	Toluene	**23**, **54** [21b]

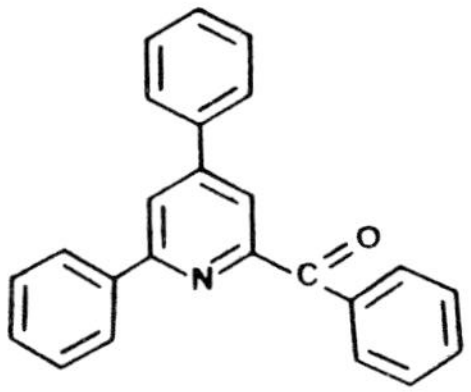 N, C=O	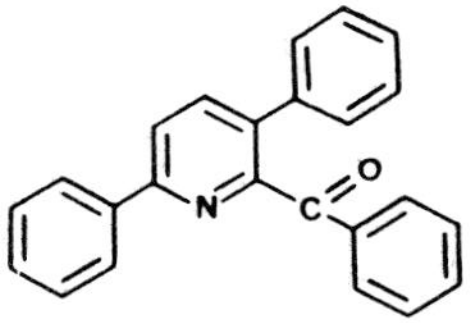 N, C=O	Phenyl-acetylene	Benzoylcyanide	Toluene	**23**
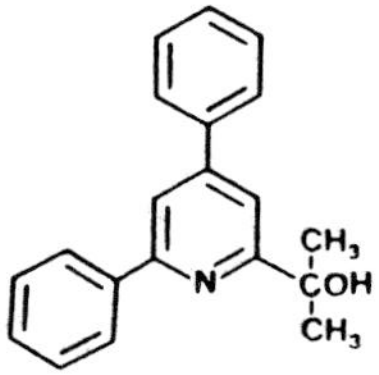 N, CH_3–COH–CH_3	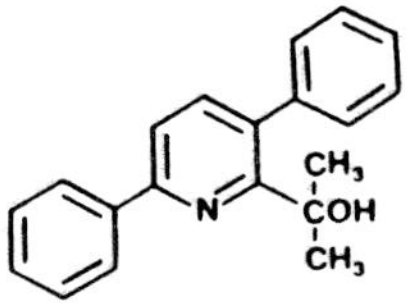 N, CH_3–COH–CH_3	Phenyl-acetylene	Acetonecyanohydrin	Toluene	**23**
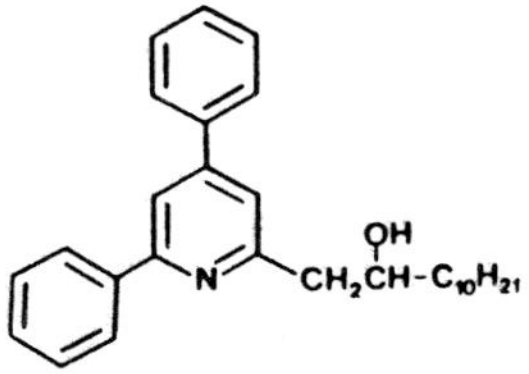 N, $CH_2CH(OH)$–$C_{10}H_{21}$	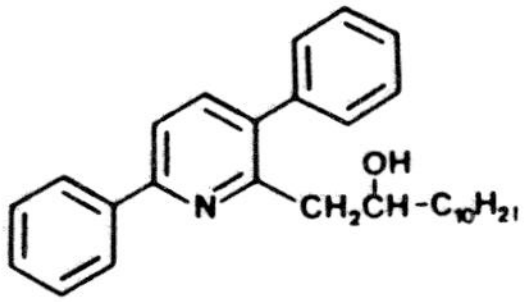 N, $CH_2CH(OH)$–$C_{10}H_{21}$	Phenyl-acetylene	2-Hydroxy-Dodecyl-cyanide	Toluene	**23**
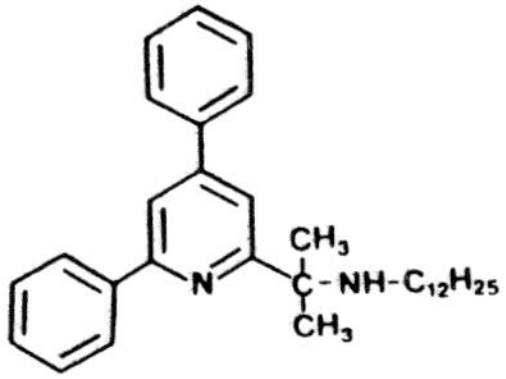 N, CH_3–C(CH_3)–NH–$C_{12}H_{25}$	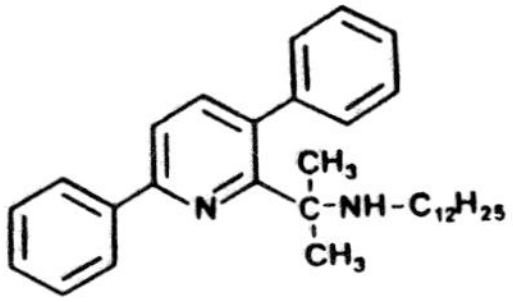 N, CH_3–C(CH_3)–NH–$C_{12}H_{25}$	Phenyl-acetylene	Dodecyl-Amino-*i*-Butyronitrile	Toluene	**23**

Table 17 (continued)

Product	Alkyne	Nitrile	Solvent	Catalyst [a]
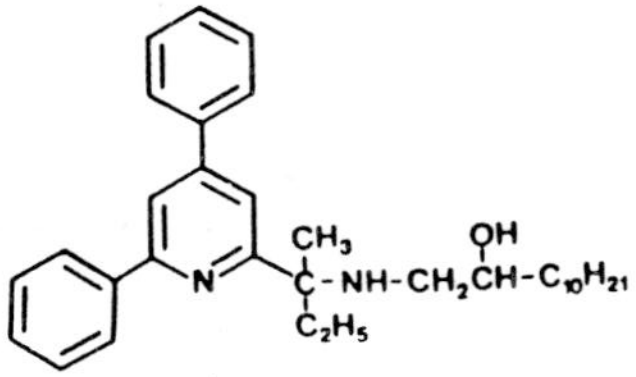 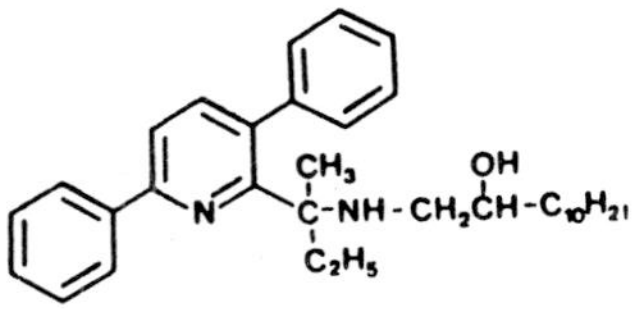	Phenyl-acetylene	α-(2-Hydroxy-Dodecyl-amino)-2-Methyl-butyronitrile	Toluene	**23**
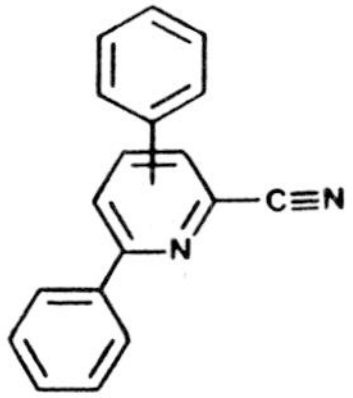 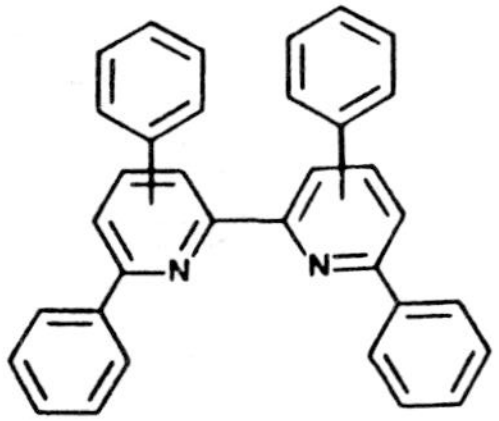	Phenyl-acetylene	Cyanogen	Toluene	**23**

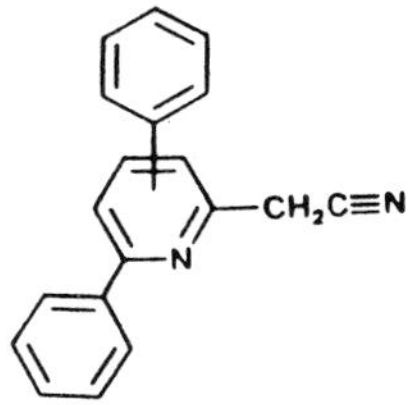 $-CH_2C\equiv N$	Phenyl-acetylene	Dicyanomethane	Toluene	20
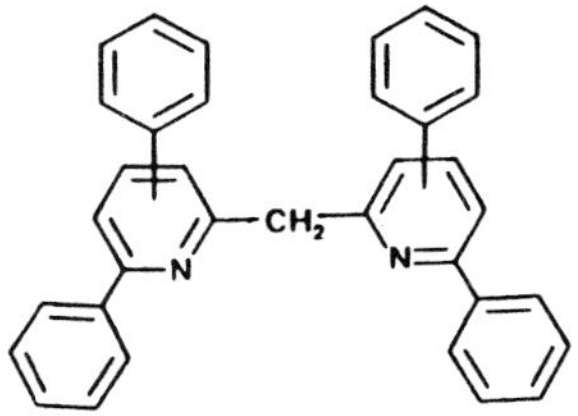 $-CH_2-$				
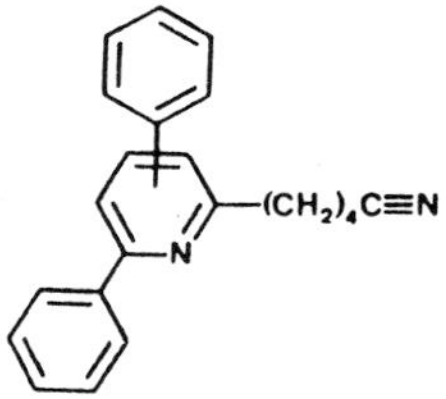 $-(CH_2)_4C\equiv N$	Phenyl-acetylene	1,4-Dicyanobutane	Toluene	20
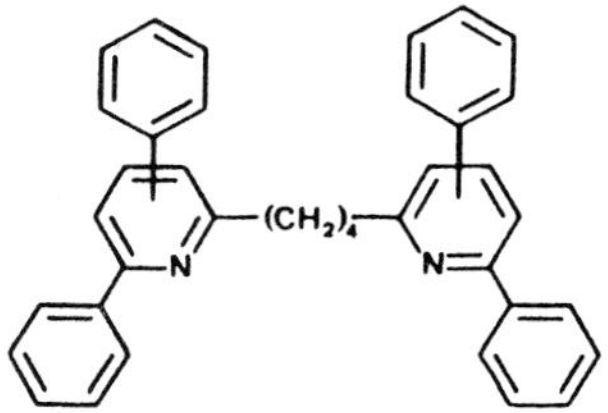 $-(CH_2)_4-$				

Table 17 (continued)

Product	Alkyne	Nitrile	Solvent	Catalyst [a]
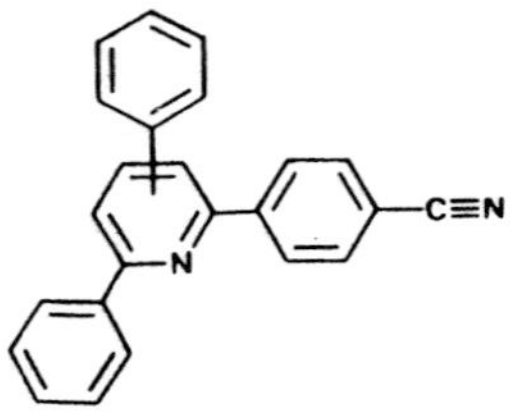	Phenyl-acetylene	1,4-Dicyanobenzene	Toluene	**20**
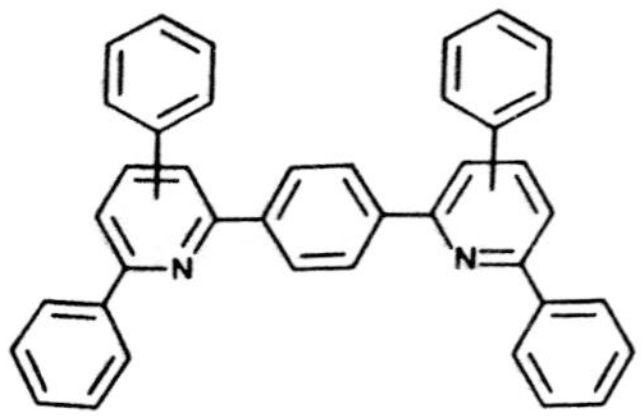				
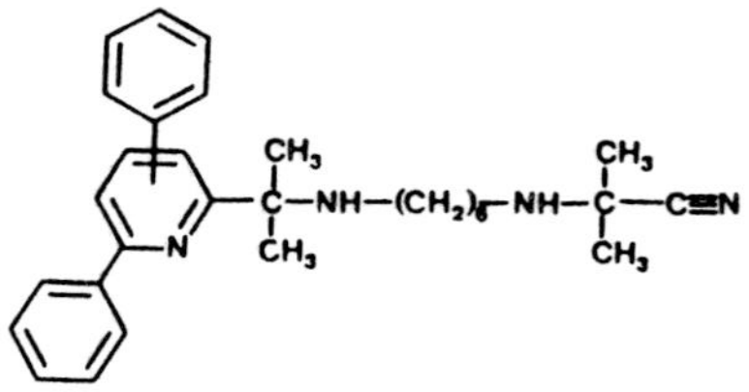	Phenyl-acetylene	1,6-Di-(2-Cyano-*i*-Propyl)-Amino-Hexane	Toluene	**26**

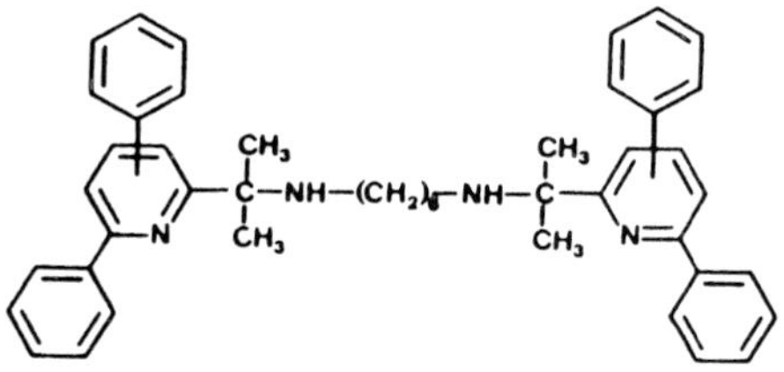 $C(CH_3)_2$–NH–$(CH_2)_6$–NH–$C(CH_3)_2$				
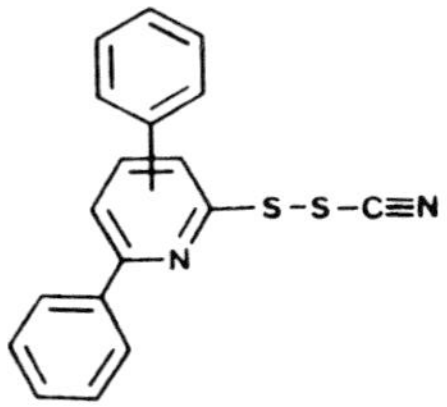 –S–S–C≡N	Phenyl-acetylene	Thiocyanogen	Toluene	23
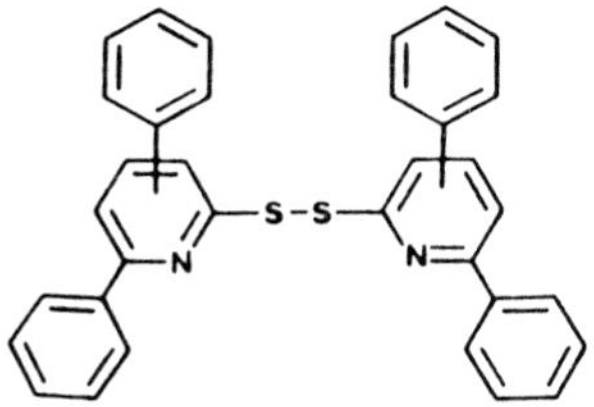 –S–S–				
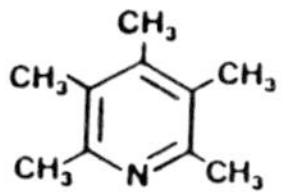 CH_3, CH_3, CH_3, CH_3, CH_3, N	2-Butyne	Acetonitrile		23

Table 17 (continued)

Product	Alkyne	Nitrile	Solvent	Catalyst [a]
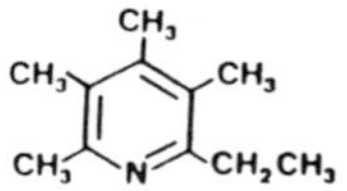 CH_3, CH_3, CH_3, CH_3, N, CH_2CH_3	2-Butyne	Propionitrile		20, 23
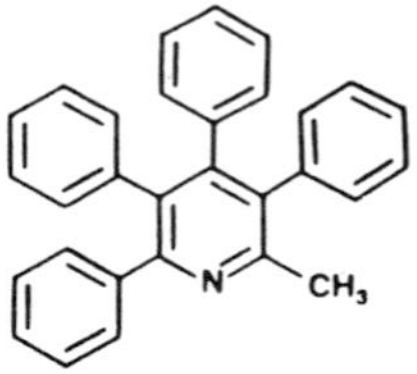 N, CH_3	Diphenyl-acetylene	Acetonitrile		23
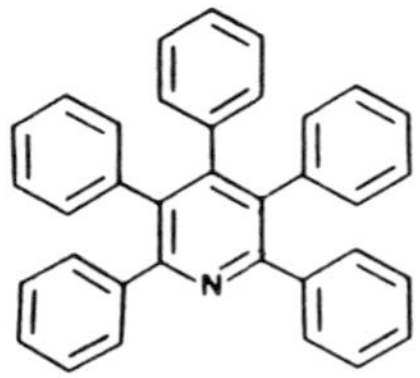 N	Diphenyl-acetylene	Benzonitrile		20, 23
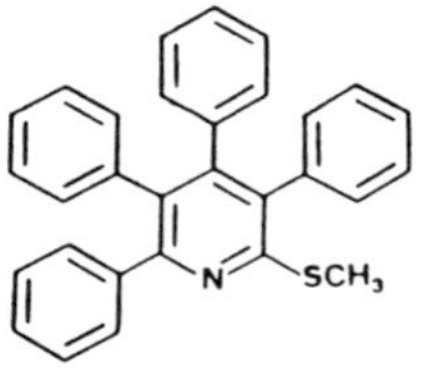 N, SCH_3	Diphenyl-acetylene	Methylthiocyanate	Toluene	23

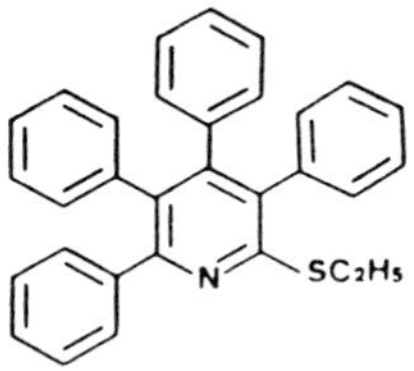	Diphenyl-acetylene	Ethylthiocyanate	Toluene	23
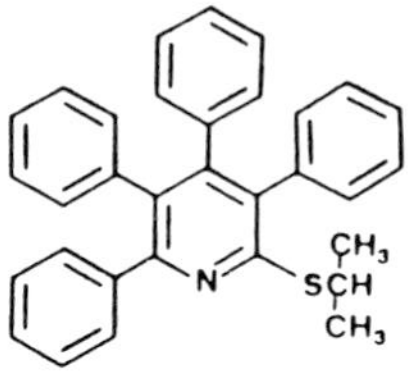	Diphenyl-acetylene	*i*-Propylthiocyanate	Toluene	23
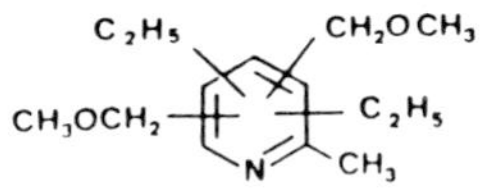	1-Methoxy-pentyne-2	Acetonitrile	Toluene	20
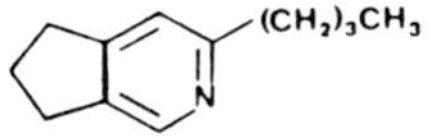	1,6-Hepta-diyne	*n*-Butylcyanide		50 [18b]
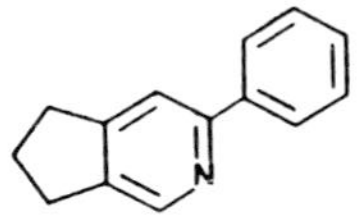	1,6-Hepta-diyne	Benzonitrile		50 [18b]
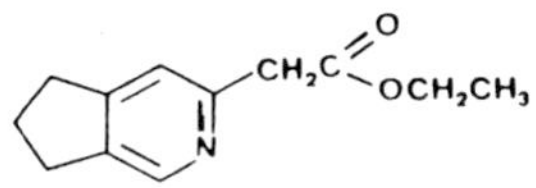	1,6-Hepta-diyne	$C_2H_5-O-C(=O)-CH_2-CN$		50 [18b]

Table 17 (continued)

Product	Alkyne	Nitrile	Solvent	Catalyst [a]
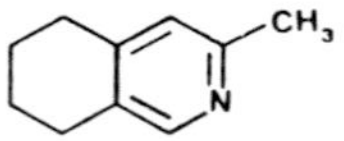 CH_3, N	1,7-Octa-diyne	Acetonitrile		**23, 50** [18b]
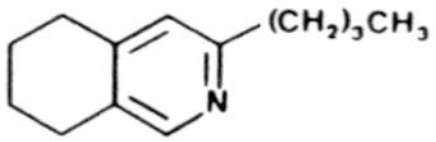 $(CH_2)_3CH_3$, N	1,7-Octa-diyne	*n*-Butylcyanide		**50** [18b]
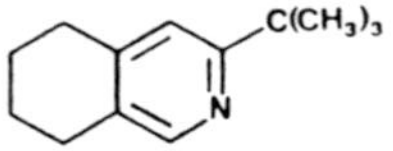 $C(CH_3)_3$, N	1,7-Octa-diyne	*t*-Butylcyanide		**50** [18b]
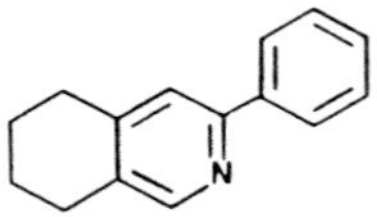 N	1,7-Octa-diyne	Benzonitrile		**50** [18b]
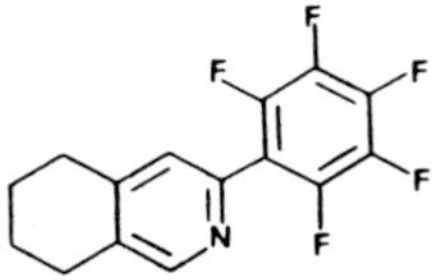 F, F, F, F, F, N	1,7-Octa-diyne	C_6F_5—CN		**50** [18b]
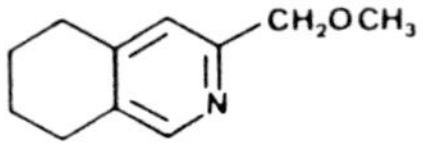 CH_2OCH_3, N	1,7-Octa-diyne	Methoxyacetonitrile		**50** [18b]
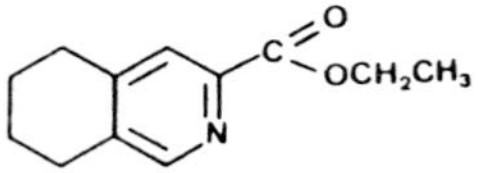 $C(=O)OCH_2CH_3$, N	1,7-Octa-diyne	C_2H_5—O—C(=O)—CN		**50** [18b]

Table 17 (continued)

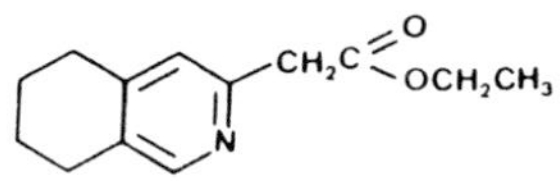 $CH_2C(=O)OCH_2CH_3$; N	1,7-Octa-diyne	$C_2H_5-O-C(=O)-CH_2-CN$	50 [18b]
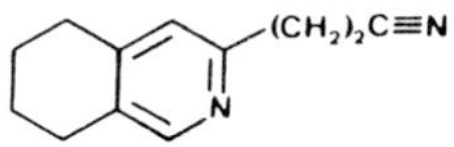 $(CH_2)_2C{\equiv}N$; N	1,7-Octa-diyne	1,2-Dicyanoethane	50 [18b]
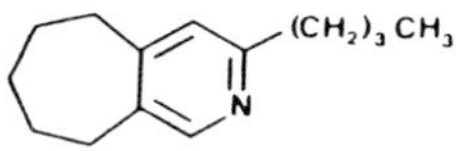 $(CH_2)_3CH_3$; N	1,8-Nona-diyne	*n*-Butylcyanide	50 [18b]
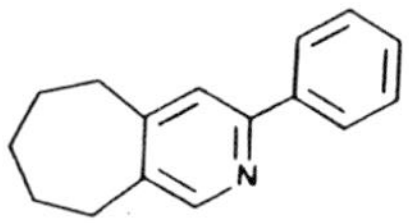 N	1,8-Nona-diyne	Benzonitrile	50 [18b]
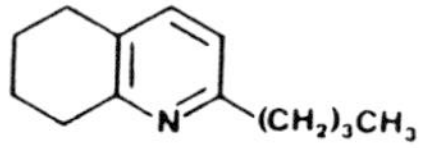 N; $(CH_2)_3CH_3$	1-Hexyne	$HC{\equiv}C-(CH_2)_4-CN$	50 [18h]
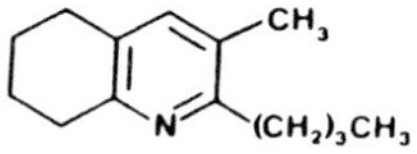 CH_3; N; $(CH_2)_3CH_3$	2-Heptyne	$HC{\equiv}C-(CH_2)_4-CN$	50 [18h]
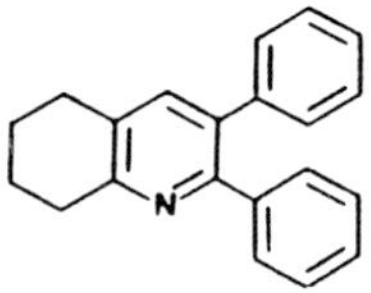 N	Diphenyl-acetylene	$HC{\equiv}C-(CH_2)_4-CN$	50 [18h]

Table 17 (continued)

Product	Alkyne	Nitrile	Solvent	Catalyst [a]
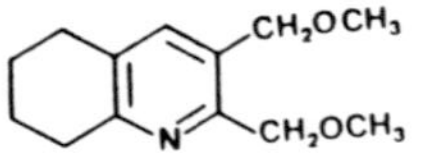 CH_2OCH_3, CH_2OCH_3, N	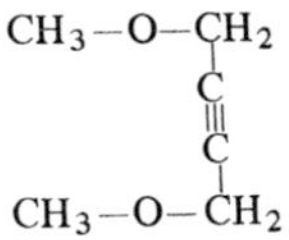 $CH_3-O-CH_2-C\equiv C-CH_2-O-CH_3$	$HC\equiv C-(CH_2)_4-CN$		**50** [18h]
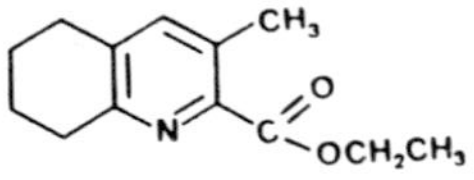 CH_3, $C(=O)OCH_2CH_3$, N	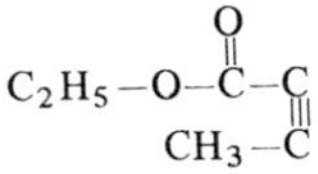 $C_2H_5-O-C(=O)-C\equiv C-CH_3$	$HC\equiv C-(CH_2)_4-CN$		**50** [18h]
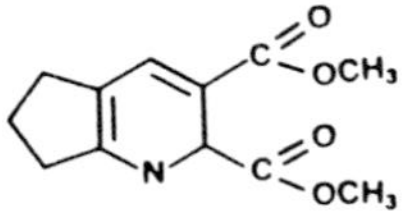 $C(=O)OCH_3$, $C(=O)OCH_3$, N	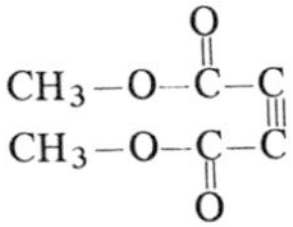 $CH_3-O-C(=O)-C\equiv C-C(=O)-O-CH_3$	$HC\equiv C-(CH_2)_3-CN$		**50** [18h]
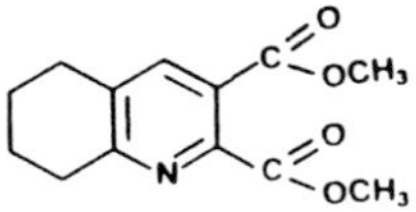 $C(=O)OCH_3$, $C(=O)OCH_3$, N	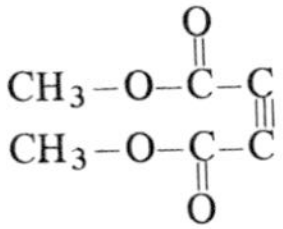 $CH_3-O-C(=O)-C\equiv C-C(=O)-O-CH_3$	$HC\equiv C-(CH_2)_4-CN$		**50** [18h]
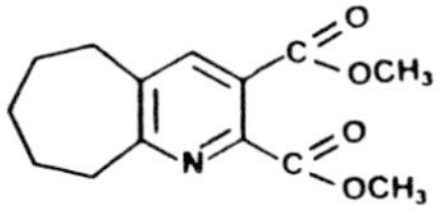 $C(=O)OCH_3$, $C(=O)OCH_3$, N	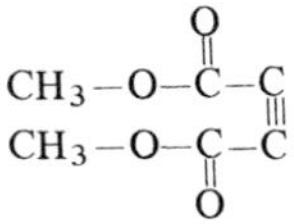 $CH_3-O-C(=O)-C\equiv C-C(=O)-O-CH_3$	$HC\equiv C-(CH_2)_5-CN$		**50** [18h]

Product	Reactants		Yield
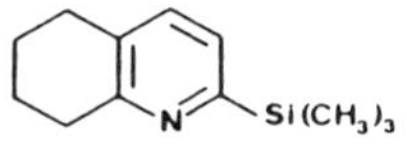 N; $Si(CH_3)_3$	$Me_3Si-C{\equiv}CH$	$HC{\equiv}C-(CH_2)_4-CN$	**50** [18h]
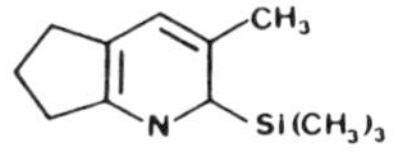 CH_3; N; $Si(CH_3)_3$	$Me_3Si-C{\equiv}C-CH_3$	$HC{\equiv}C-(CH_2)_3-CN$	**50** [18h]
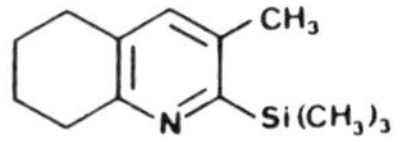 CH_3; N; $Si(CH_3)_3$	$Me_3Si-C{\equiv}C-CH_3$	$HC{\equiv}C-(CH_2)_4-CN$	**50** [18h]
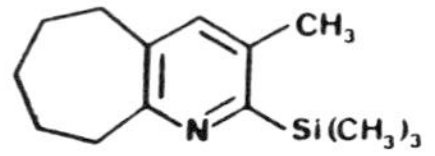 CH_3; N; $Si(CH_3)_3$	$Me_3Si-C{\equiv}C-CH_3$	$HC{\equiv}C-(CH_2)_5-CN$	**50** [18h]
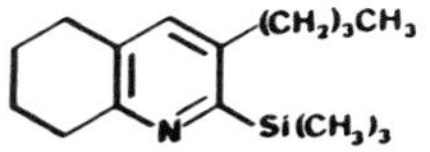 $(CH_2)_3CH_3$; N; $Si(CH_3)_3$	$Me_3Si-C{\equiv}C-C_4H_9$	$HC{\equiv}C-(CH_2)_4-CN$	**50** [18h]
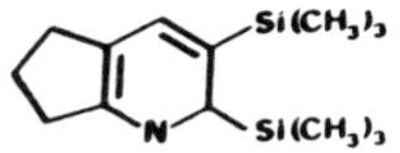 $Si(CH_3)_3$; N; $Si(CH_3)_3$	$Me_3Si-C{\equiv}C-SiMe_3$	$HC{\equiv}C-(CH_2)_3-CN$	**50** [18h]
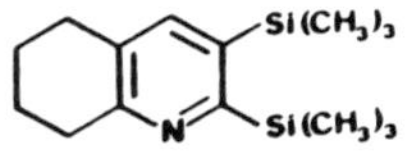 $Si(CH_3)_3$; N; $Si(CH_3)_3$	$Me_3Si-C{\equiv}C-SiMe_3$	$HC{\equiv}C-(CH_2)_4-CN$	**50** [18h]
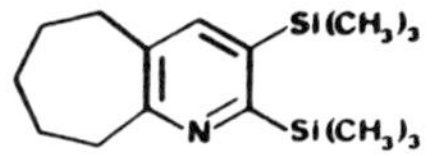 $Si(CH_3)_3$; N; $Si(CH_3)_3$	$Me_3Si-C{\equiv}C-SiMe_3$	$HC{\equiv}C-(CH_2)_5-CN$	**50** [18h]

[a] The most effective catalyst for a given reaction is shown in bold face, with underlining.

9. ACKNOWLEDGEMENT

The authors are pleased to thank Dr P. W. Jolly for transforming this article into jolly good English.

10. GLOSSARY OF ABBREVIATIONS

a = year(s)
Ac = acetyl-
acac = acetylacetonato
bp = boiling point
br = boiling range
Bu = butyl-
c = concentration; cyclo-
cat = catalyst
cod = cyclooctadiene-1,5
Cp = cyclopentadienyl-
CpH = cyclopentadiene
d = day(s)
DMF = dimethylformamide
Et = ethyl-
FID = flame ionization detector
GC = gas chromatography
h = hour(s)
M = mean value
max. = maximum
Me = methyl-
min = minute(s)
mp = melting point
MS = mass spectrometry
n = normal
NMP = *N*-methyl-pyrrolidone
NMR = nuclear magnetic resonance
Pa = Pascal (= 1 N/m^2 = 10^{-5} bar)
Ph = phenyl-
PROD = productivity (*t* product/kg cobalt)
R = organyl group
t = tertiary
T = temperature
THF = tetrahydrofurane
TLC = thin layer chromatography
TON = turnover number

UV = ultra violet
2-VP = 2-vinylpyridine
X = halogen
Z_R = response value (of the regression equation)

11. REFERENCES

1. *Comprehensive Organometallic Chemistry*, v. 7 and 8, Ed. G. Wilkinson, F. G. A. Stone, E. W. Abel. Pergamon Press (1982).
2. C. W. Bird: *J. Organomet. Chem.* **47**, 281–309 (1973).
3. R. P. A. Sneeden, in *Comprehensive Organometallic Chemistry*, v. 8, p. 275 f. Pergamon Press (1982).
4. (a) S. F. A. Kettle and L. E. Orgel: *Proc. Chem. Soc. 1959*, 307.
   (b) W. Z. Heldt: *J. Organomet. Chem.* **6**, 292 (1966).
   (c) G. Oehme and H. Pracejus: *Z. Chem.* **9**, 140 (1969).
5. K. Ohno and J. Tsuji: *J. Chem. Soc., Chem. Commun. 1971*, 247.
6. J. Kiji *et al.*: *J. Chem. Soc., Chem. Commun. 1974*, 506.
7. P. Heimbach, B. Hugelin, H. Peter, A. Roloff and E. Troxler: *Angew. Chem.* **88**, 29 (1976); *Angew. Chem. Int. Ed.* **15**, 49 (1976).
8. T. Joh and N. Hagihara:
   (a) *Tetrahedron Lett. 1967*, 4199.
   (b) *Kogyo Kagaku Zasshi* **91**, 378 (1970); *C. A.* **73**, 45294 (1970).
   (c) *Kogyo Kagaku Zasshi* **91**, 373 (1970); *C. A.* **73**, 45295 (1970).
9. Y. Watanabe, J. Tsuji and Y. Ohsugi: *Tetrahedron Lett.* **22**, 2667 (1981).
10. H. Alper *et al.*: *J. Am. Chem. Soc.* **103**, 1289 (1981); *Organometallics*, v. 1, 70 (1982); *ibid.* **1**, 332 (1982).
11. J. E. Bäckvall and J. E. Nyströn: *J. Chem. Soc., Chem. Commun. 1981*, 59.
12. S.-I. Murahashi, T. Shimamura and I. Moritani: *J. Chem. Soc., Chem. Commun. 1974*, 931.
13. K. Utimoto, H. Miwa and H. Nozaki: *Tetrahedron Lett.* **22**, 4277 (1981).
14. H. Bönnemann: *Angew. Chem.* **85**, 1024 (1973); *Angew. Chem. Int. Ed.* **12**, 964 (1973).
15. P. W. Jolly and G. Wilke: *The Organic Chemistry of Nickel*, v. 1 and 2, Academic Press, New York (1974/1975).
16. M. Berthelot: *Annalen* **141**, 173 (1866).
17. (a) W. Reppe: *Chemie und Technik der Acetylen-Druck-Reaktionen*, 2. Auflage, Weinheim, Verlag Chemie (1952).
    (b) Copenhaver and Bigelow: *Acetylene and Carbon Monoxide Chemistry*, Reinhold, New York (1949).
    (c) C. W. Bird: *Transition Metal Intermediates in Organic Synthesis*, Academic Press, New York, London (1967).
18. (a) K. P. C. Vollhardt: 'Transition-metal catalyzed acetylene cyclisation in organic synthesis', *Acc. Chem. Res.* **10**, 1 (1977).
    (b) A. Naiman and K. P. C. Vollhardt: *Angew. Chem.* **89**, 758–759 (1977); *Angew. Chem. Int. Ed.* **16**, 708 (1977).
    (c) J. R. Fritch and K. P. C. Vollhardt: *Angew. Chem.* **92**, 570–572 (1980); *Angew. Chem. Int. Ed.* **19**, 559 (1980).
    (d) G. Ville, K. P. C. Vollhardt and M. J. Winter: *J. Am. Chem. Soc.* **103**, 5267–5269 (1981).

(e) J. P. Tane and K. P. C. Vollhardt: *Angew. Chem.* **94**, 642–643 (1982).
(f) J. R. Fritch and K. P. C. Vollhardt: *Organometallics* 1, 590–602 (1982).
(g) J. S. Drage and K. P. C. Vollhardt: *Organometallics* 1, 1545–1547 (1982).
(h) D. J. Brien, A. Naiman and K. P. C. Vollhardt: *J. Chem. Soc., Chem. Commun. 1982*, 133–134.
(i) B. C. Berris, Y.-H. Lai and K. P. C. Vollhardt: *J. Chem. Soc., Chem. Commun. 1982*, 953–954.

19. (a) H. Bönnemann, R. Brinkmann and H. Schenkluhn: *Synthesis 1974*, 575.
(b) H. Bönnemann and H. Schenkluhn: DBP 2416295 (4.4.1974); U.S.P. 4,006,149.
20. (a) Y. Wakatsuki and H. Yamazaki: *J. Chem. Soc., Chem. Commun. 1973*, 280.
(b) H. Yamazaki and Y. Wakatsuki: *Tetrahedron Lett. 1973*, 3383.
(c) Rikagaku Kenkyusho, Saitama-Ken: JAP–OS 126680/74 (1974).
(d) H. Yamazaki and Y. Wakatsuki: JAP–OS 7725780 (1975); *C. A.* **87**, 68168n. (1977).
21. (a) Y. Wakatsuki and H. Yamazaki: *Synthesis 1976*, 26.
(b) H. Yamazaki and Y. Wakatsuki: JAP–OS 7725780 (1975); *C. A.* **87**, 68168n. (1977).
22. (a) W. Reppe, N. von Kutepow and A. Magin: *Angew. Chem.* **81**, 717 (1969); *Angew. Chem. Int. Ed. (Engl.)* **8**, 727 (1969).
(b) F. L. Bowden and A. P. B. Lever: *J. Organomet. Chem. (Rev.)* **3**, 227 (1968).
(c) F. R. Hartley: *Chem. Rev.* **69**, 799 (1969).
23. H. Hogeveen, R. F. Kingma and D. M. Kok: *J. Org. Chem.* **47**, 989–997 (1982).
24. H. Bönnemann, *Angew. Chem.* **90**, 517 (1978); *Angew. Chem. Int. Ed.* **17**, 505 (1978).
25. P. Hardt:
(a) DOS 2615309 (08.04.1976), Swiss Appl. 12139–75 (18.09.1975).
(b) DOS 2742541 (20.04.1078), Swiss Appl. 76/13,079 (15.10.1976).
(c) U.S.P. 4,196,387 (01.04.1980), to Lonza AG.
(d) DOS 2742542 (15.02.1979), Swiss Appl. 9471–77 (02.08.1977).
26. (a) G. Natta, U. Giannini, P. Pino and A. Cassata: *Chim. Ind.* (Milan) **47**, 524 (1965).
(b) G. Natta and P. Pino *et al.*: *J. Chem. Soc., Chem. Commun. 1967*, 1263.
27. (a) Ch. Grard: Dissertation, Universität Bochum, 1967; G. Wilke: *Kagaku Kogyo* **20**, 1308, p. 1310 (1967).
(b) S. Otsuka and M. Rossi: *J. Chem. Soc.* (A), *1968*, 2630.
(c) S. Koda, A. Tanaka and T. Watanabe: *J. Chem. Soc., Chem. Commun. 1969*, 1293.
28. S. Otsuka and M. Rossi: *J. Chem. Soc.* (A), *1969*, 497–500.
29. H. Lehmkuhl, W. Leuchte and E. Janssen: *J. Organomet. Chem.* **30**, 407–409 (1971).
30. G. E. Herberich, W. Koch and H. Lueken: *J. Organomet. Chem.* **160**, 17–23 (1978).
31. (a) D. Habermann: Dissertation, Bochum, 1980, pp. 58 and 70.
(b) K. Jonas and C. Krüger: *Angew. Chem.* **92**, 513–531 (1980); *Angew. Chem. Int. Ed.* **19**, 520 (1980).
(c) K. Jonas: *Advances in Organometallic Chemistry* **19**, p. 112 f. (1981).
32. J. L. Spencer, R. G. Beever and S. A. Frith: *J. Organomet. Chem.* **221**, 1196 (1980).
33. R. L. Pruett and W. R. Myers: U.S.P. 3.159.659 (1962/65); *Chem. Abstr.* **62**, 7800 (1965).
34. M. L. H. Green, L. Pratt and G. Wilkinson: *J. Chem. Soc. 1959*, 3753–3767.
35. H. Bönnemann, M. Radermacher, C. Krüger and H.-J. Kraus: *Helv. Chim. Acta* **66**, 185–191 (1983).
36. M. Radermacher: Diplomarbeit, RWTH, Aachen, 1982.
37. H. Kojima, S. Takahashi, H. Yamazaki and N. Hagihara: *Bull. Chem. Soc. Jap.* **43**, 2272 (1970).
38. H. H. Hoehn, L. Pratt, K. F. Watterson and G. Wilkinson: *J. Chem. Soc. 1961*, 2738–2745.
39. (a) K. Jonas, E. Deffense and D. Habermann: *Angew. Chem.*, in preparation.
(b) D. Habermann: Dissertation, Bochum, 1980, p. 75.

40. (a) *Gmelins Handbuch der anorg. Chemie*, v. 5, 'Cobalt-organische Verbindungen', Part 1, 291 ff.
(b) *Comprehensive Organometallic Chemistry*, v. 5, 248–252.
41. H. Bönnemann and M. Samson: Europ. Patent Appl. 009685 (1980); U.S.P. 4,266,061 (05.05.1981) to Studiengesellschaft Kohle m.b.H.
42. (a) *Gmelins Handbuch der anorg. Chemie*, v. 5, 'Cobalt-organische Verbindungen', Part 1, 359 ff.
(b) *Comprehensive Organometallic Chemistry*, v. 5, 244–247.
43. G. E. Herberich and G. Greiss: *Chem. Ber.* **105**, 3413–3423 (1972).
44. M. Rosenblum, B. North, D. Wells and W. P. Giering: *J. Am. Chem. Soc.* **94**, 1239–1246 (1972).
45. W.-S. Lee and H. H. Brintzinger: *J. Organomet. Chem.* **209**, 401–406 (1981).
46. J. A. Connor: *Top. Curr. Chem.* **71**, 71 (1976).
47. (a) T. S. Pieper, F. A. Cotton and G. Wilkinson: *J. Inorg. Nucl. Chem.* **1**, 165–174 (1955).
(b) A. Nakamura and N. Hagihara: *Bull. Chem. Soc. Japan* **33**, 425 (1960).
(c) H. Bönnemann and B. Bogdanović: Ger. Pat. Appl. 31 0550.1 (17.02.1982), Europ. Patent Appl. 83101246.3 (10.02.1983) to Studiengesellschaft Kohle m.b.H.
(d) T. R. Engelmann: Dissertation, Southern Illinois Univ. 1970; Diss. Abstr. Intern. **B 31**, 5859 (1971).
(e) J. Lewis and A. W. Parkins: *J. Chem. Soc.* (A) *1967*, 1150.
(f) A. Nakamura and N. Hagihara: *Nippon Kagaku Zasshi* **82**, 1392–1394 (1961).
(g) R. B. King: *Organometallic Synthesis*, v. 1; *Transition Metal Compounds*, New York (1965), 115–119 and 131–132.
(h) K. Yasufuku and H. Yamazaki: *Org. Mass. Spectrom.* **3**, 23–29 (1970).
(i) A. Nakamura: *Mem. Inst. Sci. Res. Osaka Univ.* **19**, 81/95 and 84/88 (1962).
48. R. D. Rausch and R. A. Genetti: *J. Org. Chem.* **35**, 3888 (1970).
49. (a) H. W. Sternberg and I. Wender: *Intern. Conf. Coord. Chem.*, London (1959), *Spec. Publ.* **13**, 35–55.
(b) H. W. Sternberg, R. Markby and I. Wender: *Chim. Ind.* (Milan) **42**, 41–51 (1960).
50. (a) A. Nakamura and N. Hagihara: *Bull. Chem. Soc. Japan* **34**, 452–453 (1961).
(b) R. B. King, P. M. Treichel and F. G. A. Stone: *J. Am. Chem. Soc.* **83**, 3593–3597 (1961).
(c) W. McFarlane, L. Pratt and G. Wilkinson: *J. Chem. Soc. 1963*, 2162.
51. H. P. Fritz and H. Keller: *Z. Naturforsch.* **16b**, 348 (1961).
52. R. B. King and A. Efraty: *J. Am. Chem. Soc.* **93**, 4950 (1971).
53. A. Greco, M. Green and F. G. A. Stone: *J. Chem. Soc.* (A) *1971*, 285–288.
54. H. Bönnemann, M. Samson, C. Krüger and L.-K. Liu (in preparation).
55. H. Bönnemann and M. Samson: DBP 2840460 (16.09.1978) to Studiengesellschaft Kohle m.b.H.
56. W. Meurers: Diplomarbeit, RWTH, Aachen, 1982.
57. J. C. Wollensack: U.S.P. 3 088 960 (1959/63) to Ethyl Corp.
58. A. Budzinski: *Chem. Ind.* (London) **23**, Sept. 1981.
59. R. A. Abramovitch: *Chemistry of Heterocyclic Compounds*, v. 14, 'Pyridine and Its Derivatives', Suppl. Parts 1–4, Wiley, New York (1974–1975).
60. *Ullmanns Encyklopädie der Technischen Chemie* (4. Auflage) **19**, 591–617.
61. Kirk-Othmer: *Encyclopedia of Chemical Technology* **19**, 454–483.
62. H. Beschke and H. Friedrich: *Chem. Ztg.* **101**, 377–384 (1977).
63. Rütgerswerke, DOS 20 51 316 (1970).
64. (a) Y. Kusunoki and H. Okazeku: *Hydrocarbon Process.* **53** (11), 129–131 (1974).
(b) *C. A.* **96**, 52147e (1982).
(c) *C. A.* **96**, 52148f (1982).

65. *Chem. Mark. Rep.* (Oct. 17, 1977); U.S. Pat. 3,780,082 (Dec. 18, 1973), J. M. Deumens and S. H. Groen (to Stamicarbon N. V.); Brit. Pat. 1,378,464 (Dec. 27, 1974) (to Stamicarbon B. V.); Brit. Pat. 1,304,155 (Jan. 24, 1973) (to Stamicarbon N. V.).
66. (a) G. M. Badger and W. H. F. Sasse: *Adv. Heterocycl. Chem.* **2**, 179 (1963).
    (b) M. A. E. Hodgson, in *Modern Chemistry in Industry*, Society of Chemical Industry, London, UK (1968), p. 49; L. A. Summers: *The Bipyridinium Herbicides*, Academic Press, Inc., New York (1980).
67. *Chemical Age*, July 25, *1964*, p. 168.
68. D. B. Wootton: *Dev. Adhes.* **1**, 181 (1977).
69. P. Hardt: DOS 2,751,072 (1978), Swiss Appl. 76/14,399 (15.11.1976); *C. A.* **89**, 14770, to Lonza A. G.
70. M. J. Birchenough: *J. Chem. Soc. 1951*, 1263–1266.
71. Schering: DT 66 38 91 (1936).
72. P. Arnall and N. R. Clark: *Chem. Process.* (London) **17** (10), 9, 11–13, 15 (1971).
73. (a) M. M. Boudakian: *Chem. Heterocycl. Compd.: Pyridine and Its Derivatives*, Suppl. Part II (14), p. 407 ff.
    (b) Degussa: US 39 20 657 (1972).
    (c) Olin Mathieson Chem. Corp.: US 31 53 044 (1963).
    (d) K. Thomas and D. Jerdiel: *Angew. Chem.* **70**, 737 (1958).
    (e) Dow Chem. Co.: US 32 06 358 (1965).
74. E. Shaw, J. Bernstein, K. Losse and W. A. Lott: *J. Am. Chem. Soc.* **72**, 4362 (1950); U.S. Pat. 2,745,826 (May 15, 1956); S. Semenoff and M. A. Dolliver (to Olin Mathieson Chemical Corp.).
75. W. Ramsay: *Ber. Dtsch. Chem. Ges.* **10**, 736 (1877).
76. R. Meyer and A. Tanzen: *Ber. Dtsch. Chem. Ges.* **46**, 3186 (1913).
77. H. Bönnemann: DOS 3117363.2 (02.05.1981); Europ. Pat. Appl. 064268 (28.04.1982) to Studiengesellschaft Kohle m.b.H.
78. Y. Wakatsuki and H. Yamazaki: *J. Organomet. Chem.* **149**, 385 (1978).
79. St. Goldschmidt and M. Minsinger: Ger. 952,807 (22.11.1956); *C. A.* **53**, 16160 (1959).
80. J. P. Wibaut and C. Hoogzand: *Chem. Weekblad* **52**, 357–9 (1956); *C. A.* **51**, 3593 (1957).
81. (a) D. Tatone, Trane Cong Dich, R. Nacco and C. Botteghi: *J. Org. Chem.* **40**, No. 20, 2987–2990 (1975).
    (b) *C. A.* **97**, 109819c (1982).
82. C. Botteghi: private communication.
83. In 1978–1979 we worked together with Lonza AG. on a technical process for the manufacture of vinylpyridine. The authors want to thank Drs. Peter Hardt and Colm O'Murchü for their fruitful collaboration.
84. H. Kojima, H. Yamazaki and N. Hagihara: *Mem. Inst. Sci. Ind. Res.* (Osaka Univ.) **28**, 113–20 (1971); *C. A.* **75**, 49706 (1971).
85. H. Bönnemann and M. Samson: U.S.P. 4,266,061 (14.09.1979 – 05.05.1981).
86. H. Bönnemann: U.S. Pat. Appl. 371,872 (26.04.1982).
87. H. Bönnemann and G. S. Natarajan: *Erdöl und Kohle – Erdgas – Petrochemie vereinigt mit Brennstoffchemie* **33**, 328 (1980).
88. R. A. Abramovitch: *Heterocyclic Compounds*, v. 14, 'Pyridine and Its Derivatives', Wiley, New York (1975), Part 4, Chap. 15, p. 189.
89. (a) TRAC (Technische Regeln Acetylen), Carl Heymanns Verlag K. G., D–5000 Köln; TRAC 203: compressors; TRAC 204: capillaries; TRAC 206: cylinders; TRAC 207: safety installations and recoil-guards.
    (b) B. A. Ivanov and S. M. Kogarko: *The Upper Concentration Limit for the Propagation of Flame in Mixtures of Acetylene with Oxygen and Air*, Engl. from Doklady Akademii Nauk SSSR, v. 142, No. 3, pp. 637–638, January 1962.

(c) H. B. Sargent: 'How to design a hazard-free system to handle acetylene', *Chemical Engineering* **64**, No. 2, 250–254 (1957).

90. H. Yamazaki and Y. Wakatsuki: *J. Organomet. Chem.* **139**, 157–167 (1977).
91. Y. Wakatsuki and H. Yamazaki: *J. Chem. Soc. Dalton* (1978), 1278–1282.
92. H. Bönnemann, W. Brijoux and K. H. Simmrock: *Erdöl und Kohle – Erdgas – Petrochemie vereinigt mit Brennstoffchemie* **33**, 476–479 (1980).
93. W. Brijoux: Dissertation, Universität Dortmund (1979), S. 17, 23, 24.
94. D. R. McAlister, J. E. Bercaw and R. G. Bergman: *J. Am. Chem. Soc.* **99**, 1666–1668 (1977).
95. Y. Wakatsuki and H. Yamazaki: *J. Organomet. Chem.* **139**, 167–177 (1977).
96. Y. Wakatsuki, Y. Aoki and H. Yamazaki: *J. Am. Chem. Soc.* **101**, 1123–1130 (1979).
97. J. Browning, M. Green, A. Lagura, L. E. Smart, J. L. Spencer and F. G. A. Stone: *J. Chem. Soc., Chem. Commun. 1975*, 723–724.
98. A. Stockis and R. Hoffmann: *J. Am. Chem. Soc.* **102**, 2952–2962 (1980).
99. D. L. Thorn and R. Hoffmann: *Nouveau Journal de Chimie* **3**, 39–45 (1979).
100. H. C. L. Linder: *Chemie-Ingenieur-Technik* **40**, 18, 873–875 (1968).
101. K. H. Simmrock: *Chemie-Ingenieur-Technik* **40**, 18, 875–883 (1968).
102. (a) U. Hoffmann and H. Hofmann: *Einführung in die Optimierung*, Verlag Chemie GmbH, Weinheim (1971).
    (b) A. J. Bojarinow and W. W. Kafarow: *Optimierungsmethoden in der chemischen Technologie*, Verlag Chemie GmbH, Weinheim (1972).
    (c) E. Hofer and R. Lunderstädt: *Numerische Methoden der Optimierung*, R. Oldenbourg Verlag, München, Wien (1975).
    (d) H. Erfurth and G. Biess: *Verfahrenstechnik: 'Optimierungsmethoden'*, VEB – Deutscher Verlag für Grundstoffindustrie, Leipzig (1975).
    (e) W. Entenmann: *Optimierungsverfahren*, A. Hüthig Verlag, Heidelberg (1976).
    (f) F. Bandermann: 'Optimierung chemischer Reaktionen', *Ullmanns Encyklopädie der Technischen Chemie* **1**, 362–418.
103. (a) I. Newton: 'Principiae' (1687), in J. Ph. Wolfers: *Sir Isaac Newtons Mathematische Prinzipien der Naturlehre*, II. Buch, VII. Abschnitt, Paragraph 46, 323 f., Verlag von Robert Oppenheimer.
    (b) L. Euler: *Methodus inveniendi lineas curvas maximi minimive proprietate gaudentes sive solutio problematis isoperimetrici latissimo sensu accepti*, Lausannae et Genevae (1744).
104. A. Capelli and P. Trombouze: *Chemie-Ingenieur-Technik* **49**, 1, 5–12 (1977).
105. K. H. Simmrock: *Chemie-Ingenieur-Technik* **43**, 9, 571–583 (1971).
106. K. J. Hinger and H. Blenke: *Chemie-Ingenieur-Technik* **47**, 23, 976–981 (1975).
107. L. Rockstroh and K. Hartmann: *Chemische Technik* **28**, 3, 134–138 (1976).
108. F. Bandermann: 'Statistische Methoden beim Planen und Auswerten von Versuchen', *Ullmanns Encyklopädie der Technischen Chemie* **1**, 293–360.
109. G. Schomburg, H. Husmann and F. Weeke: *Journal of Chromatography* **112**, 205–217 (1975).
110. O. L. Davies: *Statistical Methods in Research and Production*, 3rd edition, Oliver and Boyd, London (1967).
111. F. Yates: *Design and Analysis of Factorial Experiments*, Imperial Bureau of Soil Science, London (1937).
112. N. R. Draper and H. Smith: *Applied Regression Analysis*, Wiley (1967), 478.
113. W. Spendley, G. R. Hext and F. R. Himsworth: *Technometrics* **4**, 441–461 (1962).
114. J. R. Blackborrow and D. Young: *Metal Vapour Synthesis in Organometallic Chemistry*, Springer-Verlag (1979), 179.
115. Y.-H. Lai: *Synthesis 1981*, 586.

116. (a) H. Bönnemann, B. Bogdanović, W. Brijoux and R. Mynott: *Catalysis in Organic Chemistry*, Marcel Dekker Inc., New York (1983).
(b) H. Bönnemann, W. Brijoux, R. Brinkmann, M. Kajitani, G. S. Natarajan and M. Samson: *Catalysis of Organic Reactions*, Marcel Dekker Inc., New York (1984).
117. R. B. King *et al.*: *Org. Synth.* **56**, 1 (1977).
118. (a) A. J. Ashe and P. Shu: *J. Am. Chem. Soc.* **93**, 1804 (1971).
(b) E. Raabe: Dissertation, RWTH Aachen (1984).
119. (a) P. Hong and H. Yamazaki: *Tetrahedron Lett. 1977*, 1333.
(b) H. Yamazaki and Y. Wakatsuki: *Kagaku Sosetsu* **32**, 161–201 (1981).
120. H. Bönnemann, R. Brinkmann, M. Kajitani and G. S. Natarajan: to be published.
121. (a) H. Bönnemann, R. Brinkmann, C. Krüger and R. Goddard: in preparation.
(b) H. Bönnemann, R. Brinkmann, C. Krüger and J. Sekutowski: in preparation.
122. I. A. Abronin, L. G. Gorb, D. Z. Levin, N. K. Demidova and E. S. Mortikov: *Bull. Acad. Sci. U.S.S.R.* (Div. Chem. Sci.) **31**, No. 11, 2317 (1983).

Homogeneous Catalysis Using Iodide-Promoted Rhodium Catalysts

D. J. DRURY

Research and Development Department, BP Chemicals Ltd., Belgrave House, 76 Buckingham Palace Road, London SWIW OSU, U.K.

R. Ugo (ed.), Aspects of Homogeneous Catalysis, Vol. 5, 197–216.

1. INTRODUCTION

Recent years have seen an upsurge in interest in catalytic conversions involving carbon monoxide and other one carbon molecules. The impetus for such studies has come from the desirability of basing chemical and liquid fuels production on feedstocks other than petroleum. Homogeneous catalysis has a major part to play in this field and, in particular, homogeneous rhodium catalysts combined with an iodide promoter have proved to be especially important.

The first reported use of iodide-promoted rhodium catalysts came in 1968 [1] with the disclosure of the Monsanto process for carbonylation of methanol to acetic acid. Since that time the same catalyst system has been used for a wide range of carbonylation and related reactions. A number of these reactions have not yet attracted academic interest, and some of the more interesting observations are confined to the patent literature. The purpose of this review is to illustrate the range of reactions that can be catalysed with iodide-promoted rhodium catalysts and to suggest the mechanistic relationships that might exist between the various reactions.

2. RHODIUM IODOCARBONYLS

The reactions to be discussed are generally carried out in the presence of carbon monoxide and with a considerable molar excess of iodide over rhodium. Under these circumstances rhodium iodocarbonyls are likely to predominate and, indeed, to be the active catalytic species. A brief summary of the chemistry of such species is therefore a worthwhile preliminary to the discussion of the catalytic reactions.

Table 1 lists the known simple iodocarbonyl derivatives of rhodium. Apart

Table 1

SIMPLE IODOCARBONYL DERIVATIVES OF RHODIUM

Species	Reference Number
$Rh_2(CO)_4I_2$	2
$Rh(CO)_2I_2^-$	3
$Rh_2(CO)_2I_4^{2-}$ [a]	3
$Rh(CO)_3I$	4
$Rh(CO)I_4^-$ x (solvent)	3, 5
cis and *trans*-$Rh(CO)_2I_4^-$	5, 6
$Rh(CO)I_5^{2-}$	5, 7

[a] Existence doubtful: see Ref. 5.

from $[Rh(CO)_2I]_2$ and $Rh(CO)_3I$, all are anionic. These iodocarbonyl rhodates are particularly stable in polar solvents in the presence of iodide and CO. It is a feature of Rh/I-catalysed carbonylation reactions that a wide range of soluble rhodium sources often give identical catalytic results: this can be attributed to a rapid conversion of the rhodium to the relevant iodocarbonyl form. Thus, for example, although the rhodium(I) species $Rh(CO)_2I_2^-$ is the predominant rhodium species in solution during methanol carbonylation [8], a range of rhodium (I) and rhodium (III) complexes give similar reaction rates [9]. The reduction of rhodium (III) halides to the dicarbonyl dihalogenorhodates by carbon monoxide in alcoholic [5] and aqueous [10] media is known to proceed according to:

$$RhX_3 + 3CO + H_2O \longrightarrow Rh(CO)_2X_2^- + CO_2 + 2H^+ \tag{1}$$

The mechanism of this reduction is not known, but probably involves nucleophilic attack of water on a carbonyl ligand to generate a metallo-carboxylic acid which eliminates carbon dioxide:

$$I{-}\overset{III}{Rh}{-}CO \xrightarrow[-H^+]{H_2O} I{-}\overset{III}{Rh}{-}CO_2H \xrightarrow{-CO_2} I{-}\overset{III}{Rh}{-}H \xrightarrow{-HI} Rh^I \tag{2}$$

The oxidation of rhodium (I) iodocarbonyl derivatives is also important in catalytic reactions. Hydrogen iodide – a common ingredient in Rh/I catalysed reactions – can carry out such oxidations [5].

$$Rh(CO)_2I_2^- + 2HI \longrightarrow Rh(CO)I_4^- + H_2 + CO \tag{3}$$

It seems reasonable to assume that such oxidations occur via oxidative addition of HI to give a – so far undetected – hydride.

$$Rh(CO)_2I_2^- \xrightarrow{HI} HRh(CO)_2I_3^- \xrightarrow[-H_2]{HI} Rh(CO)I_4^- + CO \tag{4}$$

Oxidative addition of I_2 is another pathway for conversion of rhodium (I) to rhodium (III) [5], e.g.:

$$Rh(CO)_2I_2^- + I_2 \rightleftharpoons Rh(CO)_2I_4^- \tag{5}$$

The reversibility of this transformation is shown by the fact that amines which are capable of forming charge-transfer complexes with I_2 will cause the reduction of $Rh(CO)_2I_4^-$ to $Rh(CO)_2I_2^-$ [11].

3. CATALYSIS OF THE WATER GAS SHIFT REACTION

Simplifying and combining Reactions 1 and 2 gives the following:

$$Rh(III) + CO + H_2O \longrightarrow Rh(I) + 2H^+ + CO_2 \quad (6)$$

$$Rh(I) + 2HI \longrightarrow Rh(III) + H_2 + 2I^- \quad (7)$$

$$CO + H_2O \longrightarrow CO_2 + H_2 \quad (8)$$

This amounts to catalysis of the Water Gas Shift reaction (WGSR). Rh/I catalysis of this reaction was first demonstrated by Eisenberg and co-workers [12, 13] and has also been studied by a group of Monsanto workers [14]. Aqueous acidic conditions are needed for the catalysis, together with a large molar excess of iodide over rhodium. Carboxylic acid solvents appear to enhance the rate. The Rh/I catalyst is a moderately active one for this reaction, and turn-over numbers of around 500 mol CO_2/mol Rh/h are reported at 185°C/27 bar [14].

The kinetics of the Rh/I-catalysed WGSR are rather complex, and there are some discrepancies in the observations of the two groups of workers (but these may be attributed to the very different reaction conditions used). Both sets of workers are, however, agreed that the reaction consists of a Rh(III) reduction step (Equation 5) and a Rh(I) oxidation step (Equation 6). Which of these steps is rate determining depends, however, on the reaction conditions.

For example, Eisenberg [13] (working at a CO pressure of around 0.66 bar) finds a segmented Arrhenius plot implying that Rh(III) reduction is rate determining above 80°C (Ea = 9.3 kcal mol^{-1}) while Rh(I) oxidation is rate-determining below 70°C (Ea = 25.8 kcal mol^{-1}). In accordance with this, solutions from the high-temperature regime are found to contain only Rh(III) species, such as $Rh(CO)_2I_4^-$, but $Rh(CO)_2I_2^-$ is predominant in the lower temperature regime. The Rh(I) regime is also favoured at low acid levels. Increasing the acidity increases the reaction rate through its effect on Reaction 7, but Rh(III) reduction eventually becomes rate limiting and further increase in acidity then inhibits the reaction.

The precise mechanism of the Rh/I-catalysed WGSR is not known, but that shown in Figure has been proposed by Eisenberg [14] as being consistent with his results. It should, however, be noted that, of the species shown in the scheme, only $Rh(CO)_2I_2^-$ and $Rh(CO)_2I_4^-$ have actually been detected.

The dotted line in Figure 1 shows an alternative pathway noted by Eisenberg, involving direct conversion of $Rh(CO)_2I_3CO_2H^-$ to $HRh(CO)_2I_3^-$. In this scheme, the actual catalysis involves only Rh(III) species, and $Rh(CO)_2I_2^-$ emerges as a catalytically inactive species. Insufficient data are presently available to allow an evaluation of this possibility.

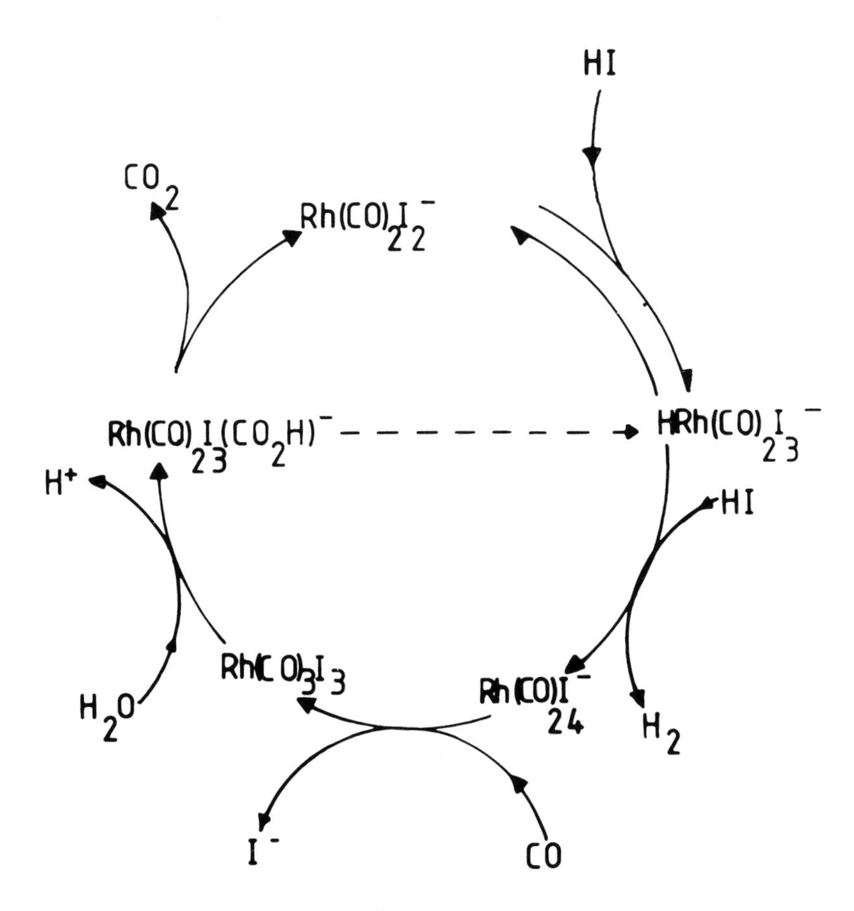

Fig. 1. Mechanism proposed for Rh/I-catalysed water gas shift reaction.

4. ALCOHOL CARBONYLATION

4.1. Methanol Carbonylation

The best-known application of the Rh/I catalyst system is in the production of acetic acid by carbonylation of methanol:

$$CH_3OH + CO \longrightarrow CH_3CO_2H \tag{8}$$

In 1968 Paulik and Roth of Monsanto [1] reported that iodide-promoted rhodium catalysts enabled this reaction to be carried out under mild conditions, constituting a very significant advance on the previously-used Co/I catalyst system, which requires severe conditions (210°C, 500 bar) [15]. Since that time a number of plants have been built using the Monsanto technology, and more are planned. At the time of writing, installed and planned capacity using the Monsanto process amounts to ca. 500 000 tn/year of acetic acid. In addition to its industrial importance, Rh/I-catalysed methanol carbonylation has considerable scientific importance, as its simple kinetics, and the ease with which reaction intermediates can be isolated and studied, have made this one of the best-understood of homogeneous catalytic processes.

The reaction may be carried out using a variety of Rh(I) or Rh(III) complexes. A large molar excess of iodide over Rh is usually used: I_2, HI or CH_3I are suitable iodide sources [9]. A wide variety of materials, including methanol itself, may be used as solvents for the reaction. Reactions run in methanol suffer, however, from the disadvantage that quantities of dimethyl ether are formed [16, 17]. Acetophenone has been claimed [17] to be a particularly advantageous solvent.

The reaction is practically quantitative in its selectivity [9]. An interesting feature of the reaction is that its rate is independent of either methanol concentration or carbon monoxide pressure, but the reaction is first order in both rhodium and iodine. Only the catalyst, and neither of the reactants therefore appear in the rate equation which is of the form [9, 16]:

$$-d[CH_3OH]/dt = k\,[Rh]\,[I]$$

with the kinetic parameters shown in Table 3, below.

These very simple kinetics are explained by the mechanistic scheme proposed by Forster [8, 18, 19] (Figure 2). In this scheme methyl iodide is formed by the reaction:

$$CH_3OH + HI \rightleftharpoons CH_3I + H_2O \qquad (9)$$

and oxidative addition of CH_3I to the anion $Rh(CO)_2I_2^-$ is the first, and rate-determining, step in the cycle. The methyl rhodium species (1) undergoes a rapid methyl migration to give the formally five-coordinate acetyl species (2), which rapidly takes up CO to give (3). Reductive elimination of acetyl iodide regenerates $Rh(CO)_2I_2^-$ and solvolysis of acetyl iodide gives acetic acid or methyl acetate, depending on the reaction conditions.

A range of observations [8, 18, 19] point to the basic correctness of this mechanism:

1. It has been demonstrated that the mechanism will reproduce the observed kinetics, if it is assumed that the equilibrium (8) lies far to the right and that the oxidative addition step is much slower than any of the other steps [16].

2. *In situ* infrared spectroscopy has shown that $Rh(CO)_2I_2^-$ is the only detectable rhodium species under catalytic conditions [8].

3. The oxidative addition of CH_3I to $Rh(CO)_2I_2^-$ has been followed spectroscopically [8]. An acetyl species is formed directly, showing that the methyl migration is very fast.

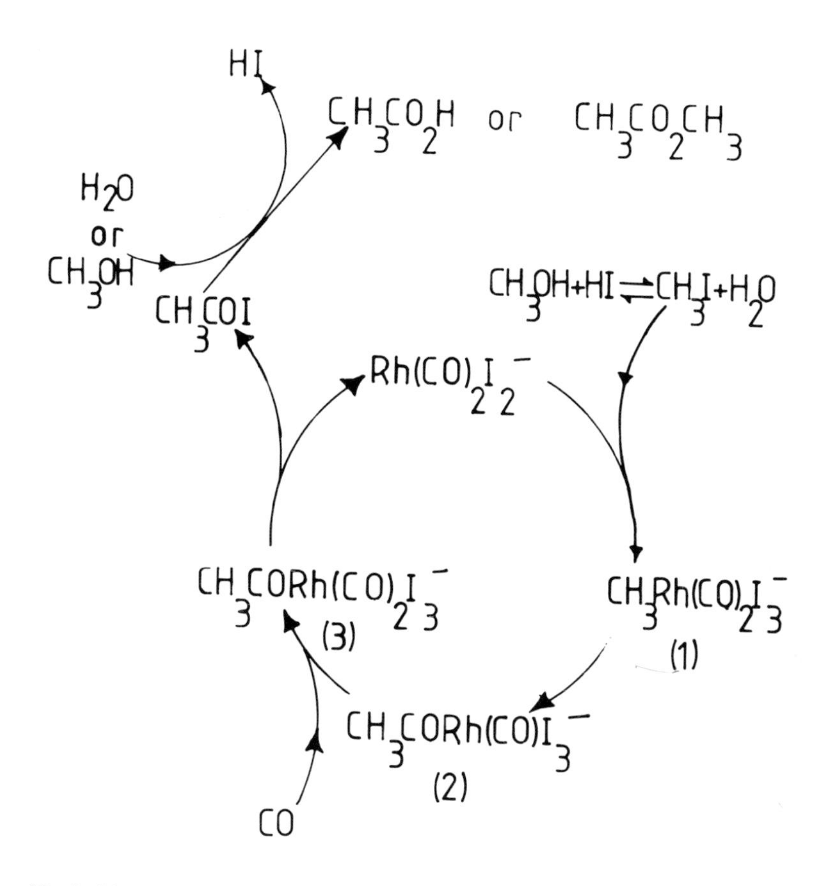

Fig. 2. Mechanism proposed for Rh/I-catalysed methanol carbonylation.

4. Species (2) has been isolated [8] and characterised [20] in the solid-state as an iodide-bridged dimer. The infrared spectrum of this material in solution is identical with that of the product of the reaction of CH_3I with $Rh(CO)_2I_2^-$.

5. Species (2) reacts very rapidly with CO under ambient conditions to give a material with one acetyl and two CO stretching frequencies [8]. Although not definitively characterised, this is reasonably ascribed as species (3).

6. The material postulated to be species (3) slowly decomposes at room temperature to give $Rh(CO)_2I_2^-$ and acetyl iodide [8]. This is good evidence that a reductive elimination step is possible, although a solvolytic mechanism involving direct attack of methanol or H_2O on the acetyl rhodium species (3) is a possibility under actual reaction conditions. The fact that addition of chelating bisphosphines blocks the catalysis [21] has also been adduced as evidence of a reductive elimination step on the theory that the intermediates then formed cannot undergo a reductive trans-elimination.

The carbonylation of methanol is thus a very rapid and generally highly selective reaction. Although the presence of hydrogen does not usually affect the reaction [9], substantial hydrogen partial pressure can, in certain circumstances, lead to formation of methane [22] through the reaction:

$$CH_3OH + H_2 \longrightarrow CH_4 + H_2O \tag{10}$$

Methane production is considerably enhanced if the catalyst is 'pre-conditioned' by prior operation of the WGSR in the Rh(III) regime. The following sequence of reactions has been proposed [22] for catalysis of methane formation.

$$Rh(CO)_2I_4^- + H_2 \rightleftharpoons HRh(CO)_2I_3^- + HI$$

$$HRh(CO)_2I_3^- + CH_3I \longrightarrow CH_4 + Rh(CO)_2I_4^-$$

$$CH_3OH + HI \rightleftharpoons CH_3I + H_2O$$

$$CH_3OH + H_2 \longrightarrow CH_4 + H_2O$$

4.2. Carbonylation of Other Alcohols

Alcohols other than methanol may be carbonylated using the Rh/I catalyst system, but these reactions are of less industrial significance. Table 2 lists some of the known reactions. Detailed data are, however, only available for ethanol [23] and isopropanol [24]. Table 3 lists the kinetics parameters for carbonylation of these alcohols.

Table 2

CARBONYLATION OF HIGHER ALCOHOLS

Alcohol	Product(s)	Reference No.
C_2H_5OH	$C_2H_5CO_2H$	23
$CH_3CH(OH)CH_3$	$C_3H_7CO_2H$; $CH_3CH(CO_2H)CH_3$	24, 25
$HO(CH_2)_4OH$	$HO_2C(CH_2)_4CO_2H$	25
$CH_3(CH_2)_7OH$	$CH_3(CH_2)_7CO_2H$	25
$(CH_3)_3COH$	$(CH_3)_3CCO_2H$; $CH_3CH(CH_3)CH_2CO_2H$	25
$C_6H_5CH_2OH$	$C_6H_5CH_2CO_2H$	26

Table 3

KINETIC PARAMETERS FOR ALCOHOL CARBONYLATION

	Methanol	Ethanol	*iso*-Propanol [a]
Ref. no.	23	23	24
k ($mol^{-1} l s^{-1}$) [b]	0.645	0.035	0.185
Ea (kcal mol^{-1})	14.7	14.6	25.9
$\Delta H^{\pm}$ (kcal mol^{-1})	13.6	13.6	24.9
$\Delta S^{\pm}$ (e.u.)	−32.2	−37.3	− 9.8

[a] Figures refer to combined production of *n*- and *iso*-butyric acids.
[b] At 473°K.

Ethanol carbonylation [23] yields propionic acid, with kinetic dependencies similar to those found for methanol carbonylation. The reaction rate is, however, only ca. 5% of that of methanol carbonylation. The mechanism of ethanol carbonylation is probably broadly similar to that described for methanol carbonylation (see Figure 2), except that the oxidative addition of ethyl iodide in the rate determining step is much slower than the oxidative addition of methyl iodide.

The carbonylation of isopropanol [24] is more complicated. Both *n*- and *iso*-butyric acids are formed, since the reaction of HI with isopropanol can yield both *n*- and *iso*-propyl iodides. The ratio of the two acid products is, moreover, dependent on temperature, and on alcohol and rhodium concentrations. A further complication is that the overall reaction rate is not linearly dependent on iodide concentration over a wide range (as it is with methanol and ethanol carbonylation). These observations cannot be completely rationalised with the data presently available, but the reaction probably involves a competitive reaction of the *n*- and *iso*-propyl iodides. It has been deduced [24] that the activation energy for formation of *n*-butyric acid is similar to that for acetic and propionic acid formation, but that *iso*-butyric acid formation has a much higher activation energy. Overall, the rate of isopropanol carbonylation is intermediate between those of methanol and ethanol carbonylation.

5. CARBONYLATION OF ESTERS AND ETHERS

5.1. Ester Carbonylation

Esters may be carbonylated under conditions similar to those used for alcohols: the products in this case are anhydrides rather than acids. A reaction of particular industrial importance is the production of acetic anhydride from methyl acetate [27, 28]:

$$CH_3CO_2CH_3 + CO \longrightarrow (CH_3CO)_2O \qquad (11)$$

The mechanism of Figure 1 can be modified to encompass this reaction if we assume that the acetyl iodide formed by reductive elimination reacts with methyl acetate to form the product anhydride and regenerate methyl iodide:

$$CH_3COI + CH_3CO_2CH_3 \longrightarrow (CH_3CO)_2O + CH_3I \qquad (12)$$

Thus, it is not surprising to find that methyl acetate carbonylation obeys broadly similar kinetics to methanol carbonylation [28]. Methyl acetate carbonylation is, however, a rather slower reaction than methanol carbonylation, although the rate may be enhanced by the addition of certain co-promoters. These co-promoters are typically metal salts, phosphines or nitrogen bases [28, 29]. It has been suggested [28] that the latter two types of materials act as ligands to the rhodium. In the presence of methyl iodide, these materials are, however, likely to be present as quaternary phosphonium or ammonium iodides (Q^+I^-). An alternative explanation [30] for the promotional effect involves catalysis of Reaction 12:

$$Q^+I^- + CH_3CO_2CH_3 \rightleftharpoons Q^+O_2CCH_3^- + CH_3I \qquad (13)$$

$$CH_3COI + Q^+O_2CCH_3^- \longrightarrow Q^+I^- + (CH_3CO)_2O \qquad (14)$$

The effect of this catalysis is to lower the standing concentration of acetyl iodide. Acetyl iodide can cause deactivation of the catalyst [19] by its thermal decomposition to give iodine, which removes part of the rhodium from the catalytic cycle by Reaction 5.

Relatively little work has been done on the carbonylation of other esters, but similar anhydride products would be expected. A notable exception is methyl formate. Under typical carbonylation conditions, methyl formate is converted selectively to acetic acid without net consumption of carbon monoxide [30].

$$HCO_2CH_3 \longrightarrow CH_3CO_2H \qquad (15)$$

The mechanism of this curious reaction is not known, but the results of labelling studies [30] are consistent with an effective decarbonylation of the methyl formate to give methanol and carbon monoxide, followed by carbonylation of the methanol. A reasonable theory would be that the mixed acetoformic anhydride is first formed by a mechanism similar to that discussed for acetic anhydride formation. The mixed anhydride would be unstable under the reaction conditions, decomposing to acetic acid and carbon monoxide [32]

5.2. Ether Carbonylation

Ether carbonylation with the rhodium/iodide catalyst system takes place in two stages, forming first esters and then anhydrides [27], e.g.:

$$(CH_3)_2O \xrightarrow{CO} CH_3CO_2CH_3 \xrightarrow{CO} (CH_3CO)_2O \qquad (16)$$

Industrially, dimethyl ether carbonylation is a possible alternative to methyl acetate carbonylation for acetic anhydride production [27].

No studies of the mechanism of dimethyl ether carbonylation have been reported, but it is likely that the mechanism is again similar to that of methanol carbonylation, with the modification that the acetyl iodide in this case reacts with the ether.

$$CH_3COI + (CH_3)_2O \longrightarrow CH_3CO_2CH_3 + CH_3I \tag{17}$$

6. REDUCTIVE CARBONYLATION OF ESTERS AND ETHERS

As noted earlier, the presence of hydrogen has relatively little effect on the Rh/I-catalysed methanol carbonylation reaction, except under conditions where methane may be formed. The course of ester and ether carbonylations is, however, radically modified when hydrogen is present.

For example, if methyl acetate carbonylation is carried out using mixtures of carbon monoxide and hydrogen, ethylidene diacetate (EDA) and acetic acid are co-produced along with acetic anhydride [30, 33, 34]. The stoichiometry of EDA formation is

$$2CH_2CO_2CH_3 + 2CO + H_2 \longrightarrow CH_3CH(OAc)_2 + CH_3CO_2H \tag{18}$$

The ratio of EDA to acetic anhydride produced depends on the hydrogen partial pressure used: higher partial pressures favouring EDA.

EDA has considerable industrial importance as it can be decomposed thermally [35] to give vinyl acetate (a precursor of a range of vinyl polymers):

$$CH_3CH(OAc)_2 \longrightarrow CH_2{=}CH(OAc) + CH_3CO_2H \tag{19}$$

Acetaldehyde – another material of considerable industrial importance – can also be produced from EDA [35]:

$$CH_3CH(OAc)_2 \rightleftharpoons CH_3CHO + (CH_3CO)_2O \tag{20}$$

When vinyl acetate is made by Reactions 18 and 19, the by-product acetic acid from both stages may be recycled to form the methyl acetate feedstock. Overall, the process can be represented by:

$$2CH_3CO_2CH_3 + 2CO + H_2 \longrightarrow CH_3CH(OAc)_2 + CH_3CO_2H \tag{18}$$

$$CH_3CH(OAc)_2 \longrightarrow CH_2{=}CH(OAc) + CH_3CO_2H \tag{19}$$

$$2CH_3CO_2H + 2CH_3OH \longrightarrow 2CH_3CO_2CH_3 + 2H_2O \tag{21}$$

$$2CH_3OH + 2CO + H_2 \longrightarrow CH_2{=}CH(OAc) + 2H_2O \tag{22}$$

The production of what is, conventionally, an ethylene derivative from the C_1 building blocks methanol and carbon monoxide is a potentially very important advance. At the time of writing this technology is not yet commercialised, but holds considerable promise for the future.

Despite the industrial potential of the EDA synthesis reaction, no mechanistic details are known. The following observations may, however, point to some of the key steps in the reaction.

1. The reaction conditions for EDA formation are very similar to those for acetic anhydride formation except for the presence of hydrogen. Indeed, the flexibility of the process to produce either EDA or acetic anhydride has been emphasised [34].

2. EDA can be formed from acetaldehyde and acetic anhydride by the reverse of Reaction 20.

3. Acetic anhydride may itself be converted to EDA using a similar catalyst system to that used for production of EDA from methyl acetate [36]:

$$2(CH_3CO)_2O + H_2 \longrightarrow CH_3CH(OAc)_2 + CH_3CO_2H \qquad (23)$$

These observations are consistent with the theory that acetic anhydride is the first-formed product, that acetaldehyde is next produced by hydrogenolysis of the anhydride, and that EDA is finally formed by Reaction 20:

$$2CH_3CO_2CH_3 + 2CO \longrightarrow 2(CH_3CO)_2O \qquad (11)$$

$$(CH_3CO)_2O + H_2 \longrightarrow CH_3CHO + CH_3CO_2H \qquad (24)$$

$$CH_3CHO + (CH_3CO)_2O \longrightarrow CH_3CH(OAc)_2 \qquad (20)$$

We do not, however, know what sort of mechanism might be operating in Reaction 23. One possibility is that a key step is the thermal decomposition of acetic anhydride to give ketene [37], which is then hydrogenated to acetaldehyde:

$$(CH_3CO)_2O \rightleftharpoons CH_2{=}C{=}O + CH_3CO_2H \qquad (25)$$

$$CH_2{=}C{=}O + H_2 \longrightarrow CH_3CHO \qquad (26)$$

In support of this, there are indications [38] that the presence of a large excess of acetic acid suppresses EDA formation. This could be explained as being due to suppression of equilibrium (25).

7. HYDROCARBONYLATION OF OLEFINS

As noted in Section 4.2 above, the carbonylation of ethanol is a considerably slower reaction than the carbonylation of methanol: rates of reaction with higher primary alcohols would be expected to be still lower. Fortunately, however, there is a facile route to the higher carboxylic acids from the corresponding olefins via the so-called hydrocarboxylation reaction. This reaction, like hydroformylation, produces a mixture of branched and linear isomers:

$$RCH{=}CH_2 + CO + H_2O \longrightarrow RCH_2CH_2CO_2H \text{ and } \underset{\displaystyle CO_2H}{RCHCH_3} \qquad (27)$$

A variety of homogeneous transition-metal catalyst systems will effect the hydrocarboxylation reaction [39], with the Rh/I system being amongst the most effective [40].

A wide variety of olefins may be hydrocarboxylated using the Rh/I catalyst. Reaction rates generally decrease as the length of the carbon chain increases, and internal olefins are less reactive than terminal olefins. Of particular industrial interest is the preparation of propionic acid – widely used as a grain preservative – by hydrocarboxylation of ethylene. Monsanto [41] have demonstrated the co-production of acetic and propionic acids by conducting the carbonylation of methanol and the hydrocarboxylation of ethylene in the same reactor. Another reaction of potential industrial interest is the production of adipic acid from 1,3-butadiene, which has been studied by BASF [42]:

$$CH_2{=}CH{-}CH{=}CH_2 + 2CO + 2H_2O \longrightarrow HO_2C(CH_2)_4CO_2H \qquad (28)$$

The Rh/I catalyst system has the considerable advantage of being operable under relatively mild conditions (typically 150–220°C, 30–75 bar). As with methanol carbonylation, a wide variety of soluble rhodium sources may be used as the catalyst precursor. In this case, however, the iodide source can be important, and very different reaction rates may be observed depending on whether HI or an alkyl iodide promoter is used (see below). The *n/iso* ratio in the products is rather lower with the Rh/I catalyst system than with some other hydrocarboxylation catalysts. Values of 1.0–1.5 have been claimed [40] for Rh/I, while a $Pd/PPh_3/HCl$ system can attain up to 4.0 [43]. This failing has, of course, no significance in the case of propionic acid synthesis.

Mechanistic and kinetic studies of the hydrocarboxylation reaction have been few, but the Monsanto group [40, 44] have outlined some of the major features of the mechanism of ethylene hydrocarboxylation. Initial observations showed that ethyl iodide was a more effective promoter than HI for this reaction. This suggested a mechanism similar to that proposed above for ethanol carbonylation, except that the ethyl iodide is, in this case, formed by reaction of HI with ethylene:

$$C_2H_4 + HI \rightleftharpoons C_2H_5I \qquad (29)$$

Experiments with ^{14}C labelled ethylene precluded this possibility, however, as activity accumulated in the product propionic acid more quickly than in the ethyl iodide. These observations were interpreted as meaning that, although a part of the propionic acid is formed via oxidative addition of ethyl iodide to a rhodium (I) complex, the majority is formed *via* insertion of ethylene into the Rh—H bond of a rhodium (III) hydride. In other respects the reaction mechanism is probably analogous to that of methanol homologation. The scheme of Figure 3 has been drawn on this basis.

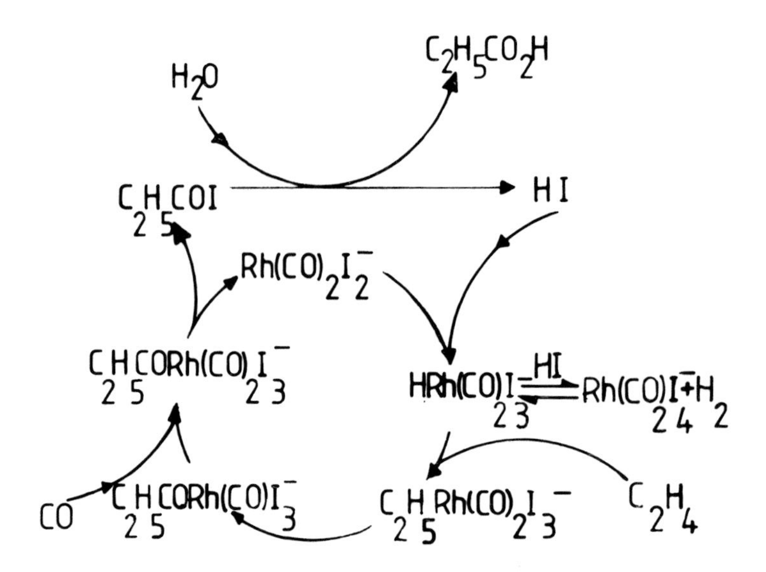

Fig. 3. Mechanism proposed for Rh/I-catalysed hydrocarboxylation of ethylene.

The explanation for the differing activities of HI and ethyl iodide as promoters seems to lie in the mode of formation of the active catalytic species.

The rhodium (III) hydride is probably formed by oxidative addition of HI to a rhodium (I) species along the lines of Equation 4 above. Too high a level of HI will, however, result in removal of the hydride as shown in Equation 4. In support of this, spectroscopic studies [40, 44] have shown that the majority of the rhodium in this reaction is present as $Rh(CO)_2I_4^-$. Rhodium in this form may be restored to the main catalytic cycle by a reduction sequence such as that shown in Equation 2.

The mechanism of hydrocarboxylation of higher olefins is probably similar to that for ethylene. It is noteworthy, however [40], that the form of the iodide promoter is less critical with the higher olefins than with ethylene. Presumably the alkyl iodide-forming equilibrium is rapidly established in these cases.

A variant of the hydrocarboxylation reaction occurs when the reaction is carried out under anhydrous conditions with a carboxylic acid as solvent [45]. In this case the acyl iodide intermediate reacts with the carboxylic acid to generate an anhydride in a manner analogous to that involved in the acetic anhydride synthesis described above (Section 5.1). The overall reaction is then of the form:

$$RCH{=}CH_2 + CO + R'CO_2H \longrightarrow RCH_2CH_2CO.O.COR' \text{ (+ isomers)} \quad (30)$$

8. HYDROCARBOXYLATION OF FORMALDEHYDE

The hydrocarboxylation of formaldehyde yields glycolic acid, which can, in turn, be hydrogenated to ethylene glycol:

$$CH_2O + CO + H_2O \longrightarrow HOCH_2CO_2H \quad (31)$$

$$HOCH_2CO_2H + 2H_2 \longrightarrow HOCH_2CH_2OH + H_2O \quad (32)$$

The C_1-based route to ethylene glycol was operated commercially by du Pont until around 1968 [46]. The hydrocarboxylation step was carried out with a concentrated sulphuric acid catalyst. In recent years attempts have been made to improve on the costly sulphuric acid-based process by use of transition metal catalysts. The Rh/I catalyst system has been patented for this purpose [47].

Unfortunately, however, the Rh/I-catalysed reaction is rather unselective. In a typical example [47], equimolar quantities of water and formaldehyde (added as trioxane) are heated for 20.5 h at 150°C/68 bar CO in the presence of a $RhCl_3$/HI catalyst mixture. The glycolic acid yield is only 23% and acetic and formic acids (Yields of 63% and 14% respectively) are also formed.

In aqueous solution formaldehyde is largely present as methylene glycol [48]. This would react with HI to give hydroxymethyl iodide:

$$CH_2O + H_2O \rightleftharpoons H_2C(OH)_2 \quad (33)$$

$$H_2C(OH)_2 + HI \rightleftharpoons HOCH_2I + H_2O \quad (34)$$

A reasonable mechanism for formaldehyde hydrocarboxylation would therefore be a similar cycle to that proposed for methanol carbonylation (Figure 2) but with hydroxymethyl iodide substituted for methyl iodide. This scheme does not, however, explain the formation of acetic and formic acids. The acetic acid could be produced by decarboxylation [49] of malonic acid, itself produced by carbonylation of glycolic acid. The origin of the formic acid is less easy to explain.

9. HYDROGENOLYSIS REACTIONS

The hydrogenolysis of methanol to give methane was mentioned in Section 4.1 above as being a competing reaction with methanol carbonylation under certain circumstances. There are indications in the patent literature [50] that other alcohol hydrogenolysis reactions may be carried out using the Rh/I catalyst; e.g.:

$$C_6H_5CH(OH)CH_3 + H_2 \longrightarrow C_6H_5CH_2CH_3 + H_2O \quad (35)$$

When 1,2-diols or epoxides are carbonylated with the Rh/I catalyst, this hydrogenolysis reaction seems to proceed at the same time as the carbonylation reaction, so that monocarboxylic acids are formed rather than the expected dicarboxylic acids [51] (see Table 4). The mechanism of this curious reaction has not been studied.

Table 4

CARBONYLATION OF 1,2-DIOL DERIVATIVES [51] [a]

Substrate	Product	Selectivity (mol %)
$HO(CH_2)_2OH$	$C_2H_5CO_2H$	95
$CH_3O(CH_2)_2OH$	$C_2H_5CO_2H$	47
CH_2—$CHCH_3$ (epoxide, bridged by O)	n-$C_3H_7CO_2H$ and $CH_3CH(CH_3)CO_2H$	88
$Cl(CH_2)_2OH$	$C_2H_5CO_2H$	93

[a] Reaction Conditions: 175–215°C, 48–75 bar CO, 17 h, 1mM Rh, 0.6 MI.

A simpler reaction, which is also formally a hydrogenolysis is the reduction of I_2 to HI. This reaction, catalysed by rhodium, has been the subject of a patent [52]. The reaction can be carried out in aqueous acidic solution with carbon monoxide as the reducing agent:

$$I_2 + H_2O + CO \longrightarrow 2HI + CO_2 \quad (36)$$

Mechanistic studies have not been reported, but it is not difficult to formulate a reasonable mechanism by combining Equation 5 with elements of the WGSR mechanism (Figure 1), e.g.:

$$Rh(CO)_2I_4^- + H_2O \longrightarrow H^+ + CO_2 + HRh(CO)I_4^{2-}$$

$$HRh(CO)I_4^{2-} + CO \longrightarrow HRh(CO)_2I_3^- + I^-$$

$$HRh(CO)_2I_3^- \longrightarrow HI + Rh(CO)_2I_2^-$$

$$Rh(CO)_2I_2^- + I_2 \longrightarrow Rh(CO)_2I_4^-$$

In non-aqueous media the reduction of iodine to iodide can be carried out using hydrogen [53]:

$$I_2 + 2CH_3CO_2CH_3 + H_2 \longrightarrow 2CH_3I + 2CH_3CO_2H \quad (37)$$

10. HOMOLOGATION OF CARBOXYLIC ACIDS

Texaco workers have recently developed catalyst systems for the conversion of carboxylic acids to higher homologues using synthesis gas:

$$RCO_2H + CO + 2H_2 \longrightarrow RCH_2CO_2H + H_2O \tag{38}$$

Although iodide-promoted ruthenium catalysts seem to be preferred for this reaction [54], the use of Rh/I catalysts has also been patented [55]. In a typical example, acetic acid is treated with CO/H_2 at 429 bar/220°C in the presence of a Rh_2O_3/CH_3I catalyst to give a mixture containing 29.1% propionic acid, 14.7% butyric acid and 3% valeric acid. No indication is given of the rate of reaction, however, and it is too early to say whether this type of reaction is a realistic alternative to olefin hydrocarboxylation (Section 7 above) for the production of higher carboxylic acids.

11. HETEROGENISATION OF THE Rh/I CATALYST

The carbonylation of methanol to give acetic acid is readily carried out in the gas phase by feeding a mixture of methanol, methyl iodide and carbon monoxide over a solid rhodium-containing catalyst. While heterogeneous catalysis is, strictly speaking, outside the scope of the present review, these reactions show sufficient similarity to the homogeneous variants to warrant some comment.

A number of approaches to heterogenising the Rh/I catalyst have been tried. These include impregnating charcoal with a pre-formed rhodium iodocarbonyl solution [56], preparation of rhodium-exchanged zeolites [57] and use of polymer-based phosphines to complex the rhodium [58].

In all cases the reaction is found to be first order in methyl iodide partial pressure, and zero order in methanol and carbon monoxide pressures. The order on rhodium has only been determined in the case of the polymer-based catalyst. Perhaps surprisingly, the reaction appears to be second-order in rhodium.

While these gas-phase reactions show broadly similar kinetics to the liquid-phase variants, there are also indications that the active rhodium species are similar in the two cases. Originally [58] the polymer-bound catalyst was thought to contain $RhCl(CO)(PPh_3)_2$ as the active catalytic species but subsequent studies [19] suggest that the surface phosphine groups are, in fact, quaternarised by methyl iodide and that iodocarbonyl rhodate anions are ionically bound to these phosphonium groups. $Rh(CO)_2I_2^-$, $Rh(CO)I_5^{2-}$ and $Rh(CO)_2I_4^-$ can be eluted from the catalyst by solutions containing iodide ion.

An infrared spectroscopic study [59] of a rhodium-exchanged zeolite-X catalyst (RhNaX) suggests that rhodium iodocarbonyls are the predominant catalytic species for this catalyst also. Thus, treatment of the rhodium (III)-exchanged zeolite with carbon monoxide led to accumulation of surface carbon dioxide, and the appearance of bands ascribed to $Rh(CO)_2I_2^-$. This parallels the homogeneous reduction of rhodium (III) species by carbon monoxide (see Section 2 above). Treatment of the reduced catalyst with methyl iodide led to the formation of an acetyl-rhodium species, again mirroring homogeneous behaviour. Under actual reaction conditions $Rh(CO)_2I_2^-$ appeared to be the major surface species, but rhodium (III) species were also detected.

As well as these gas-phase examples, there is also a report [60] of the heterogenisation of the Rh/I catalyst for liquid-phase methanol carbonylation. The approach adopted was to attach ionically the $Rh(CO)_2I_2^-$ to an ion-exchange resin. Reaction rates were similar to those found with the homogeneous catalyst and, under suitable conditions, leaching of the rhodium from the resin could be minimised.

12. CONCLUSIONS

The commercial implementation of the Monsanto acetic acid process was a considerable chemical engineering achievement in containing the corrosive iodide-containing catalyst system and avoiding losses of the very expensive rhodium catalyst. Now that it has been shown that such problems can be solved, other Rh/I-catalysed reactions – such as acetic anhydride and vinyl acetate production – have become more realistic possibilities. A number of other reactions of both industrial and academic interest are still at an early stage of development.

We now have a good working knowledge of the key mechanistic steps involved in some reactions – such as the WGSR, methanol carbonylation and hydrocarboxylation – due largely to the work of the Monsanto Group. There are, however, many details of these reactions which are unclear and also reactions where mechanistic data is wholly absent. In time, no doubt, our understanding of this catalyst system will improve, and it is possible that it may prove to be even more versatile than it appears at present.

13. REFERENCES

1. F. E. Paulik and J. F. Roth: *J. Chem. Soc., Chem. Comm.*, 1578 (1968).
2. W. Hicher and H. Lagally: *Z. Anorg. Allg. Chem.* **98**, 251 (1943).
3. L. M. Vallarino: *Inorg. Chem.* **4**, 161 (1965).
4. D. E. Morris and H. B. Tinker: *J. Organomet. Chem.* **49**, C53 (1873).
5. D. Forster: *Inorg. Chem.* **8**, 1556 (1969).
6. J. J. Daly, F. Sanz and D. Forster: *J. Amer. Chem. Soc.* **97**, 2551 (1975).
7. D. J. Dahm and D. Forster: *Inorg. Nucl. Chem. Letts.* **6**, 15 (1970).
8. D. Forster: *J. Amer. Chem. Soc.* **98**, 846 (1976).
9. J. F. Roth, J. H. Craddock, A. Hershman and F. E. Paulik: *Chemtech.* **600** (1970).
10. B. R. James and G. L. Rempel: *J. Chem. Soc. Comm.* **158**, (1967).

11. D. J. Drury and A. G. Kent: unpublished observations.
12. C. H. Cheng, D. E. Hendriksen and R. Eisenberg: *J. Amer. Chem. Soc.* **99**, 2791 (1977).
13. E. C. Baker, D. E. Henriksen, and R. Eisenberg: *J. Amer. Chem. Soc.* **102**, 1020 (1980).
14. T. C. Singleton, L. J. Park, J. L. Price and D. Forster: *Prepr. Div. Pet. Chem. Amer. Chem. Soc.* **24**, 329 (1979).
15. N. von Kutepow, W. Himmele and H. Hohenschutz: *Chem. Ing. Tech.* **37**, 383 (1965).
16. J. Hjortkjaer and V. W. Jensen: *Ind. Eng. Chem. Prod. Res. Dev.* **15**, 46 (1976).
17. T. Matsumoto, K. Mori, T. Mizoroki and A. Ozaki: *Bull. Chem. Soc. Japan* **50**, 2337 (1977).
18. D. Forster: *Ann, N. Y. Acad. Sci* **295**, 79 (1977).
19. D. Forster: *Adv. Organomet. Chem.* **17**, 255 (1979).
20. G. W. Adamson, J. J. Daly and D. Forster: *J. Organomet. Chem.* **71**, C17 (1974).
21. D. Brodzki, C. Leclere, B. Denise and G. Pannetier: *Bull. Soc. Chim. Fr.* **61** (1976).
22. D. J. Drury, M. J. Green, D. J. M. Ray and A. J. Stevenson: *J. Organomet.* **236**, C23 (1982).
23. J. Hjortkjaer and J. C. Jorgensen: *J. Mol. Catal.* **4**, 199 (1978).
24. J. Hjortkjaer and J. C. E. Jorgensen: *J. Chem. Soc., Perkin II*, 763, (1978).
25. U.S. Patent No. 3,769,329 (1969) (Monsanto); *C.A.* **71**, 12573 (1969).
26. U.K. Patent No. 1,276,326 (1970) (Monsanto); *C.A.* **72**, 110807 (1970).
27. U.K. Patent No. 1,468,940 (1974) (Halcon Corp.); *C.A.* **83**, 96448 (1975).
U.S. Patent No. 4,046,807 (1974) (Hoechst A.G.); *C.A.* **85**, 62675 (1976).
28. M. Schrod and G. Luft: *Ind. Eng. Chem. Prod. Res. Dev.* **20** 649 (1981).
29. U.S. Patent No. 4,115,444 (1975) (Halcon Corp.); *C.A.* **85**, 159463 (1976).
30. D. J. Drury: *Ind. Chem. Bull.* **177** (1982).
31. F. J. Bryant, W. R. Johnson and T. C. Singleton: *Prep. Div. Pet. Chem. Amer. Chem. Soc.* **18**, 193 (1973).
32. *Beilsteins Handbuch der Organischen Chemie*, vol. 2, p. 166, Springer Verlag (1920).
33. U.K. Patent No. 1,538,782 (1975) (Halcon Corp.); *C.A.* **85**, 176870 (1976).
34. J. L. Ehrler and B. Juran: *Hydrocarbon Process.* **109** (Feb. 1982).
35. *Kirk-Othmer Encyclopaedia of Chemical Technology*, 2nd edn., vol. 21, p. 327, John Wiley and Sons (1970).
36. European Patent Appl. No. 34,062 (1981) (Mitsubishi Gas Chemical); *C.A.* **96**, 34596 (1982).
37. *Kirk-Othmer Encyclopaedia of Chemical Technology*, 3rd edn., vol. 1, p. 155, John Wiley and Sons (1978).
38. Belgian Patent No. 891,211 (1982) (Eastman Kodak).
39. P. Pino, F. Piacenti and M. Bianchi: in *Organic Synthesis via Metal Carbonyls*, vol. II, ed. I. Wedneer and P. Pino, John Wiley and Sons (1977).
40. D. Forster, A. Hershman and D. E. Morris: *Catal. Rev. Sci. Eng.* **23**, 89 (1981).
41. U.K. Patent No. 4,111,982 (1978) (Monsanto); *C.A.* **90**, 103429 (1979).
42. U.K. Patent No. 1,348,800 (1972) (BASF); *C.A.* **76**, 154404 (1972).
43. K. Bittler, N. V. Kutepow, D. Neubauer and H. Reis: *Angew. Chem. Int. Ed. Engl.* 7, 329 (1968).
44. D. E. Morris and G. V. Johnson: *Proc. Symp. Rhodium Homogeneous Catal.* **113** (1978).
45. U.K. Patent No. 1,448,010 (1974) (Monsanto); *C.A.* **81**, 119316 (1974).
46. *Kirk-Othmer Encyclopaedia of Chemical Technology*, 3rd. edn., vol. 11, p. 939, John Wiley and Sons (1978).
47. U.S. Patent No. 3,754,028, (1973) (Chevron); *C.A.* **79**, 104766 (1973).
48. J. F. Walker: *Formaldehyde*, Reinhold (1944).
49. L. W. Clark: in *The Chemistry of Carboxylic Acids and Esters*, p. 589, ed. S. Patai, Interscience (1969).
50. U.S. Patent No. 4,067,900, (1978) (Monsanto); *C.A.* **89**, 23930 (1978).
51. U.K. Patent No. 1,276,326 (1970) (Monsanto); *C.A.* **72**, 110807 (1970).
52. U.K. Patent No. 1,361,302 (1973) (Monsanto); *C.A.* **78**, 138415 (1973).

53. European Patent Appln. No. 46,870 (1982) (Hoechst A. G.); *C.A.* **97**, 5788 (1982).
54. J. F. Knifton: *Chemtech.* **609** (1981).
55. U.K. Patent No. 2,078,722 (1980) (Texaco Dev. Corp.); *C.A.* **96**, 180797 (1982).
56. K. K. Robinson, A. Hershman, J. H. Craddock, and J. F. Roth: *J. Catal.* **27**, 389 (1972).
57. N. Takahashi, Y. Orisaka and T. Yashima: *J. Catal.* **59**, 61 (1979).
58. M. S. Jarrell and B. C. Gates: *J. Catal.* **40**, 255 (1975).
59. J. Yamanis and K.-C. Yang: *J. Catal.* **69**, 498 (1981).
60. R. S. Drago, E. D. Nyberg, A. E. A'mma and Z. Zombeck: *Inorg. Chem.* **20**, 641 (1981).

Recent Developments in the Homogeneous Catalysis of the Water-Gas Shift Reaction

RICHARD M. LAINE AND ROBERT B. WILSON, JR.

Contribution from the Physical Organic Chemistry Department, SRI International, Menlo Park, CA 94025, U.S.A.

1. INTRODUCTION

The major portion of the world's hydrogen supply is derived from steam cracking of hydrocarbons, particularly methane and naphtha, which produces water-gas:

$$C_x + H_2O \xrightarrow{\text{catalyst}} CO + H_2 + CO_2 \tag{1}$$

R. Ugo (ed.), Aspects of Homogeneous Catalysis, Vol. 5, 217–240.

Stripping the water-gas mixture of CO_2 and any residual water gives syngas (CO/H_2). If a particular industrial process using syngas requires that the ratio of CO to H_2 in the syngas coming from (1) be adjusted in favor of H_2 or if only pure H_2 is required, as it would be in ammonia synthesis, then it is possible to catalytically convert the reducing equivalents of CO to H_2 via the water-gas shift reaction (WGSR):

$$H_2O + CO \xrightarrow{\text{catalyst}} H_2 + CO_2 \tag{2}$$

$\Delta G = -6.82$ kcal/mol, $\Delta H = -9.84$, $\Delta S = -10.1$ eu for $H_2O(g)$ at 298°C

$\Delta G = -4.76$ kcal/mol, $\Delta H = 0.68$, $\Delta S = +18.3$ eu for $H_2O(\ell)$ at 298°C

In effect, the WGSR is another major source of the world's hydrogen.

On an industrial scale [1, 2] catalysis of (2) for ammonia synthesis is achieved using heterogeneous catalysts based on mixtures of iron and chromium oxides at temperatures above 350°C [high temperature (HT) shift catalysts]. A second type of catalyst, often used in series with HT shift catalysts, consists of mixtures of zinc and copper oxides [low temperature (LT) shift catalysts]. The LT shift catalysts operate at above 200°C. The need to use two catalyst systems is mandated by the thermodynamics of (2) as well as by the sensitivity of the LT shift catalysts to poisoning by sulfur compounds.

The thermodynamics of (2) are such that the HT shift catalysts can convert only 90%–98% of the CO to H_2, depending on the operating conditions. Therefore, the partially shifted gas must be further treated with LT shift catalysts to "complete" the conversion to H_2. In addition, the HT catalysts must be used before the LT catalysts becaue the feed gases coming from (1) contain trace amounts of sulfur that can be removed (as H_2S) only after the gas has been processed with the HT shift catalysts.

Thus the currently used industrial WGSR process for producing H_2 from CO, suffers from being multistep, thermodynamically inefficient, and sensitive to poisoning. Consequently, the search for catalysts that overcome these inherent problems, that is, the search for single-step LT shift catalysts that are efficient and sulfur tolerant, is a continuing effort for catalyst researchers.

One domain of this continuing search has been the exploration of homogeneous catalysts as catalysts for the WGSR. The impetus for this exploration comes from the knowledge that homogeneous catalysts are (in general) more active and more selective than their heterogeneous counterparts. Therefore, the development of highly active homogeneous catalysts for the WGSR was viewed as a potential solution to the need for better LT shift catalysts.

In the last decade, several research groups have demonstrated, with varying degrees of success, that it is possible to homogeneously catalyze the WGSR. However, except for a brief account of the early work in this area [3], a comprehensive report of this work has not been attempted. The purpose of the present paper is to describe in detail the recent progress (through 1982) in homogeneous catalysis of the WGSR.

Another very closely related area of research (generally identified with its discoverer, Reppe [4]) concerns the catalytic reactions of CO and H_2O with organic substrates, as generally illustrated by reaction (3) and specifically by (4):

$$15\,CO + H_2O + S \xrightarrow{\text{catalyst}} SH_2 + CO_2 \quad (S = \text{Substrate}) \tag{3}$$

$$C_3H_7CH{=}CH_2 + 2CO + H_2O \xrightarrow[\text{MeOH/KOH}]{Ru_3(CO)_{12}/150°} C_3H_7CH_2CH_2CHO + CO_2 \tag{4}$$

Although, it would be appropriate to discuss the recent developments in the area of Reppe catalysis chemistry along with recent developments in WGSR catalysis chemistry, the extent of recent contributions to Reppe catalysis chemistry dictates that our discussions of this area be published separately.

2. BACKGROUND

The first researcher to speculate, in the literature, on the possibility of homogeneous catalysis of the WGSR was Reppe. His work with catalytic reactions of the general form (3) led him to postulate that a process akin to WGSR catalysis was occurring [4]. Yet there is no evidence from his published work in 1953 that he attempted to validate his proposal through experimentation. It was almost 20 years later that the first examples of homogeneous catalysis of the WGSR appeared in the patent literature and then in the open literature. In the early seventies, a series of patents, issued to Fenton [5–7], described the homogeneous catalysis of the WGSR using a variety of group 8 metals in conjunction with phosphine, arsine, or stibine ligands and amine or inorganic bases.

In 1977, the first three reports on the homogeneous catalysis of the WGSR appeared in the open literature. Each report described a different catalyst system. For example, Laine *et al.* [8] reported that aqueous alcoholic solutions of ruthenium carbonyl (made basic with KOH) gave active WGSR catalyst systems. Kang *et al.* [9] were able to demonstrate that a variety of group 8 metals were active catalysts for both the WGSR and Reppe type hydroformylation [e.g. (4)] when dissolved in aqueous THF and made basic with trimethylamine. In contrast, Cheng *et al.* [10] were able to promote catalysis of the WGSR in acidic solution. They were able to show that rhodium carbonyl complexes, when dissolved in aqueous acetic acid containing iodide, gave active WGSR catalysts. The initial evidence suggested that the catalytic cycles in each system were different.

From these original reports, as well as from the mechanistic evidence presented later by other research groups, it is evident that there are several different types of catalytic cycles by which homogeneous catalysis of the WGSR occurs. In general, the various types of catalytic cycles proposed for the WGSR can be distinguished according to whether CO or H_2O activation occurs in the primary step and whether catalysis occurs under acidic, basic, or neutral conditions. Other distinguishing features include the presence or absence of ligands other than CO or H_2O and the use of group 6 or 8 metal complexes. Because so many mechanisms for catalysis of the WGSR have been proposed in the literature, we have organized our discussions in terms of the types of mechanisms proposed to date.

3. CHEMISTRY OF THE WGSR CATALYTIC CYCLES

As a prelude to the mechanistic discussions, it would be advantageous to run through a simple exercise that will be of value to the reader in comparing and contrasting the chemistry involved in the various catalytic cycles.

Our original conceptual approach for developing homogeneous WGSR catalysts was to propose a series of hypothetical catalytic cycles in which each step in the catalytic cycle could be justified according to literature precedent and then test these potential catalytic cycles experimentally. The approach actually involved starting with the products, H_2 and CO_2, and working backwards stepwise to the reactants, using examples found in the literature. Given the considerable progress made in the area of the organometallic chemistry of H_2O and CO_2, it would be instructive to repeat this exercise, including the recent developments, in order to realize just how many mechanisms are possible for the WGSR.

3.1. Sources of H_2

We can begin by examining the types of organometallic compounds that react to eliminate H_2. The simplest source of H_2 is one in which a metal dihydride undergoes unimolecular reductive elimination to produce H_2:

$$MH_2 \longrightarrow H_2 + \text{"M"} \tag{5}$$

Examples of complexes that eliminate H_2 by this pathway include $H_2Fe(CO)_4$ [11], $H_2RhCl(PPh_3)_3$ [12], $H_2IrCl(CO)(PPh_3)_2$ [13], and $H_2Co[P(OR)_3]_4^+$ [14]. Kinetic studies have shown that these complexes eliminate H_2 via unimolecular processes. H_2 can also be formed through bimolecular elimination reactions [e.g. (6)] as shown by Marko and Ungvary [15] for $HCo(CO)_4$ and as recently shown by Evans and Norton [16] for $H_2Os(CO)_4$ [17]:

$$MH + MH \longrightarrow M_2 + H_2 \tag{6}$$

$$2HCo(CO)_4 \rightleftharpoons Co_2(CO)_8 + H_2 \tag{7}$$

$$2H_2Os(CO)_4 \longrightarrow H_2Os_2(CO)_8 + H_2 \tag{8}$$

The mechanisms by which reactions (6)–(8) proceed are somewhat more complicated than shown above because dissociation evidently occurs before elimination of H_2. In the case of $H_2Fe(CO)_4$, the process is further complicated by competing multinuclear processes, which provide additional routes for reductive elimination of H_2, as demonstrated by Collman *et al.* [18]:

$$H_2Fe(CO)_4 + Fe(CO)_4 \longrightarrow H_2Fe_2(CO)_8 \quad (9)$$

$$H_2Fe_2(CO)_8 + HFe_2(CO)_8^- \longrightarrow HFe_3(CO)_{11}^- + Fe(CO)_5 + H_2 \quad (10)$$

In the analogous ruthenium systems, it appears that multinuclear H_2 elimination processes are the only pathways available (11), but see below [19].

$$3[H_3Ru_4(CO)_{12}]^- + 9CO \longrightarrow 3[HRu_3(CO)_{11}]^- + Ru_3(CO)_{12} + 3H_2 \quad (11)$$

From these examples, it becomes obvious that there are several pathways available for H_2 formation via reductive elimination from metal hydrides. However, there are still other routes available for H_2 production that only partially involve metal hydrides. Consider, for example, the reactions of alkali metal hydrides or $LiAlH_4$ or $NaBH_4$ with HX:

$$MH + HX \longrightarrow MX + H_2 \quad (12)$$

Transition metal hydrides can also react via similar pathways as exemplified by reactions (13)–(16) [20]:

$$HFe(\eta^5Cp)(CO)_2 + HCl \longrightarrow FeCl(\eta^5Cp)(CO)_2 + H_2 \quad (13)$$

$$H_2Zr(\eta^5Cp)_2 + HOAc \longrightarrow Zr(OAc)_2(\eta^5Cp)_2 + H_2 \quad (14)$$

$$\text{mer-}HIrCl_2(PPh_3)_3 + HCl \longrightarrow \text{mer-}IrCl_3(PPh_3)_3 + H_2 \quad (15)$$

$$Pd[P(Ph)_3]_4 + 2HX + H_2O \longrightarrow Pd(H_2O)[P(Ph)_3]_3X_2 + P(Ph)_3 + H_2 \quad (16)$$

As discussed below, it is likely that transition metal hydrides will undergo similar reactions when $HX = H_2O$.

A third alternative for H_2 production involves an organometallic complex, but does not result from the immediate participation of a metal hydride, although metal hydrides are likely to be precursors. Recent work by a number of researchers has demonstrated that metal formyl complexes are very efficient hydride donors [21]. As such, they should react readily with H_2O to produce H_2:

$$MC(=O)H + H_2O \longrightarrow M(CO)^+ + OH^- + H_2 \quad (17)$$

Although there are as yet no examples of reactions such as (17), evidence is now available that supports the existence of such reactions (see below).

Finally, there is one reaction of importance to the present discussion in which H_2 equivalents are produced in the absence of transition metal species:

$$HCO_2^- + H_2O \longrightarrow CO_2 + OH^- + H_2 \quad (18)$$

Reaction (18) has not been observed directly, but analogous reactions (for example, the Leuckart reaction [22] are fairly common when there is a substrate to receive the H_2.

3.2. Sources of CO_2

Carbon dioxide, the other product in the WGSR, can be produced via several well-known reactions. The most common source of CO_2 is carbonate decomposition, which can be observed in several forms:

$$M(O_2COH) \longrightarrow M(OH) + CO_2 \tag{19}$$

$$2M(HCO_3) \longrightarrow M_2(CO_3) + H_2O + CO_2 \tag{20}$$

$$M(CO_3) \longrightarrow M(O) + CO_2 \tag{21}$$

Most transition and nontransition metal carbonates and bicarbonates will decompose according to reactions (19)–(21).

Less common, but still well-known sources of CO_2, are the formate and formic acid decomposition reactions:

$$M(O_2CH) \longrightarrow MH + CO_2 \tag{22}$$

$$M + HCO_2H \longrightarrow [HM(O_2CH] \longrightarrow M + H_2 + CO_2 \tag{23}$$

Species such as the intermediate shown in (23) have been isolated in the reversible reaction of $PtH_2(PEt_3)_2$ with CO_2; moreover, this complex has been shown to act as a catalyst for (23) [23]. In addition to the $PtH_2(PEt_3)_2$ complex, the decomposition of transition metal formate complexes as shown in reaction (22) has been discussed in a review by Eisenberg and Hendriksen [24]. More recently, Darensbourg *et al.* reported that complexes such as (η^5Cp) $Fe(CO)_2(O_2CH)$ [25], and the group 6 complexes $(HCO_2)M(CO)_5^-$ (where M = Cr, Mo, W [26]) also decompose readily as shown in (22) to produce CO_2. The mechanism of η^1 or η^2 formamato decomposition is generally assumed to be one that involves a β elimination process (see Scheme 1).

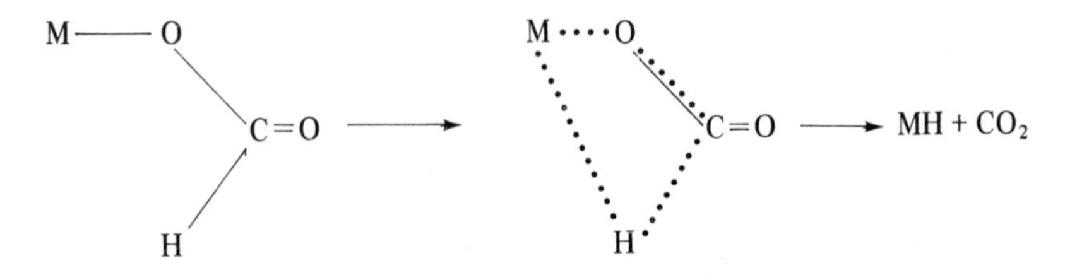

Scheme 1

Many metal complexes can act as efficient catalysts for formic acid decomposition, as shown in (23). Strauss *et al.* [27] report that $Rh(C_6H_4PPh_2)(PPh_3)_2$ catalyzes (23), and they compare the activity of this catalyst with the activities of $Pd(HCO_2)_2$ [28] $PtCl_2(PBu_3)_2$ [29] $IrH_2Cl(PPh_3)_3$ [29], and $Pt[P(iPr)_3]_3$ [30], which have previously been shown to catalyze (23). The platinum isopropylphosphine complex appears to have the highest activity of the catalysts compared. Other compounds including $Ru_3(CO)_{12}$ have also been found to be active catalysts for (23) [31]. In most cases, it is assumed that the pathway for production of CO_2 in formic acid decomposition proceeds as shown in Scheme 2, in analogy to Scheme 1.

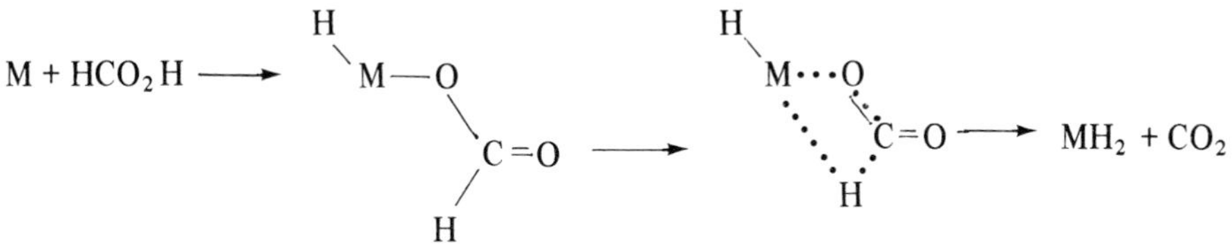

$$M + HCO_2H \longrightarrow H{-}M{-}O{-}C({=}O){-}H \longrightarrow H{-}M\cdots O{\cdots}C({=}O)\cdots H\cdots M \longrightarrow MH_2 + CO_2$$

Scheme 2

However an alternative β elimination pathway involving initiation by C—H insertion is also possible:

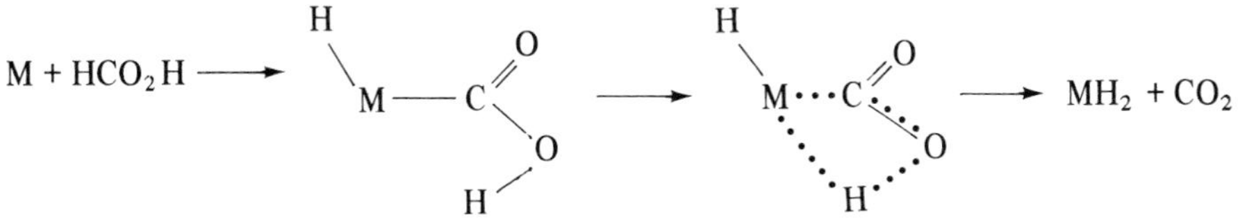

$$M + HCO_2H \longrightarrow H{-}M{-}C({=}O){-}O{-}H \longrightarrow H{-}M\cdots C({=}O)\cdots O\cdots H\cdots M \longrightarrow MH_2 + CO_2$$

Scheme 3

Support for the reactions shown in Scheme 3 comes from the work of Grey *et al.* [32] in which these researchers find that $H_2Ru(C_6H_4PPh_2)(PPh_3)_2$ reacts with formate esters to give decarbonylation reactions. Presumably these reactions are initiated by C—H insertion.

The latter part of Scheme 3 provides another possible route by which CO_2 is formed. This route proceeds via the decomposition of a metallocarboxylic acid such as shown in reaction (24):

$$M(CO_2H) \longrightarrow MH + CO_2 \quad (24)$$

The decomposition of metallocarboxylic acids to give CO_2 was first reported in 1969 by Deeming and Shaw [33] for $IrCl_2(CO_2H)(CO)(PMe_2Ph)_2$:

$$IrCl_2(CO_2H)(CO)(PMe_2Ph)_2 \longrightarrow IrHCl_2(CO)(PMe_2Ph)_2 + CO_2 \quad (25)$$

Other examples of metallocarboxylic acids that decompose to release CO_2 include (η^5-Cp)Fe(CO)(PPh_3)CO_2H [34], $(Et_3P)_2$Pt(Cl)CO_2H [35], *trans*-PtH(CO_2K)[P($i$$Pr_3$)$_3$]$_2$, and ($\eta^5$-Cp)Re(NO)(CO)$CO_2$H [36]. In some some instances the presence of excess base promoted decomposition of the metallocarboxylic acid, suggesting that the metallocarboxylate, MCO_2^-, is the species that actually decomposes with the release of CO_2.

$$M(CO_2)^- \longrightarrow CO_2 + M^- + H_2O \longrightarrow MH + OH^- \qquad (26)$$

Unfortunately, no one has as yet examined the decomposition kinetics for these compounds; therefore, the exact mechanisms of decomposition are not known. It is generally assumed that the reaction is unimolecular and that, when the metallocarboxylic acid is involved, the hydrogen is transferred to the metal center through a β-elimination mechanism.

3.3. Intermediates

The next step backward in the WGSR catalytic cycle is to look for intermediates that tie either the process of CO_2 elimination to the formation of a metal hydride (eventual source of H_2) or the process of H_2 elimination to the formation of a species that can release CO_2. The general reactions (22)–(24) and (26) all result in the coproduction of CO_2 and a metal hydride. Intermediates of the general form **1**–**3**

1 $M(O_2CH)$ **2** $M(CO_2H)$ **3** $M(CO_2^-)$

are thus the most likely sources of CO_2 in many WGSR catalytic cycles. Darensbourg has recently suggested one alternative intermediate, **4**, that could also result in CO_2 production [37, 38] (as carbonate):

4 $M{-}C(OH)_2O^-$

The only evidence, to our knowledge, of the possibility of the alternative route, where H_2 elimination precedes CO_2 elimination, comes from work by Yoshida *et al.* [39], who report that the reaction of rhodium phosphine hydride complexes with H_2CO_3 eliminates H_2 with the concomitant formation of a bicarbonato complex. This observation when coupled with the Darensbourg proposal of intermediate **4**, gives Scheme 4:

$$MCO + HO^- \rightleftharpoons MC(O)OH^-$$

$$MC(O)OH^- + H_2O \rightleftharpoons HM^- + H_2CO_3$$

$$HM^- + H_2CO_3 \longrightarrow H_2 + M(O_2COH)^-$$

$$M(O_2COH)^- \rightleftharpoons M + OH^- + CO_2$$

Scheme 4.

The next step backward must account for the formation of intermediates like **1**–**4**. As illustrated in Scheme 4, the formation of **4** could readily proceed through nucleophilic attack of hydroxide on any of the complexes **1**–**3**. Furthermore, it is likely that the simplest route to **3** is through deprotonation of **2**. Thus, our main concern is to develop intermediates that react to produce **1** and **2**. Moreover, **1** and **2** must be formed starting with CO and H_2O, which limits the choices.

For example, literature evidence suggests that there are only two potential routes to **1**. One route involves the reaction of CO_2 with a metal hydride, essentially the reverse of (22). This reaction is a dead end in that CO_2 is a desired product and cannot be a reactant. The alternative route to **1** is through reaction of the coordinately unsaturated metal with formic acid or formate ion:

$$M + HCO_2H \longrightarrow HM(O_2CH) \quad (27)$$

$$M + HCO_2^- \longrightarrow [M(O_2CH)]^- \quad (28)$$

Because the formate ion and its conjugate acid are readily available from the reaction of hydroxide with CO, (29) and (30),

$$CO + OH^- \longrightarrow HCO_2^- \quad (29)$$

$$HCO_2^- + H_2O \rightleftharpoons HCO_2H + OH^- \quad (30)$$

we have defined one possible pathway for the WGSR proceeding through formate intermediates [reactions (29), (28), (22), and (6) or (30), (27), and (23)], although another route is possible as will be described shortly.

Unlike the formate routes through **1**, the potential pathways for the WGSR that pass through metallocarboxylic acids such as **2** are numerous. As in the formate routes, working backward brings us to the reactants, CO and H_2O. However, unlike the formate precursors, there are several approaches to **2** that differ in the initial activation step. That is, some approaches begin with CO activation while others begin with H_2O activation. These differences are more readily explained through illustration rather than description:

$$MCO + OH^- \longrightarrow M(CO_2H)^- \quad (31)$$

$$MCO^+ + H_2O \longrightarrow M(CO_2H) + H^+ \quad (32)$$

$$MCO^+ + H_2O \longrightarrow HM(CO_2H)^+ \quad (33)$$

Reactions (31)–(33) obtain metallocarboxylic acids through nucleophilic attack of free OH^- or H_2O on metal complexed, metal activated CO. The differences in the reactions are determined by the extent to which CO is activated by the metal through backbonding and charge and by the extent to which the metal can stabilize the various possible intermediates. When the metal is neutral and supports several good electron donor ligands or is negatively charged, it will backdonate considerable electron density to the coordinated CO group. Thus, a strong nucleophile, such as OH^-, will be required to react with the coordinated CO to obtain the metallocarboxylic acid. In contrast, when the metal has little electron density to donate to the coordinated CO, as occurs when the metal is in a high oxidation state, then even weak nucleophiles such as H_2O can react with the CO to produce a metallocarboxylic acid. This reaction has been used by several researchers as an easy way of labeling the carbonyl oxygen [40–42]:

$$Re(CO)_6^+ + H_2{*}O \longrightarrow (CO)_5Re[C(=O){*}OH] + H^+$$

$$(CO)_5Re[C(={*}O)OH] + H^+ \longrightarrow (CO)_5Re(C{*}O)^+ + H_2O \qquad (34)$$

For reactions (32) and (33), the question of when the complex retains the additional hydrogen or releases it is probably dictated by the charge on the metal: the higher the charge, the more likely that the complex will release a proton.

Reactions (35)–(37) illustrate the possible pathways for the formation of metallocarboxylic acids that begin with activation of H_2O.

$$M + H_2O \longrightarrow HM(OH) \qquad (35)$$

$$HM(OH) + (CO) \longrightarrow [HM(CO)(OH)] \longrightarrow HM(CO_2H) \qquad (36)$$

$$HM(OH) + CO \longrightarrow HM(CO) + OH^- \longrightarrow HM(CO_2H) \qquad (37)$$

At present the compounds *trans*-$PtH(OH)[P(i\mathrm{Pr})_3]$ [43] $HRh(OH)[P(i\mathrm{Pr})_3]_3$ [44] and $HW(OH)(CO)_3(PCy_3)$ [45] appear to be the only isolated examples of H_2O activation by organometallic complexes. Only the first two complexes have been shown to react with CO to form metallocarboxylic acids, although the tungsten compound may eventually be shown to function analogously. Mason and Ibers [45] have shown that the tungsten compound reacts reversibly with CO_2 to give the carbonato complex as in reaction (19).

Finally, the formation of metallocarboxylic acids could conceivably serve as an additional pathway for the formation of formate complexes as shown in reaction (38).

$$HM(CO_2H) \longrightarrow M + HCO_2H \longrightarrow HM(O_2CH) \qquad (38)$$

Keeping the concepts and examples of the above exercise in mind, we can now begin to discuss the various reported examples of homogeneous catalysis of the WGSR. These examples will be separated into two groups: catalysis of the WGSR under basic conditions and catalysis under acidic or neutral conditions.

4. CATALYSIS OF THE WGSR UNDER BASIC CONDITIONS

4.1. Ruthenium Carbonyl Complexes

Since the original reports on ruthenium catalysis of the WGSR in 1977, this system has been the subject of considerable study by the original discoverers as well as many other groups in the field. Early work by Ford *et al.* [3, 46] on the hydroxide-promoted system resulted in the identification of two complexes present in the active catalyst solution. The two complexes, $H_3Ru_4(CO)_{12}^-$ and $HRu_3(CO)_{11}^-$, justified the cluster catalysis mechanism proposed in Scheme 5:

$$HRu_4(CO)_{13}^- + H_2O \longrightarrow H_2Ru_4(CO)_{13} + OH^-$$

$$H_2Ru_4(CO)_{13} + OH^- \longrightarrow H_2Ru_4(CO)_{12}(CO_2H)^-$$

$$H_2Ru_4(CO)_{12}(CO_2H)^- \longrightarrow H_3Ru_4(CO)_{12}^- + CO_2$$

$$H_3Ru_4(CO)_{12}^- + CO \longrightarrow HRu_4(CO)_{13}^- + H_2$$

Scheme 5.

An alternative mechanism involving the trinuclear cluster was also formulated as shown in Scheme 6:

$$HRu_3(CO)_{11}^- + H_2O \longrightarrow H_2Ru_3(CO)_{11} + OH^-$$

$$H_2Ru_3(CO)_{11} + CO \longrightarrow Ru_3(CO)_{12} + H_2$$

$$Ru_3(CO)_{12} + OH^- \longrightarrow HRu_3(CO)_{11}^- + CO_2$$

Scheme 6.

In recent work, Gross and Ford [47] present concrete evidence for the intermediacy of the trinuclear cluster metallocarboxylic acid formed in Scheme 6. This work also gives credence to the formation of the tetranuclear metallocarboxylic acid in Scheme 5. In related work, Bricker *et al.* [19] and Suss-Fink [48] described the equilibrium shown in (39):

$$HRu_3(CO)_{11}^- + H_2 \rightleftharpoons H_3Ru_4(CO)_{12}^- + CO \tag{39}$$

Bricker *et al.* [19] in studies of (39), concluded that the actual mechanism for ruthenium catalysis of the WGSR centers on the trinuclear species rather than the tetranuclear complex and emphasized a new twist to the originally proposed mechanism in which H_2 generation requires that $HRu_3(CO)_{11}^-$ serve as a source of free hydride as shown in Scheme 7:

$$Ru_3(CO)_{12} + OH^- \longrightarrow HRu_3(CO)_{11}^- + CO_2$$

$$HRu_3(CO)_{11}^- + CO \longrightarrow Ru_3(CO)_{12} + H^-$$

$$H^- + H_2O \longrightarrow H_2 + OH^-$$

Scheme 7.

Although no mechanism for hydride generation is offered, the choices available, as illustrated above [e.g., reactions (11) and (16)] together with the work of Pearson *et al.* [49] suggest that a formyl intermediate may serve as the source of hydride [see reaction (17)]. The rate-limiting step would then be the migration of hydride from metal to CO.

4.2. Iron Carbonyl Catalysis of the WGSR

Perhaps the simplest WGSR catalyst system studied to date is the system based on $Fe(CO)_5$. The first report on a hydroxide-promoted catalyst system was made by King *et al.* in 1978 [50]. They have continued their studies in this area and, based on high temperature IR data [51] and previous work by Kang *et al.* [9] (see below), propose the mechanism shown in Scheme 8:

$$Fe(CO)_5 + OH^- \longrightarrow Fe(CO)_4C(=O)OH^-$$

$$Fe(CO)_4C(=O)OH^- \longrightarrow HFe(CO)_4^- + CO_2$$

$$HFe(CO)_4^- + H_2O \longrightarrow H_2Fe(CO)_4 + OH^-$$

$$H_2Fe(CO)_4 \longrightarrow H_2 + Fe(CO)_4$$

$$Fe(CO)_4 + CO \longrightarrow Fe(CO)_5$$

Scheme 8.

King *et al.* report that the rate determining step in the catalytic cycle is the reaction of base with $Fe(CO)_5$, which was recently confirmed by Pearson *et al.* [49]. From kinetic studies of these same systems, Pearson *et al.* contribute the fact that loss of CO_2 proceeds via deprotonation of the metallocarboxylic acid rather than directly from the metallocarboxylic acid:

$$Fe(CO)_4C(=O)OH^- + OH^- \longrightarrow [Fe(CO)_4(CO_2)]^{2-} + H_2O \tag{40}$$

$$[Fe(CO)_4(CO_2)]^{2-} + H_2O \longrightarrow HFe(CO)_4^- + HCO_3^- \tag{41}$$

4.3. Group 6 Metal Complexes as Catalysts for the WGSR

King *et al.* [50, 51] were also the first group of researchers to report the use of group 6 metal carbonyls [$Cr(CO)_6$, $Mo(CO)_6$, $W(CO)_6$] as catalyst precursors for catalysis of the WGSR in the presence of OH^-. Kinetic studies of WGSR catalyst solutions generated from these three carbonyl complexes implicate the following catalytic cycle:

$$CO + OH^- \longrightarrow HCO_2^-$$

$$M(CO)_6 \rightleftharpoons M(CO)_5 + CO$$

$$M(CO)_5 + HCO_2^- \longrightarrow M(CO)_5(O_2CH)^-$$

$$M(CO)_5(O_2CH)^- \longrightarrow HM(CO)_5^- + CO_2$$

$$HM(CO)_5^- + H_2O \longrightarrow H_2M(CO)_5 + OH^-$$

$$H_2M(CO)_5 \longrightarrow M(CO)_5 + H_2$$

Scheme 9.

A contrasting argument has been presented by Darensbourg and Rokicki [38, 52] in which they point out that the reaction that produces the anionic metal hydride and CO_2 is an equilibrium that favors the formato complex

$$M(CO)_5(O_2CH)^- \rightleftharpoons HM(CO)_5^- + CO_2 \tag{42}$$

and thus is not a particularly viable intermediate. In addition, from kinetic studies of the reaction of $M(CO)_6$ with base, they find that a metallocarboxylic acid intermediate is energetically a more favorable intermediate. Moreover, their results indicate that group 6 metal carbonyl catalysis of the WGSR exhibits second order dependence on OH^-, a feature that cannot be reconciled with the mechanism proposed in Scheme 9. Thus, Darensbourg and Rokicki propose an alternative mechanism [38, 52] that takes into account these differences and in which (42) is a deadend side reaction:

$$M(CO)_6 + OH^- \longrightarrow M(CO)_5CO_2H^-$$

$$M(CO)_5CO_2H^- + OH^- \longrightarrow M(CO)_5H^- + HCO_3^-$$

$$M(CO)_5H^- + H_2O \rightleftharpoons M(CO)_5H_2 + OH^-$$

$$M(CO)_5H_2 \longrightarrow M(CO)_5 + H_2$$

$$M(CO)_5 + CO \longrightarrow M(CO)_6$$

Scheme 10.

4.4. Group 8 Metal Catalysis of the WGSR in the Presence of Amine Bases

Pettit *et al.* reported [53, 54] the first examples of amine-promoted group 8 metal catalysis of the WGSR in 1977–79. Since then, only one follow-up report [55] has appeared that attempts to elucidate the mechanisms by which WGSR catalysis occurs in these systems. The original reports, of Pettit *et al.* proposed a mechanism for OH^--promoted iron carbonyl catalysis of the WGSR quite similar to the mechanism proposed by King *et al.* above, although not in as great detail. At that time, they concluded that this proposed mechanism was a general mechanism applicable to all the systems studied whether they were promoted by OH^- or by amine bases. As can be seen from the data in Tables 1–3, there are considerable differences in relative activities of the various catalysts depending on the type of base/catalyst solution studied. In particular, the relative activities of ruthenium and rhodium OH^--promoted

Table 1

HYDROXIDE PROMOTED CATALYSIS OF THE WGSR

Catalyst	Solvent	Pressure CO (atm)	Temp. (°C)	Turnovers (/24 h)	Ref.
$Fe(CO)_5$	*n*-BuOH	28.2	137	16	50
$Fe(CO)_5$	*n*-BuOH	28.2	181	72	50
$Ru_3(CO)_{12}$	$EtOCH_2CH_2OH$	1	100	2.2	46
$Ru_3(CO)_{12}/Fe_3(CO)_{12}$	$EtOCH_2CH_2OH$	1	100	7.4	46
$Ru_3(CO)_{12}$	MeOH	75	135	53	46
$Os_3(CO)_{12}$	MeOH	75	135	12	46
$Rh_6(CO)_{16}$	MeOH	75	136	110	46
$Ir_4(CO)_{12}$	MeOH	75	135	17	46
$Cr(CO)_6$	MeOH	7.8	140	280	50
$Mo(CO)_6$	MeOH	11	140	31	50
$W(CO)_6$	MeOH	7.8	130	140	50

Table 2

AMINE PROMOTED CATALYSIS OF THE WGSR

Catalyst	Solvent/amine	Pressure CO (atm)	Temp. (°C)	Turnovers (/10 h)	Ref.
$Fe(CO)_5$	$THF/(CH_3)_3N$	23.8	110	5	54
$Ru_3(CO)_{12}$	$THF/(CH_3)_3N$	23.8	100	3300	54
$Os_3(CO)_{12}$	$THF/(CH_3)_3N$	23.8	180	270	54
$Rh_6(CO)_{16}$	$THF/(CH_3)_3N$	23.8	126	1700	54
$Ir_4(CO)_{12}$	$THF/(CH_3)_3N$	23.8	126	300	54
$[Pt_3(CO)_6]_2^+$	$THF/(CH_3)_3N$	23.8	126	700	54
$Rh_6(CO)_{16}$	$EtOCH_2CH_2OH/NH_2CH_2CH_2NH_2$	0.8	100	250	59
$Rh_6(CO)_{16}$	$EtOCH_2CH_2OH/NH_2(CH_2)_3NH_2$	0.8	100	76	59
$Rh_6(CO)_{16}$	$EtOCH_2CH_2OH/NH_2(CH_2)_4NH_2$	0.8	100	15	59

Table 3

AMINE PROMOTED RUTHENIUM CATALYSIS OF THE WGSR

Catalyst	Solvent/amine	Pressure CO (atm)	Temp. (°C)	Turnovers (/10 h)	Ref.
$Ru_3(CO)_{12}$	Diglyme/$(CH_3)_3N$	51	100	5740	55
$Ru_3(CO)_{12}$	Diglyme/E_3N	51	100	860	55
$Ru_3(CO)_{12}$	Diglyme/Bu_3N	51	100	540	55
$Ru_3(CO)_{12}$	Diglyme/pyridine	51	100	300	55
$Ru_3(CO)_{12}$	Diglyme/$NH(CH_3)_2$	51	100	2200	55

and amine promoted catalysis of the WGSR are reversed, with amine enhanced ruthenium catalysis of the WGSR having the highest activity of all the systems studied. These observations do not support Pettit's conclusion about the generality of the iron carbonyl mechanism.

Slegeir *et al.* [55] briefly studied several aspects of ruthenium catalysis of the WGSR in the $(CH_3)_3N/THF/H_2O$ system and proposed a mechanism for amine promoted WGSR catalysis as illustrated in Scheme 11:

$$Ru_3(CO)_{12} + 3CO \rightleftharpoons 3Ru(CO)_5$$

$$Ru(CO)_5 + (CH_3)_3N \rightleftharpoons (CO)_4Ru^-C(=O)N^+(CH_3)_3$$

$$(CO)_4Ru^-C(=O)N^+(CH_3)_3 + H_2O \longrightarrow HRu(CO)_4^- + CO_2 + (CH_3)_3NH^+$$

$$HRu(CO)_4^- + H_2O \rightleftharpoons H_2Ru(CO)_4 + OH^-$$

$$H_2Ru(CO)_4 \rightleftharpoons H_2 + Ru(CO)_4$$

$$Ru(CO)_4 + CO \rightleftharpoons Ru(CO)_5$$

Scheme 11.

The evidence in support of this mechanism is contrary. At high initial $Ru_3(CO)_{12}$ concentrations, Slegeir *et al.* [55] present CO pressure dependence data that, as the authors state, support cluster catalysis. Moreover, the authors isolate $H_4Ru_4(CO)_{12}$ from acidification of the reaction solution. This also supports cluster catalysis. At lower CO pressures and $Ru_3(CO)_{12}$ concentrations, they observe higher WGSR catalysis activity and conclude that mononuclear species are involved. Unfortunately, the authors do not consider the equilibrium shown in (39), which would account for their observations, retain a cluster catalyzed mechanism and correlate with the observations of Ford *et al.* and Shore, that in the OH^- promoted ruthenium system, the true active catalyst is the $HRu_3(CO)_{11}^-$ species. However, these similarities do not explain the disparity in the catalyst activities between the amine and OH^- promoted WGSR as seen from a comparison of the data recorded in Table 1 through 3.

The work of Laine *et al.* [56–58] provides at least a partial explanation of these differences. In an effort to explain the reversal in relative activities of the two catalysts, Laine *et al.* presented evidence that a majority of the second and third row group 8 metals interact strongly with tertiary amines as evidenced by reaction (43) as catalyzed by ruthenium, osmium, or rhodium.

$$Et_3N + D_2O + CO \longrightarrow Et_2NCHDCD_3 + CO_2 + HDO \qquad (43)$$

The possibility exists that amine-metal complexes play an active role in the various WGSR catalyst solutions. At a minimum these results suggest that the catalyst system is extremely complex. Therefore, a simple explanation of the mechanisms of amine promoted group 8 metal catalysis of the WGSR is not as yet possible.

Kaneda *et al.* [59] also studied amine promoted rhodium catalysis of the WGSR and reported that the use of diamines (see Table 3) such as ethylene diamine considerably enhances the activity of the catalyst. One major conclusion from their work is that amine configuration plays a more important role in determining catalyst activity than amine basicity. Their work provides further evidence for the participation of amines as ligands in the WGSR catalytic cycle.

One major drawback to using primary or secondary amines as cocatalysts is that the well-known formamidation reaction, (44), irreversibly consumes the amine:

$$R_2NH + CO \longrightarrow R_2NC(=O)H \qquad (44)$$

4.5. Mixed-Metal Catalysis of the WGSR

Only three examples of mixed-metal catalysis of the WGSR have been reported in the literature. One system, described by Ford *et al.* [60] involves the use of iron/ruthenium mixtures to catalyze the WGSR. The other two reports concern the use of iron/ruthenium or iron/rhodium mixture for Reppe reactions and will be discussed elsewhere.

As seen in Table 1, the use of mixtures of iron and ruthenium in place of the individual metals in conjunction with OH^- gives catalyst solutions that are more active than identical catalysts solutions made up with the individual metals. At present there is no firm evidence to provide a rationale for these observations. However, the two most reasonable explanations, which are based on the formation of a mixed-metal cluster, are that this cluster is either more susceptible to OH^- attack or undergoes more facile reductive elimination of H_2 than the single metal catalyst intermediates. Knox *et al.* [61] report that the cluster $H_4FeRu_3(CO)_{12}$ loses H_2 more readily than the all ruthenium analog. In the related Reppe hydroformylations, mixed-metal rate enhancement is observed where essentially no H_2

is produced, thus making the first conclusion unlikely. Gross and Ford [47] find that the order of reactivity for the iron triad clusters, for nucleophilic attack by methoxide, is $Fe_3(CO)_{12} > Ru_3(CO)_{12} > Os_3(CO)_{12}$. They propose that the iron cluster is more reactive than the ruthenium or osmium clusters because, of the three, it alone contains bridging carbonyl groups. Iron may also cause the formation of bridging carbonyls in the mixed-metal cluster making it more susceptible to nucleophilic attack by OH^-. This explanation is also reasonable for the ruthenium catalyzed Reppe hydroformylations.

As we emphasized in the introduction, LT sulfur tolerant WGSR catalysts are needed. King *et al.* [62] demonstrated that, with the exception of iron, all the group 8 metal catalysts listed in Table 4 are active catalysts for the WGSR when Na_2S is substituted for OH^- as the base. Although the rates are low, this important contribution clearly demonstrates the considerable potential available for the use of homogeneous catalysts for industrially important processes.

Table 4

SULFIDE PROMOTED CATALYSIS OF THE WGSR

Catalyst	Solvent	Pressure CO (atm)	Temp. (°C)	Turnovers (/24 h)	Ref.
$Fe(CO)_5$	MeOH	27.2	140	0	62
$Ru_3(CO)_{12}$	MeOH	27.2	160	550	62
$Os_3(CO)_{12}$	MeOH	27.2	160	200	62
$Cr(CO)_6$	MeOH	27.2	160	60	62
$Mo(CO)_6$	MeOH	27.2	160	130	62
$W(CO)_6$	MeOH	27.2	160	180	62

5. CATALYSIS OF THE WGSR UNDER ACIDIC OR NEUTRAL CONDITIONS

The first examples of homogeneous catalysis of the WGSR using acidic media were demonstrated in 1977. Despite the comments in the introduction, there were actually two systems reported that require the addition of an acid cocatalyst to effect WGSR catalysis. One system is the often cited work of Cheng *et al.* [10] using a rhodium catalyst in acetic acid/HCl solution. The second system reported by Zudin *et al.* [63] and essentially unnoticed in the literature, involves the use of palladium phosphine complexes in trifluoroacetic acid:

$$CO + H_2O \xrightarrow[CF_3CO_2H/H_2O]{Pd(PPh_3)_4/100^\circ} CO_2 + H_2 \qquad (45)$$

In continuing work on the rhodium system, Baker *et al.* [64] describe kinetic and mechanistic investigations that suggest the catalytic cycle shown in Scheme 12:

$$HRh(CO)_2I_3^- + HI \longrightarrow H_2 + Rh(CO)_2I_4^-$$

$$Rh(CO)_2I_4^- + CO \rightleftharpoons I^- + Rh(CO)_3I_3$$

$$Rh(CO)_3I_3 + H_2O \rightleftharpoons H^+ + Rh(CO)_2I_3(CO_2H)^-$$

$$Rh(CO)_2I_3(CO_2H)^- \longrightarrow HI + Rh(CO)_2I_2^- + CO_2$$

$$Rh(CO)_2I_2^- + HI \rightleftharpoons HRh(CO)_2I_3^-$$

or

$$Rh(CO)_2I_3(CO_2H)^- \longrightarrow HRh(CO)_2I_3^- + CO_2$$

and

$$Rh(CO)_2I_4^- + I^- \rightleftharpoons CO + Rh(CO)I_5^-$$

Scheme 12.

Baker *et al.* [64] have spectroscopically identified the species $Rh(CO)_2I_2^-$, $Rh(CO)I_5^-$, $Rh(CO)I_4^-$, *cis*- and *trans*-$Rh(CO)_2I_4^-$ in solution, and they isolated and characterized $Rh(CO)_2I_2^-$. An Arrhenius plot of the WGSR catalysis over the range of 55°C–100°C reveals unusual behavior, giving an Ea of 25.8 kcal/mol between 55°C–60°C and an Ea of 9.3 kcal/mol above this range. The authors argue that there is a change in the rate-limiting step at higher temperatures. Furthermore, they propose that at low temperatures the rate-limiting step is oxidation of Rh(I) by HI to Rh(III), and at high temperatures the rate-limiting step is reduction of a Rh(III) carbonyl species with concomitant release of CO_2. These observations are similar, but not completely in accord with, the observations of Singleton *et al.* [65] who studied the same system but under more forcing conditions. The differences do not warrant further discussion here.

More recently, Marnot *et al.* [66] identified another rhodium WGSR catalyst that operates under acidic conditions. These workers report that rhodium and iridium 2,2′-bipyridine (bipy) or related ligand complexes such as $Rh(bipy)_2(H_2O)_2{}^{3+}$ and $Rh(L)_2(H_2O)_2{}^{3+}$ $Ir(L)_2(H_2O)_2{}^{3+}$, where L = 4,7-diphenyl-1,10-phenanthroline disodium sulfonate (Phen-S) and 2,9-dimethyl-4,7-diphenyl-1,10-phenanthroline disodium sulfonate (2,9-dmphen-S) are active WGSR catalysts under acidic conditions. Aside from the catalyst activities (listed in Table 5), no mechanistic work has as yet been reported.

In the platinum metals group, both palladium and platinum complexes have been shown to be active WGSR catalysts in the presence of acid cocatalysts. The original report by Zudin *et al.* has been followed up by Likholobov *et al.* [67, 68] who propose the following catalytic cycle for a Ph_3P complexed palladium WGSR catalysis system run in 20% aqueous trifluoroacetic acid:

$$P_2PdX_2 + CO \rightleftharpoons [P_2Pd(X)CO]^+ + X^-$$

$$[P_2Pd(X)CO]^+ + H_2O \rightleftharpoons [P_2Pd(H_2O)CO]^{2+} + X^-$$

$$[P_2Pd(H_2O)CO]^{2+} + H_2O \longrightarrow [P_2Pd(OH)CO]^+ + H_3O^+$$

$$[P_2Pd(OH)CO]^+ + X^- \longrightarrow P_2Pd(CO_2H)X$$

$$P_2Pd(CO_2H)X \longrightarrow P_2PdHX + CO_2$$

$$P_2PdHX + HX \longrightarrow P_2PdX_2 + H_2$$

$$P = Ph_3P,\ HX = CF_3CO_2H$$

Scheme 13.

Table 5

CATALYSIS OF THE WGSR UNDER ACID OR NEUTRAL CONDITIONS

Catalyst	Solvent	Pressure CO (atm)	Temp. (°C)	Turnovers (/24 h)	Ref.
$[Rh(CO)_2Cl]_2$	HOAc/HCl	0.53	100	34	63
$RhCl_3 \cdot 3H_2O/2 \cdot$ 2,9-dmphen-S	H_2O	1.0	100	550	65
$IrCl_3 \cdot 3H_2O/2 \cdot$ 2,9-dmphen-S	H_2O	1.0	100	225	65
$IrCl_3 \cdot 3H_2O$/bipy-S	H_2O	1.0	100	9.6	65
$Pd(PPh_3)_4$	CF_3CO_2H	1.0	70	60	66
$K_2PtCl_4/SnCl_4 \cdot 5H_2O$	HOAc/HCl	0.53	88	25	68
$Pt[P(i\text{Pr})_3]_3$	Acetone	19.3	100	125	69
$RhH[P(i\text{Pr})_3]_3$	Acetone	19.3	100	672	70
$RhH[P(i\text{Pr})_3]_3$	Pyridine	19.3	100	792	70
$Rh_2(H)(CO)_3(DPPM)^+$	PrOH	1.0	90	60	71

In view of the work of Cariati *et al.* [20] reaction (16), the mechanism proposed in Scheme 13 appears quite reasonable. Although no other research groups have pursued studies of this type, there appears to be considerable potential for this type of catalyst system as exemplified by work done on Reppe type reactions using phoshine-palladium/trifluoroacetic acid [68].

Another study on WGSR catalysis in acid solutions has been reported recently. This is the work reported by Cheng and Eisenberg [69] in which they explore the use of a platinum chloride-tin chloride catalyst that is active in an acetic acid/HCl acid medium. They report that a spectroscopic analysis of the active catalyst solution shows the presence of both $PtCl(CO)(SnCl_3)_2^-$ and $PtCl_2(CO)(SnCl_3)^-$. Preliminary kinetic and mechanistic studies allow them to suggest the catalytic cycles shown in Scheme 14.

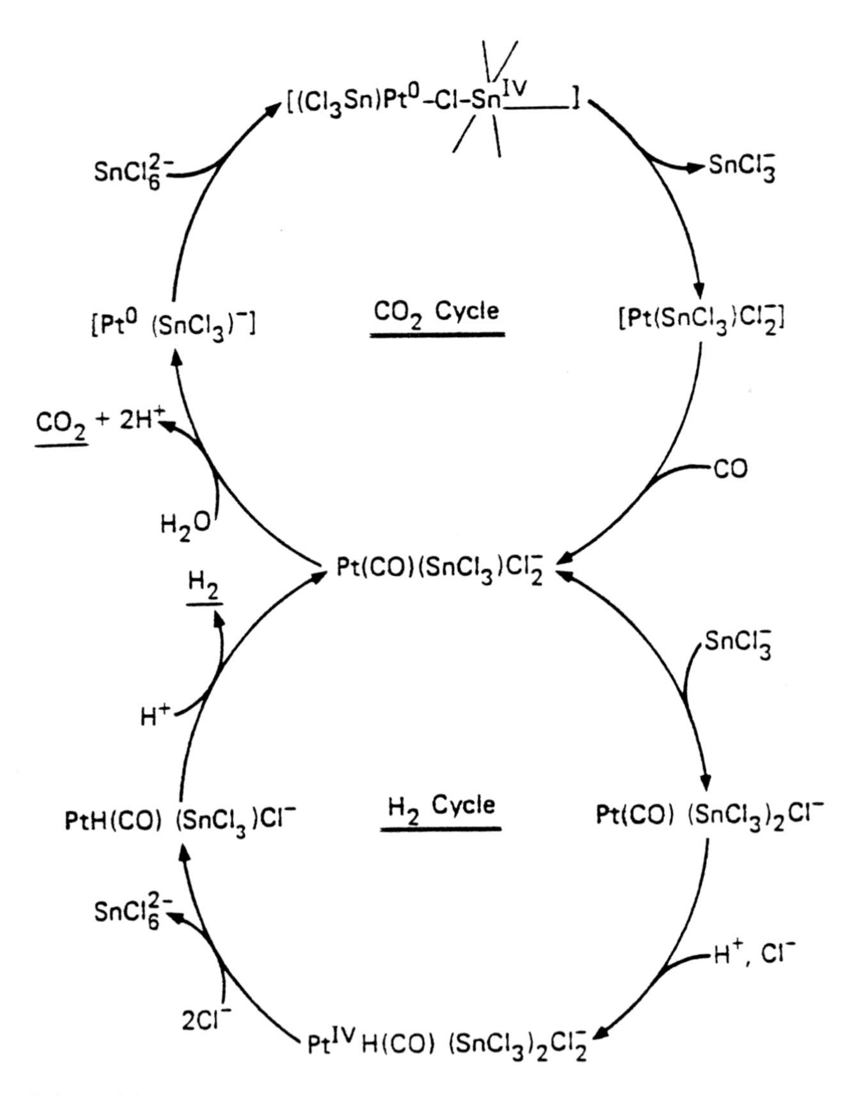

Scheme 14.

These researchers conclude that the Sn(II)/Sn(IV) redox couple is actively involved in the observed reaction chemistry. The H_2-forming catalytic cycle results in the oxidation of Sn(II) to Sn(IV) with coproduction of H_2 and, CO is oxidized to CO_2 concurrent with the reduction of Sn(IV) to Sn(II).

Ford and coworkers [71] have reported that ruthenium carbonyl will also catalyze the WGSR under acidic conditions. Thus, $Ru_3(CO)_{12}$ in a mixture of H_2SO_4/diglyme will actively catalyze the WGSR at 100°C, as shown in Table 5, with an activation energy of 14 kcal/mol. From kinetic and spectroscopic studies, these researchers have concluded that the principal species in solution is a monohydridic anionic or neutral complex. The most likely choice is $HRu_2(CO)_8^-$. They propose the catalytic cycle shown in Scheme 15.

$$HRu_2(CO)_8^- + CO \rightleftharpoons HRu(CO)_4^- + Ru(CO)_5$$

$$HRu(CO)_4^- + H_2O \rightleftharpoons H_2Ru(CO)_4 + OH^-$$

$$H_2Ru(CO)_4 \longrightarrow Ru(CO)_4 + H_2$$

$$Ru(CO)_4 + Ru(CO)_5 \longrightarrow Ru_2(CO)_9$$

$$Ru_2(CO)_9 + H_2O \rightleftharpoons HRu_2(CO)_8(CO_2H)$$

$$HRu_2(CO)_8(CO_2H) \longrightarrow HRu_2(CO)_8^- + CO_2 + H^+$$

Scheme 15.

The rationale for attack of H_2O on $Ru_2(CO)_9$ in this proposed catalytic cycle lacks precedence. It seems more reasonable to expect that a neutral carbonyl species is protonated and then reacts with H_2O, as in any of the reactions (32)–(34), to give $HRu_2(CO)_8(CO_2H)$ or a closely related intermediate. However, this proposed catalytic cycle requires much more experimental exploration before a complete anaylsis of the system is possible.

Only three WGSR catalyst systems have been described for which it has not been necessary to activate and maintain the catalytic reactivity of the system through addition of either base or acid. One reaction system uses plantinum phosphine catalysts, and the other two systems use rhodium phosphine catalysts. The one platinum system and one of the rhodium systems were reported by the Otsuka research group [30, 39]. These investigators report that the two catalyst systems function by H_2O activation rather than CO activation, an apparent requirement of all the systems described above.

Yoshida *et al.* [30] reported in 1978 that platinum phosphine complexes of the type PtL_3, where L = $P(iPr)_3$ or PEt_3, could be used to catalyze the WGSR under mild conditions in a number of solvents. They proposed the catalytic cycle shown in Scheme 16 to account for their observations:

$$PtL_3 \rightleftharpoons PtL_2 + L$$

$$PtL_2 + H_2O \rightleftharpoons PtH(OH)L_2$$

$$PtH(OH)L_2 + S \rightleftharpoons [PtH(S)L_2]OH$$

$$[PtH(S)L_2]OH + CO \rightleftharpoons [PtH(CO)L_2]OH + S$$

$$[PtH(CO)L_2]OH \longrightarrow [PtH(CO_2H)L_2]$$

$$[PtH(CO_2H)L_2] \longrightarrow PtH_2L_2 + CO_2$$

$$PtH_2L_2 + L \rightleftharpoons PtL_3 + H_2$$

Scheme 16.

Complexes such as *trans*-$PtH(CO_2K)L_2$, *trans*-$PtH(CO_2CH_3)L_2$, and *trans*-PtH_2L_2 were prepared as examples of the proposed intermediates, and the last complex was independently shown to catalyze the WGSR.

In the neutral rhodium system [39] the complex $RhHL_3$, where L = P(*i*Pr)$_3$ or P(*c*-C_6H_{11})$_3$, was found to be an active WGSR catalyst in either acetone, THF, or pyridine, with the pyridine (py) solvent system having the highest catalyst activity. The following reaction sequence is proposed for the catalytic cycle:

$$RhHL_3 + H_2O + py \rightleftharpoons [RhH_2(py)_2L_2]OH$$

$$[RhH_2(py)_2L_2]OH + CO \longrightarrow [Rh(CO)(py)_2L_2]OH + H_2$$

$$[R(CO)(py)_2L_2]OH + CO \rightleftharpoons Rh(CO)_2(py)L_2]OH + py$$

$$[Rh(CO)_2(py)L_2]OH \longrightarrow Rh(CO)py(CO_2H)L_2$$

$$Rh(CO)(py)(CO_2H)L_2 + CO \longrightarrow RhH(CO)_2L_2 + CO_2 + py$$

$$RhH(CO)_2L_2 + H_2O \rightleftharpoons [RhH_2(CO)L_2]OH$$

$$[RhH_2(CO)L_2]OH + py \rightleftharpoons [Rh(CO)(py)L_2]OH + H_2$$

Scheme 17.

The complexes, *trans*-$[Rh(CO)(py)L_2]^+$ and $[RhH_2(py)_2L_2]OH$, can be isolated when pyridine is used as solvent. When acetone is the solvent, $RhH(CO)_2L_2$ can be isolated from the reaction solutions, as can the complex $Rh_2(CO)_4L_2$. The majority of these complexes were shown spectroscopically or in separate reaction studies to react as shown in Scheme 17. In addition, the Otsuka group also demonstrated [30, 39] that both the platinum and rhodium complexes were active catalysts for formic acid and carbonic acid decomposition as explored in the exercise performed above [30, 39]. Thus, at least for these systems, formic acid decomposition must be considered as an alternative or additional route for WGSR catalysis.

The most recent entry in this area comes as part of a communication on the chemistry of "A-Frame" rhodium phosphine complexes. Kubiak and Eisenberg [70] reported that the complex, $[Rh_2(\mu\text{-}H)(\mu\text{-}CO)(CO)_2$ (bisdiphenylphosphinomethane)$_2]^+$ is an active catalyst for the WGSR under neutral conditions. Unfortunately, no further discussion of the WGSR catalysis chemistry of this complex was reported.

6. REFERENCES

1. C. L. Thomas: *Catalytic Processes and Proven Catalysts*, Academic Press, New York (1970).
2. M. I. Temkin: *Adv. Catal.* **28**, 263–267 (1979).
3. P. C. Ford: *Acc. Chem. Res.* **14**, 31–37 (1981).
4. J. W. Reppe: *Liebigs Ann. Chem.*, 582 (1953).
5. D. M. Fenton: U.S. Patent 3,490,872 (1970).
6. D. M. Fenton: U.S. Patent 3,539,298 (1970).

7. D. M. Fenton: U.S. Patent 3,781,364 (1973).
8. Richard M. Laine, Robert G. Rinker, and Peter C. Ford: *J. Am. Chem. Soc.* **99**, 252 (1977).
9. H. C. Kang, C. H. Mauldon, T. Cole, W. Slegeir, K. Cann, and R. Pettit: *J. Am. Chem. Soc.* **99**, 8323 (1977).
10. C. H. Cheng, D. E. Hendriksen, and R. Eisenberg: *J. Am. Chem. Soc.* **99**, 2791–2792 (1977).
11. R. G. Pearson and H. Mauermann: *J. Am. Chem. Soc.* **104**, 500–504 (1982).
12. J. A. Osborn, F. H. Jardine, J. F. Young, and G. Wilkinson: *J. Chem. Soc.*, Z, 1711–1732 (1966).
13. P. B. Chock and J. Halpern: *J. Am. Chem. Soc.* **88**, 3511–3514 (1966).
14. E. L. Muetterties and P. L. Watson: *J. Am. Chem. Soc.* **100**, 6978–6989 (1978).
15. L. Marko and F. Ungvary: *J. Organomet. Chem.* **20**, 205 (1969).
16. J. Evans and J. R. Norton: *J. Am. Chem. Soc.* **96**, 7577–7578 (1974).
17. K. H. Brandes and H. B. Jonassen: *Z. Anorg. Allg. Chem.* **343**, 215 (1966).
18. J. P. Collman, R. G. Finke, P. L. Matlock, R. Wahren, R. G. Komato, and J. I. Brauman: *J. Am. Chem. Soc.* **100**, 1119–1140 (1978).
19. J. C. Bricker, C. C. Nagel, and S. G. Shore: *J. Am. Chem. Soc.* **104**, 1444–1445 (1982).
20. J. P. Collman and L. S. Hegedus: *Principles and Applications of Organotransition Metal Chemistry*, pp. 69–70, Univ. Sci Books, Mill Valley, Calif. (1980).
21. C. P. Casey and S. M. Neuman: *Adv. Chem. Ser.* **173**, 131–139 (1979).
22. R. Moore: *Org. Reactions* **5**, 301–330 (1949).
23. R. S. Paonessa and W. C. Trogler: *J. Am. Chem. Soc.* **104**, 3529–3530 (1982).
24. R. Eisenberg and D. E. Hendriksen: *Adv. Catal.* **28**, 79 (1979).
25. D. J. Darensbourg, M. B. Fisher, R. E. Schmidt, Jr., and B. J. Baldwin: *J. Am. Chem. Soc.* **103**, 1297–1298 (1981).
26. D. J. Darensbourg, A. Rokicki, and M. Y. Darensbourg: *J. Am. Chem. Soc.* **103**, 3223–3224 (1981).
27. S. H. Strauss, K. H. Whitmire, and D. F. Shriver: *J. Organomet. Chem.* **174**, C59–C62 (1979).
28. A. Aguilo: *J. Catal.* **13**, 283 (1969).
29. R. S. Coffey: *J. Chem. Soc., Chem. Comm.*, 923–924 (1967).
30. T. Yoshida, Y. Ueda, and S. Otsuka: *J. Am. Chem. Soc.* **100**, 3941–3942 (1978).
31. R. M. Laine, R. G. Rinker, and P. C. Ford: unpublished work.
32. R. A. Grey, G. P. Pez, and A. Wallo: *J. Am. Chem. Soc.* **103**, 7536–7542 (1981).
33. A. J. Deeming and B. L. Shaw: *J. Chem. Soc.*, A, 443 (1969).
34. N. Grice, S. C. Kao, and R. Pettit: *J. Am. Chem. Soc.* **101**, 1627–1628 (1979).
35. M. Catellani and J. Halpern: *J. Inorg. Chem.* **19**, 566–568 (1980).
36. J. R. Sweet and W. A. G. Graham: *Organomet.* **1**, 982–986 (1982).
37. D. J. Darensbourg: *Isr. J. Chem.* **15**, 247–252 (1977).
38. D. J. Darensbourg and A. Rokicki: *Organomet.* **1**, 1685–1693 (1982).
39. T. Yoshida, T. Okano, and S. Otsuka: *ACS Symposium Ser.* **152**, 79–94 (1981).
40. E. L. Muetterties: *Inorg. Chem.* **4**, 1841 (1965).
41. D. J. Darensbourg, and J. A. Froelich: *J. Am. Chem. Soc.* **99**, 4727–4729 (1977).
42. R. L. Kump and L. J. Todd: *Inorg. Chem.* **20**, 3715–3718 (1981).
43. T. Yoshida, T. Matsuda, T. Okano, T. Kitani, and S. Otsuka: *J. Am. Chem. Soc.* **101**, 2027–2038 (1979).
44. T. Yoshida, T. Okano, K. Saito, and S. Otsuka: *Inorg. Chim. Acta* **44**, L135–L136 (1980).
45. M. G. Mason and J. A. Ibers: *J. Am. Chem. Soc.* **104**, 5153–5157 (1982).
46. C. Ungermann, R. G. Rinker, P. C. Ford, V. Landis, S. A. Moya, and R. M. Laine: *Adv. Chem. Ser.* **173**, 81–93 (1979).
47. D. C. Gross and P. C. Ford: *Inorg. Chem.* **21**, 1704–1706 (1982).
48. G. Suss-Fink: *J. Organomet. Chem.* **193**, C20–C22 (1980).

49. R. G. Pearson, H. W. Walker, H. Mauermann, and P. C. Ford: *Inorg. Chem.* **20**, 2741–2743 (1981).
50. R. B. King. C. C. Frazier, R. M. Haines, and A. D. King: *J. Am. Chem. Soc.* **100**, 2925–2927 (1978).
51. A. D. King. Jr. R. B. King, and D. B. Yang: *J. Am. Chem. Soc.* **102**, 1028–32 (1980).
52. D. J. Darensbourg and A. Rokicki: *ACS Symp. Ser.* **152**, 107–122 (1981).
53. R. Pettit, K. Cann., T. Cole, C. H. Mauldin, and W. Slegeir: *Ann. N.Y. Acad. of Sci.* **333**, 101–106 (1980).
54. R. Pettit, K. Cann, T. Cole, C. Mauldin, and W. Slegeir: *Adv. Chem. Ser.* **173**, 121 (1979).
55. W. A. R. Slegeir, R. S. Sapienza, and B. Easterling: *ACS Symp. Ser.* **152**, 325–344 (1981).
56. R. M. Laine, D. W. Thomas, S. E. Buttrill, and L. W. Cary: *J. Am. Chem. Soc.* **100**, 6527 (1978).
57. Y. Shvo, D. W. Thomas, and R. M. Laine: *J. Am. Chem. Soc.* **103**, 2461 (1981).
58. R. M. Laine, D. W. Thomas, and L. W. Cary: *J. Am. Chem. Soc.* **104**, 1763–1764 (1982).
59. K. Kaneda, M. Hiraki, K. Sano, T. Imanaka, and S. Teranishi: *J. Molec. Cat.* **9**, 227–230 (1980).
60. P. C. Ford, R. G. Rinker, C. Ungermann, R. M. Laine, V. Landis, and S. A. Moya: *J. Am. Chem. Soc.* **100**, 4595 (1978).
61. S. A. R. Knox, J. W. Koepke, M. A. Andrews, and H. D. Kaesz: *J. Am. Chem.* **97**, 3942 (1975).
62. A. D. King, R. B. King and D. B. Yang: *J. Chem. Soc., Chem Comm.* 529–30 (1980).
63. V. N. Zudin, V. A. Likholobov, Yu. I. Yermakov, and N. K. Yeremenko: *Kinet. i Katal.* **18**, 524 (1977).
64. E. C. Baker, D. E. Hendriksen, and R. Eisenberg: *J. Am. Chem. Soc.* **102**, 1020–1027 (1980).
65. T. C. Singleton, L. J. Park, J. L. Price, and D. Forster: *Prepr. Div. Pet. Chem. Am. Chem. Soc.* **24**, 329 (1979).
66. P. A. Marnot, R. R. Ruppert, and J. P. Sauvage: *Nouv. J. Chem.* **5** 543–545 (1981).
67. V. A. Likholobov, V. N. Zudin, and Yu. I. Yermakov: *Proceedings of the Fifth Japanese-Soviet Seminar on Catalysis*, Osaka, Japan, 9–18 (1980).
68. Yu. I. Yermakov: private communication.
69. C. H. Cheng and R. Eisenberg: *J. Am. Chem. Soc.* **100**, 5968–5970 (1978).
70. C. P. Kubiak and R. Eisenberg: *J. Am. Chem. Soc.* **102**, 3637–3639 (1980).
71. P. C. Ford, P. Yarrow, and H. Cohen: *ACS Symp. Ser.* **152**, 95–106 (1981).

Homologation of Alcohols, Acids and Their Derivatives by $CO + H_2$

GIUSEPPE BRACA and GLAUCO SBRANA

Institute of Industrial Organic Chemistry, University of Pisa, Via Risorgimento 35, 56100 Pisa, Italy

R. Ugo (ed.), Aspects of Homogeneous Catalysis, Vol. 5, 241–337.

1. INTRODUCTION

"Homologation", which etymologically means: "formation from a product of a new compound with the same structure and with a slight increase in the chain length", despite the formal simplicity of the transformation involved, is a very difficult reaction to carry out.

The history of this type of reaction is also unusual: in fact, the first reports by BASF researchers on the cobalt-catalyzed hydrocarbonylation of aliphatic primary alcohols into the next highest alcohols date back to 1941–1943 [1, 2] while the appearance in open literature came eight years later. Wender in 1949 [3] practically rediscovered the reaction and gave impulse to the research in this field [4, 5]. After an initial interest taken by some groups in the United States [4–7], Germany [8], Hungary [9] and Italy [10], a long period of oblivion followed (1950–1975), mainly due to the poor selectivity of this reaction and to the low activity of the catalysts which made drastic conditions, extremely high pressure and long reaction times [1, 3, 6–8]. In 1975 the sheiks magically focussed intense interest in this reaction: in the following seven years more than 40 papers and about the same number of patents were published. All the most important companies in the world have been engaged in this chemistry and appear as assignors of patents. This interest derived from the prospective of synthesizing base chemicals and liquid fuels from syngas obtained by coal gasification. So most of the work has been devoted to the improvement of catalysts performances (activity, selectivity and recycling) especially in the case of methanol homologation. As result of these efforts important improvements in selectivity and catalytic activity have been obtained and, at present, economic analysis demonstrates interesting possibilities for commercial realization [11, 12].

The difficulties encountered in carrying out this reaction may be interpreted by looking at Scheme 1 showing the main catalytic steps of the different homologation reactions. Considering as is generally accepted [13] a metal hydride as the catalytically active species, the complexity of the action of the catalyst is immediately apparent. In fact the ability to activate the oxygenated substrate must be associated with the activity in two very different reaction steps: carbonylation and hydrogenation.

Moreover these reactions must occur under the same reaction conditions without possible side-reactions which may produce hydrocarbons through the hydrogenation of the metal-alkyl intermediate, or aldehydes by a partial hydrogenation of the metal acyl derivative. Another complication arises from the possibility of successive homologations, carbonylations and hydrogenations of the products of primary reaction. Moreover, whereas for alcohols and ethers a carbonylation step must precede the hydrogenation step, for acids this latter must precede the carbonylation and for esters both possibilities can be foreseen. The difficulty of finding at the same time all these features in a simple and single catalytic system is quite clear. This condition is the main handicap of homologation reactions and the efforts of research are devoted to overcoming it: at present important results have been reached but a definite solution of the problem has not yet been found.

2. HOMOLOGATION OF ALCOHOLS

Looking at the general scheme for homologation reactions (Scheme 1), the homologation of an alcohol which proceeds through the carbonylation of an alkyl-metal derivative and the successive hydrogenation of the acyl to an alkoxy intermediate is accompanied by several side-reactions involving the same intermediates and leading to different products (Scheme 2).

The hydrocarbonylation of the alcohol to the homologous aldehyde (Equation 2) proceeds through the hydrogenolysis of the acyl-metal species. Whenever the aldehyde is formed, even hemi-acetals and acetals, arising from its reaction with the excess of the alcohol (Equations 3 and 4) and products of aldol condensation, such as unsaturated aldehydes and their hydrogenation products, are formed especially when operating at high conversion and at high temperature.

Acids and esters, products of a simple carbonylation of the substrate (Equations 5 and 6), are formed via the hydrolysis and alcoholysis of the acyl-metal intermediate.

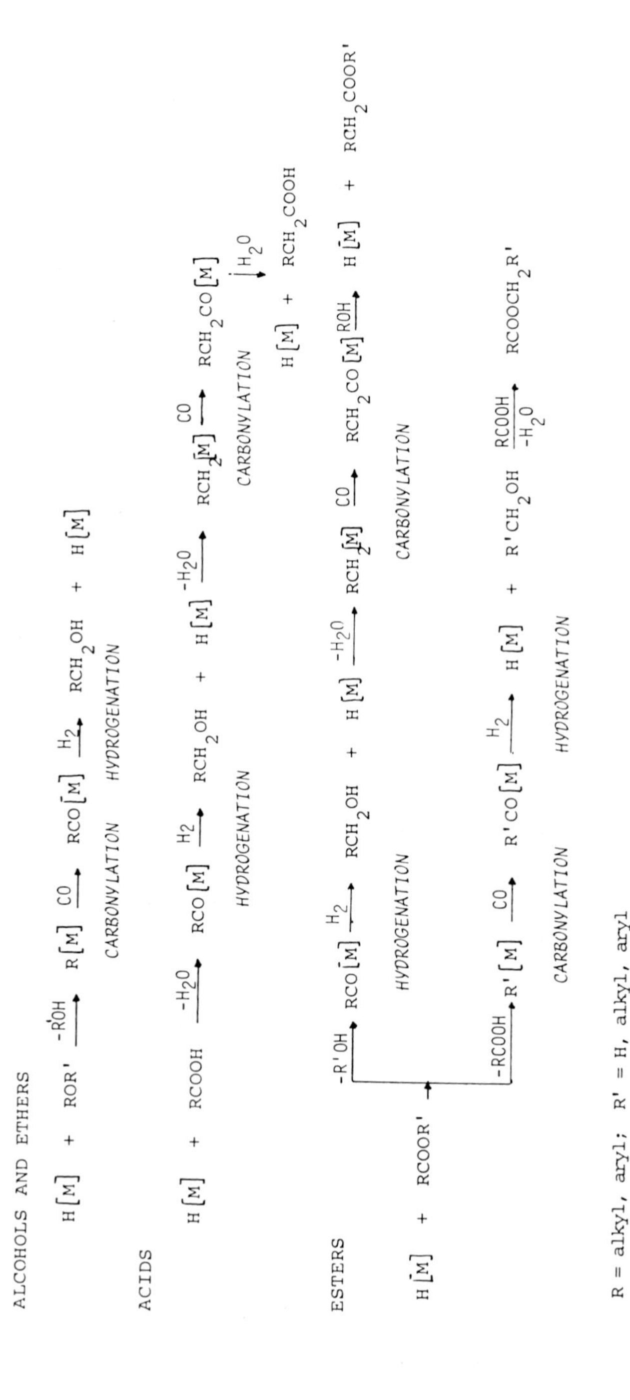

Scheme 1. Catalytic steps in the homologation reactions.

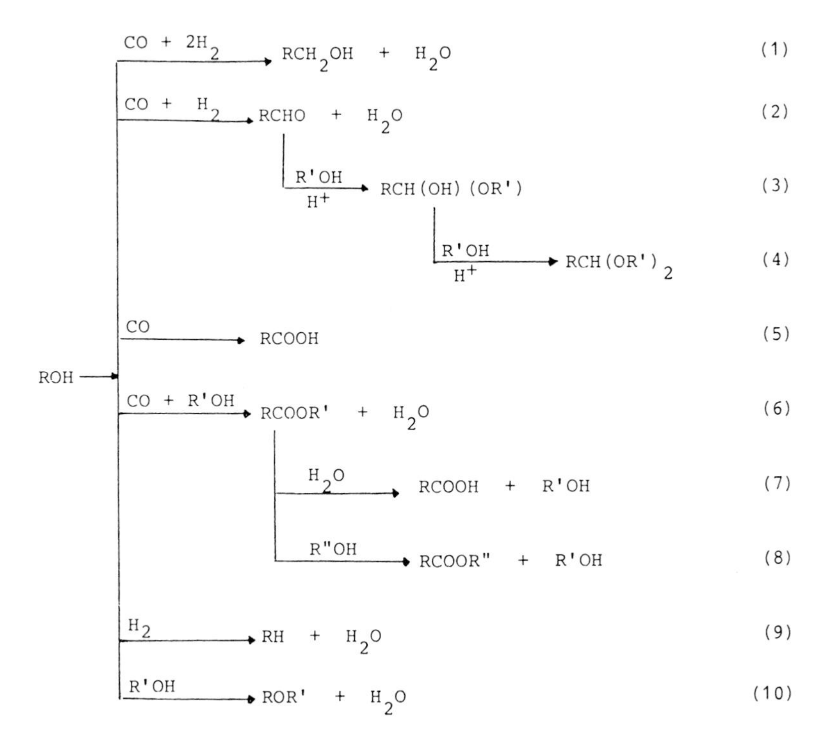

Scheme 2. Possible reactions in alcohol homologation.

Moreover all the possible products of esterification and *trans*-esterification reactions (Equations 7 and 8) may be present when the homologation process is carried out in the presence of an acid component (i.e., HI, $HCo(CO)_4$, etc.).

Hydrocarbons, products of total hydrogenation of the alcohol, come from the hydrogenolysis of the alkyl intermediate (Equation 9).

Ethers can be produced by an acid catalyzed dehydration of the alcohols (Equation 10) or via a catalytic hydrogenation of acetals (Equation 11) [14, 15].

$$RCH(OR')_2 + H_2 \xrightarrow{H[M]} RCH_2OR' + R'OH \qquad (11)$$

Formates, products of a partial hydrogenation of CO, sometimes present in low percentage, arise probably from the alcoholysis of a formyl-metal intermediate (Equation 12) or from the carbonylation and hydrogenolysis of an alkoxy-metal intermediate (Equation 13) [16].

$$HCO[M] + R'OH \longrightarrow HCOOR' + H[M] \qquad (12)$$

$$R'O[M] + CO \longrightarrow R'OCO[M] \xrightarrow{H_2} HCOOR' + H[M] \qquad (13)$$

Formation of methoxypropanol [16] may result from the interaction of acetaldehyde or its hemi-acetals and acetals with the metal hydride through carbonylation and hydrogenation steps (Equations 14 and 15).

$$CH_3CHO + H[M] \longrightarrow CH_3\underset{\displaystyle OH}{\underset{|}{C}}H[M] \xrightarrow[CH_3OH]{CO + H_2} CH_3\underset{\displaystyle OH}{\underset{|}{C}}HCH_2OCH_3 + H_2O + H[M] \tag{14}$$

$$CH_3CH(OH)OCH_3 + H[M] \xrightarrow{-H_2O} CH_3\underset{\displaystyle OCH_3}{\underset{|}{C}}H[M] \xrightarrow{CO + H_2} CH_3\underset{\displaystyle OCH_3}{\underset{|}{C}}HCH_2OH + H[M] \tag{15}$$

The water gas shift reaction (Equation 16) is also an important side reaction which can change the CO/H_2 ratio consuming carbon monoxide and enriching the gases in hydrogen.

$$CO + H_2O \rightleftharpoons CO_2 + H_2 \tag{16}$$

This very complicated distribution of products is really the most serious drawback of the alcohol homologation process; in fact some of the above mentioned side reactions may become predominant with respect to the homologation reaction. Thus, for instance, the hydrocarbonylation of methanol to acetaldehyde and its acetals is the main reaction operating with cobalt catalysts under suitable reaction conditions [17–24], and the simple carbonylation to acetic acid and its esters generally predominates with rhodium [25] and nickel [26] catalysts. The occurrence and the extent of the side reactions as well as the effect of the reaction parameters on the formation of the different products will be discussed in the following sections.

2.1. Methanol Homologation

The hydrocarbonylation of methanol can be directed towards the production of either ethanol or of acetaldehyde and its derivatives.

The second route has been pursued particularly, since the impossibility of obtaining a satisfactorily high selectivity to ethanol, with usual catalytic systems, was ascertained. In fact no great problems exist for the successive hydrogenation of acetaldehyde and its derivatives to ethanol, the main disadvantage being the formation of aldol condensation high boiling products.

A great part of the research into methanol homologation has been devoted to improving the performance of the cobalt-based catalytic system which was the oldest system proposed. Unfortunately it is at times difficult make a quantitative comparison of the data reported in the literature owing to different modes used to define ethanol selectivity (Table 1).

Thus in addition to the correct expression for evaluating the selectivity to an individual product (relation *a*, Table 1), which accounts for the consumption of more than one molecule of substrate in the formation of acetals, esters, ethers, etc., some authors report the simple molar selectivity evaluated on the basis of relation (*b*) (Table 1). In other cases the selectivity to "realisable individual product" is defined as in the relation (*c*) (Table 1). Finally some authors report the yield of an individual product defined by the relation (*d*) (Table 1).

2.1.1. COBALT CATALYSTS

Historic development of the catalytic system

Monometallic systems. The first catalyst based on cobalt oxide was of low activity and required a very high syngas pressure (~600 atm) [1, 2]. The use of $Co_2(CO)_8$ introduced by Wender [3] allowed the reduction of pressure and temperature to 300 atm and 185°C, respectively, and to obtain a $\overline{TN}$ of 2.2×10^{-3} sec^{-1} for a methanol conversion of about 30% operating in batch and of 8×10^{-3} sec^{-1} in continuous operation [16] (Table 2).

The addition of an iodide promoter, proposed firstly by Berty [9] in 1956, increased the catalytic activity but also lowered the selectivity to ethyl derivatives.

The best results seem to be those claimed by Walker [27] working in dioxane with a Co/I ratio of 1; the selectivity to ethyl derivatives was of about 66% at a methanol conversion of 93%. However, these data did not account for the formation of methane and high boiling products.

The addition of phosphine ligands to the cobalt-iodine catalysts proposed by Riley [28] and successively optimized by Slaugh (Shell) [29], Gane (BP) [30] and Pretzer (Gulf) [31] allowed selectivities to ethanol of up to 70% at methanol conversions of 30–40% to be obtained. These systems showed a lower activity and generally required temperatures $\geqslant$ 200°C.

On the other hand, the possibility of a preferential production of acetaldehyde was the second object of efforts in the study of the methanol hydrocarbonylation (Table 3). It was first recognized that operating at low cobalt concentration [32] and low temperatures [17] acetaldehyde and its acetals were favoured with respect to ethanol. Thus successive researches using cobalt catalysts with iodide promoters and phosphine ligands succeeded in making a remarkable improvement in the selectivity to acetaldehyde up to 65–70%. However, the best selectivity, $\simeq$ 80%, and productivity of about 2000 g/(h $\times$ g of Co) seems to have been attained by Rhone-Poulenc [21] using a mixture of covalent (CH_3I) and ionic (KI) iodide promoters in large excess with respect to cobalt (I/Co = 20) and or adding a small amount of a ruthenium cocatalyst (Co/Ru = 10–25) [21].

Table 1

DIFFERENT RELATIONS USED TO EVALUATE THE SELECTIVITY

$$\text{Selectivity to an individual product, \%} = \frac{\text{(moles of the individual product) (no. of methyl groups)}}{\Sigma\text{(moles of product) (no. of methyl groups)}} \times 100 \qquad (a)$$

$$\text{Molar selectivity, \%} = \frac{\text{moles of the individual product}}{\text{total moles of new products arising from the substrate}} \times 100 \qquad (b)$$

$$\text{Selectivity to realizable individual product, \%} = \frac{\text{moles of the substrate converted into individual realizable product}}{\text{total moles of substrate converted}} \times 100 \qquad (c)$$

$$\text{Yield to an individual product, \%} = \frac{\text{moles of the individual product}}{\text{moles of substrate converted}} \times 100 \qquad (d)$$

Table 2

HOMOLOGATION OF METHANOL TO ETHANOL WITH COBALT CATALYSTS

Catalytic system (solvent)	Temp. (°C)	Pressure (atm)	MeOH conv. (%)	SELECTIVITY (%) [a] EtOH (total ethyl derivatives)	AcH + acetals	AcOH + esters	Others alcohols and ethers	Hydrocarbons (unidentified products)	TN × 10^3 ($sec.^{-1}$) [b]	Ref.
$Co_2(CO)_8$, batch (*d*)	185	230–350	76	39 (45)	n.d.	15	8.0	8.5 (29.5)	2	4
$Co_2(CO)_8$, batch (*b*)	190	260	30	37.5 (41.6)	7.2	17.4	23.3	11 (3.6)	2.2	16
$Co_2(CO)_8$, continuous (*b*)	190	330	31	51.6 (55.6)	12.8	8.3	21.4	5.9	8	16
$Co_2(CO)_8$ (*b*) (oxygenated solvents)	180	300	53	69 (70)	12.7	3.0	4.5	n.d.	0.8	45
$Co(OAc)_2$ (*c*) (tetrahydrofuran)	185	200	35	n.d. (72.6)	8.6	15	45.4 (from THF)	n.d.	3.9	46
$Co(OAc)_2$ (*c*) (methyl acetate)	185	200	24	n.d. (66.4)	n.d.	1.1	32.5	n.d.	1.8	47
$Co(C_6F_5COO)_2$	190	380	58	36 (44)	2.0	14.7	39.1 (others alcohols and ethers + hydrocarbons)		3.2	35
$Co(OAc)_2/I_2$ (*b*) 1/0.065	195	180–250	46	43	traces	32	9	n.d. (16)	9	9
$Co(OAc)_2/NaI$ (*b*) (acetic acid) 1/0.6	200	400–450	46	n.d. (44.3)	8.3	27.6	9.6	10.2	20	48
$Co(OAc)_2/I_2$ (*d*) 1/0.05	190	290	30	n.d. (72.6)	3.4	3	n.d.	n.d. (19.6)	n.d.	49

Table 2 (continued)

Catalytic system (solvent)	Temp. (°C)	Pressure (atm)	MeOH conv. (%)	SELECTIVITY (%)[a] EtOH (total ethyl derivatives)	AcH + acetals	AcOH + esters	Others alcohols and ethers	Hydrocarbons (unidentified products)	TN × 10^3 ($sec.^{-1}$)[b]	Ref.
$Co(OAc)_2/I_2$ (*b*) 1/0.5	190	200	93	69.5 (74.6)	5.6	19.8	n.d.	n.d.	5.4	27
$Co(OAc)_2/I_2$ (*c*) (dioxane) 1/0.5	185	200	66	n.d. (40)	19.7	13	4.5	n.d.	7	46
CoI_2/PBu_3 (*b*) 1/3.5	200	140	11.7	35.3 (70.7)		5.1	2.6	21.6	2.8	29
$Co(OAc)_2/I_2/PR_3$ or NR_3 (*c*) 1/0.2–1/1–5 (non-polar solvents)	205	200	47	n.d. (69.8)	5.1	6.8	<4	n.d.	5.3	30
$Co(acac)_2/I_2/P(OR_3)$ (*b*) 1/1/6	200	285	38.5	70.7	13.5	10.6	1.7	n.d. (3.5)	n.d.	31
$Co(OAc)_2/I_2/PP$ (*c*) (chlorobenzene) 1/0.5/0.9	205	200	65	56.2	11	11	4	n.d. (17.8)	7.3	19
CoI_2/PP (*b*) [c] 1/3	200	180	30	84	4.5	1	1.7	5.7 (4)	2.6	50
$C_5H_5Co(CO)_2/I_2/PPh_3$	190	200	67	n.d. (30.6)	9.5	14	n.d.	n.d. (40.5)	7.4	20

[a] The superscript italic letter in parentheses on the catalytic system (Column 1) indicates the relation used to evaluate the selectivity data (see Table 1).
[b] Moles of converted methanol/(moles of Co × sec).
[c] Selectivity data do not account for methane and high molecular weight products.

Table 3

METHANOL HYDROCARBONYLATION TO ACETALDEHYDE AND ITS ACETALS WITH COBALT CATALYSTS

Catalytic system	Temp. (°C)	Pressure (atm)	MeOH conv. (%)	SELECTIVITY (%)[a] EtOH (total ethyl derivatives	AcH + acetals	AcOH + esters	Other alcohols, ethers and esters	Hydrocarbons (unidentified products)	TN × 10^3 ($sec.^{-1}$)[b]	Ref.
$Co_2(CO)_8$ *(b)*	110	260	20	traces	66	34	n.d.	n.d.	0.66	17
$Co(OAc)_2/I_2$ *(d)* 1/1	185	300–400	48	n.d.	51.5	n.d.	n.d.	n.d.	50	32
$Co(OAc)_2/I_2/SbPh_3$ *(c)* 1/0.5/1.75	185	200	46	n.d. (4.5)	56	16.8	22.7 (other alcohols, ethers and esters + hydrocarbons)		10	18
$Co_2(CO)_8/CH_3I/KI$ *(b)* 1/2/34	200	250	19	n.d.	87	n.d.	<1	n.d.	730	21
$Co(OAc)_2/HI/PPh_3$ *(b)* 1/0.6/0.66	200	285	73	34	46	7.5	13.5	n.d.	130	22
$Co(OAc)_2/HI/PPh_3$ *(a)* 1/2/2	200	300	73	4.4	66	16.6 (AcOH + esters + other alcohols, ethers and esters)		n.d.	n.d.	23

[a] The superscript italic letter in parentheses on the catalytic system (Column 1) indicates the relation used to evaluate the selectivity data (see Table 1).

[b] Moles of converted methanol/(moles of Co × sec).

Bimetallic systems. The positive effect on the catalytic activity and selectivity to ethanol adding a second metal promoter based on a ruthenium or osmium halide to the cobalt-iodine system was firstly reported in a patent of the Commerical Solvents Co. [33]. On this line, other companies, Gulf [34], BP [20, 36], Rhone-Poulenc [37], Exxon [38] and Union Carbide [39] developed bimetallic systems based preferentially on cobalt and ruthenium derivatives in conjunction with iodide promoters and phosphines, arsines or stibines as ligands.

Cobalt or ruthenium were proposed as the main component and a large variety of mononuclear and cluster Co—Ru derivatives were used [38, 40]. With an optimized Ru/Co system, selectivities to ethanol of up to 70–80% at methanol conversion of 30–40% with $\overline{TN}$ of $1-10 \times 10^{-2}$ sec^{-1} were achieved [35–39].

Other VIII group metals added to cobalt systems as soluble complexes or used as mixed cobalt clusters affect the course of the reaction without significantly improving the selectivity to ethanol. Nickel [41] and platinum [43] compounds favoured the formation of acetaldehyde whereas iron, used as ionic cluster $[(C_4H_9)_4N][FeCo_3(CO)_{12}]$ [44] gave selectivity to ethanol of 70% operating at high temperature (220°C) and selectivity to acetaldehyde of 82% operating at low temperature (140°C).

The improvements and the limits of the performance of the catalytic systems due to the addition of promoters, ligands and other metal cocatalysts, are well summarized by the data, obtained under the same conditions, by Gane [36] (Table 4).

Effect of the reaction parameters

An unusually high number of parameters affects the course of the homologation process due to the complexity of the catalytic system and of the reaction mixture. These effects will be discussed in detail and then related to the steps and to the reaction mechanism in an attempt to rationalize the process.

Nature and concentration of the cobalt compound. Firstly it must be emphasized that the effect of the nature of the cobalt precursor on the reaction course is different for unpromoted catalytic systems and for systems with iodide promoters and/or ligands.

With unpromoted cobalt catalysts, the presence of $[Co(CO)_4]^-$ and of derivatives of oxidized cobalt probably containing both methanol and CO in the coordination sphere (for example $[Co(CH_3OH)_x(CO)_y]^{n+}$) as dominant species have been evidenced in the reaction medium by high pressure I.R. spectroscopy [31]. Moreover a direct correlation between the $[Co(CO)_4]^-$ concentration and the catalytic activity under homologation conditions has been shown and the effect of the nature of the starting cobalt compound has been related to the ability of the cobalt compound to form $[Co(CO)_4]^-$ [24]. Accordingly $Co_2(CO)_8$ allows one to obtain higher methanol conversions with respect to $Co_4(CO)_{12}$ [26] (Table 5).

Table 4

IMPROVEMENTS IN THE PERFORMANCE OF COBALT CATALYSTS BY ADDITION OF IODIDE PROMOTERS AND LIGANDS [36]

Catalytic system	MeOH conv. (%)	SELECTIVITY (%) [a]					Molar yield $CO_2 + CH_4$ (%)
		EtOH realizable	AcOH realizable	AcH + acetals	Higher alcohols	Unidentified products	
$Co(OAc)_2$	36.8	39.1	16	11	3.5	30.4	5.9
$Co(OAc)_2/RuCl_3$ 1/0.125	25.2	54.4	18.6	21	6	–	11.5
$Co(OAc)_2/I_2$ 1/0.5	45.0	36.9	23.3	13.3	<1	25.5	6.8
$Co(OAc)_2/RuCl_3/I_2$ 1/0.12/0.5	39.3	25.2	11.0	23	<1	39.8	48.4
$Co(OAc)_2/I_2/PPh_3$ 1/0.5/1.7	53.6	18.5	18.0	37.5	<1	25	8.8
$Co(OAc)_2/RuCl_3/I_2/PPh_3$ 1/0.12/0.5/0.21	43.0	64.4	19.7	9	<1	5.9	7.1

Reaction conditions: Temperature, 185°C; pressure, 200 atm; time, 2 h; cobalt concentration, 0.31–0.35 mol/liter.

[a] Selectivity evaluated according to relation (*c*) (see Table 1).

Table 5

EFFECT OF THE NATURE OF THE COBALT PRECURSOR IN THE METHANOL HOMOLOGATION WITH $CO + H_2$ OR $CO + H_2O$

Cobalt precursor	Reaction conditions	Methanol conversion (%)	SELECTIVITY (%)[a] Ethanol	Methyl acetate	Ethyl acetate	Ref.
$Co_2(CO)_8$ (*b*)	temperature: 180°C	32.5	4.6	13.5	20.6	
$Co_4(CO)_{12}$ (*b*)	pressure: 300 atm H_2/CO: 2/1	9.0	10.0	13.3	67.8	26
$Co(OAc)_2$ (*b*)	Cobalt concentration:	47.5	37	36	8.5	
$Co(acac)_2$ (*b*)	7.5×10^{-3} g atom/liter of methanol	20.8	42	20	6.7	
$Co_2(CO)_8$ (*d*) *	temperature: 260°C	42	29	14	n.d.	
$Co(OAc)_2$ (*d*) **	pressure: 190 atm $MeOH/H_2O$: 0.5	49	21	16	n.d.	51
CoI_2 (*d*) **	Na_2CO_3/Co: 28	50	23	24	n.d.	
$CoCO_3$ (*d*) **	NaI/Co: 1.6 Cobalt concentration: g atom/liter of methanol * 6.2×10^{-2} ** 3.1×10^{-2}	37	25	20	n.d.	

[a] The superscript italic letter on the catalyst indicates the relation used to evaluate the selectivity data (see Table 1).

However, this seems to be true only when the cobalt precursors are Co(O) derivatives. It is likely that cobalt is catalytically more active for the homologation reaction when present in a higher oxidation state, for example as a carbonyl Co(II) intermediate. This explains the higher activity of $Co(OAc)_2$ [26], $Co(CF_3COO)_2$ [35], and CoI_2 [51] with respect to that shown by $Co_2(CO)_8$ (see Table 3 and 5) and also the promoting effect on the activity caused by the addition of iodide derivatives such as HI, I_2 etc. to $Co_2(CO)_8$ which leads to the formation of Co^{2+} species such as $[Co(CH_3COO)_{4-n}I_n]^{2-}$ [52, 53].

Also the catalytic properties of a series of clusters, i.e., $Co_4(CO)_{12}$, $[Co_4I(CO)_{11}]^-$, $Co_3(\mu_3\text{-PPh})(CO)_9$, $Co_4(\mu_4\text{-PPh})(CO)_{10}$, $Co_2(\mu\text{-PPh}_2)_2(CO)_6$ and $Co_2(\mu\text{-PPh})(CO)_8$ were studied in the absence and in the presence of iodide promoters [23]. A low activity was observed in all cases without iodide promoters and the main product was 1,1-dimethoxyethane. Moreover at the end of the runs, a substantial decomposition of the cluster was observed with production of $[Co(CO)_4]^-$ fragments. These results discouraged the assumption of clusters being active species in the homologation of methanol even if the formation of some active intermediates by cleavage of metal—metal bonds could not be excluded.

The catalyst concentration generally used was in the range 0.005–0.3 moles of Co derivative/liter of methanol. By increasing the catalyst concentration from 0.05 to 0.20 M, the reaction rate and the conversion increased, however the order with respect to Co was not determined because of the tendency to deposition of solid products at high concentration which could not be quantified [16].

Iodide promoters. The positive effect on catalytic activity of adding iodide promoters observed for some carbonylation reactions (such the synthesis of saturated or unsaturated esters from olefins, acetylenes, CO and alcohols, or the synthesis of acetic acid from methanol and CO) was discovered in the B.A.S.F. research laboratories and reported in the Reppe works [54].

Regarding methanol homologation, Berty [9] and successively Mizoroki [48] and Walker [27] showed the important promoting effect of iodine which increases the reaction rate remarkably but generally decreases the selectivity to ethanol (Table 6).

The iodine derivatives proposed as promoters can be divided between simple iodine compounds (I_2 and HI), covalent iodides (alkyl or aryl iodides: CH_3I, C_2H_5I, PhI, etc.) and ionic iodides (metallic iodides: LiI, NaI, KI, CoI_2, etc.). Under reaction conditions, I_2 is transformed into HI which, in the presence of methanol, is in equilibrium with CH_3I (Equation 17) ($\log K_{200°C} = 5.07$ [53]).

$$CH_3OH_{(g)} + HI_{(g)} \rightleftharpoons CH_3I_{(g)} + H_2O_{(g)} \qquad (17)$$

Even the ionic iodides such as NaI in the presence of methanol and acetic acid can be transformed in part into methyl iodide (Equation 18) ($\log K_{200°C} = -0.67$).

$$CH_3OH_{(g)} + NaI_{(s)} + CH_3COOH_{(g)} \rightleftharpoons CH_3I_{(g)} + CH_3COONa_{(s)} + H_2O_{(g)} \qquad (18)$$

Table 6

EFFECT OF I/Co RATIO ON METHANOL HOMOLOGATION WITH COBALT CATALYSTS WITHOUT DONOR LIGANDS

Catalytic system and reaction conditions	I/Co	Methanol conversion (%)	SELECTIVITY (%) [a] Ethanol	Acetaldehyde	Methyl acetate	Ethyl acetate	Others	Ref.
$Co(OAc)_2/I_2$ (*b*)	0	50.6	69.2	1.9	24.3	4.6	–	
Temperature: 190°C	0.5	90	68.4	2.6	22.3	6.7	–	27
Pressure: 210 atm Time: 4 h	1.0	93	69.5	5.6	19.8	5.1	–	
$Co(OAc)_2/NaI$ (*b*)	0.25	58.8	46.7	8.5	21.2	–	23.6	
Temperature: 200°C	1.0	63.7	39.2	13.6	27.5	–	19.7	
Pressure: 300 atm Time: 1.5 h	2.0	59.8	31.2	18.7	22.9	–	27.2	48
H_2/CO: 1/1	9.0	52.5	9.5	30.8	26	–	33.7	
$Co(OAc)_2$ (*b*)	0	27	7.8	52.7	39.5	–	–	
CoI_2 (*b*)	2	46.4	0	68.1	31.9	–	–	29
Temperature: 160°C Pressure: 140 atm H_2/CO: 2/1								

[a] The superscript italic letter in parentheses on the catalyst indicates the relation used to evaluate the selectivity data (see Table 1).

The contemporaneous presence in the reaction mixture of CH_3I, HI and CoI_2 in the former case and CH_3I, NaI and CoI_2 with a very low amount of HI in the latter case, was evidenced both for Co [53, 55] and Ru systems [56].

From the data reported in the literature it is not easy to recognize the real effect of the nature and concentration of the iodide promoter especially on the selectivity of the reaction, which changes by changing the methanol conversion and is also affected by the presence of a donor ligand or of a metal cocatalyst.

However, the following points seem to be acquired for cobalt systems without donor ligands (Table 6): (i) comparable activity of I_2, HI and NaI at low reaction times [48]; (ii) loss of activity of the systems with HI due to the formation of considerable amounts of dimethyl ether [48, 55] formed by an acid-catalyzed dehydration process; (iii) high selectivity to ethanol with HI or CH_3I at low methanol conversions [55] and high selectivity to acetaldehyde and its acetals with ionic iodides [21, 55] and alkyl ammonium iodides [44]; (iv) high selectivity to ethanol at low I/Co ratios (optimal values in the range 0.2–2); (v) selectivity up to 90% and hourly productivity up to 1650 g/(g Co $\times$ h) of acetaldehyde and acetals and to acetic derivatives increasing the I/Co ratio (see Table 6) and using a mixture of ionic and covalent iodides (optimal values: I/Co = 5; CH_3I/KI = 0.01–0.1) [21].

For cobalt systems with donor ligands the results obtained indicate that: (i) only I_2, HI and alkyl iodides are effective promoters for the reaction, ionic iodides are almost inactive (Table 7). The phosphines prevent the nucleophilic attack of I^- on cobalt catalyst [23]; (ii) the reaction is lower than first-order in iodide concentration (0.6 dependence on I^-) over a I/Co range of 0.5–2 [31]; (iii) optimal values of I/Co are in the range 0.2–2; larger amounts of iodine reduce the selectivity to ethanol and also to ethanol + acetaldehyde and acetals.

The same effects due to the nature and concentration of the iodide promoters have been observed and emphasized using bimetallic Co–Ru systems in the absence or in the presence of phosphine ligands.

In particular, a positive effect both on the methanol conversion and on ethanol selectivity can be obtained by increasing the iodide promoter concentration in conjunction with the phosphine ligands [39] or by using an ionic iodide in excess [I/(Co + Ru) = 10–100] with an alkyl iodide/(Co + Ru) ratio of about 2 and with Ru/Co > 2.

The function of iodine as promoter in the homologation reaction is still unclear. Iodine, in fact, present in different forms in the reaction medium, may participate in several ways in the process, either by forming Co–I species more active in the carbonylation steps of the reaction [53], or by producing reactive intermediates from the substrate [24] but even by giving inactive cobalt iodide species, thus producing an inhibitory effect.

A detailed discussion of the role of iodine will be reported in the section dealing with the reaction mechanism.

In conclusion, the simple addition of the corrosive iodide promoters to cobalt systems is not decisive for the process and only the contemporary use of iodide

Table 7

EFFECT OF I/Co RATIO ON METHANOL HOMOLOGATION WITH COBALT CATALYSTS IN THE PRESENCE OF DONOR LIGANDS

Catalytic system and reaction conditions	I/Co	Methanol conversion (%)	SELECTIVITY (%)[a] Ethanol	Acetaldehyde + acetals	Acetates	Ethers	Others	Ref.
$Co(acac)_2/I_2$/tris-*p*-tolylphosphite [(*b*)]	0	n.d.	56.6	n.d.	n.d.	n.d.	n.d.	
Co/P: 6	0.125	n.d.	58.0	n.d.	n.d.	n.d.	n.d.	
Temperature: 200°C Pressure: 286 atm	0.5	63.1	53.5	18.5	15.8	6.7	5.5	31
H_2/CO: 1/1	1.0	n.d.	36.2	n.d.	n.d.	n.d.	n.d.	
Time: 3 h	5.0	n.d.	6.3	n.d.	n.d.	n.d.	n.d.	
$Co(OAc)_2/HI/P(C_6H_5)_3$ [(*a*)]	0	7.4	1.2	42	3.4	2.3	51.1	
Co/P: 0.5	1.0	62.0	6.1	65.2	7.2	2.0	19.5	
Temperature: 200°C Pressure: 300 atm	2.0	73	4.4	66	15	1.6	13	23
H_2/CO: 1/1 Time: 2 h	4.0	67	1.6	31.5	10	5.6	51.3	

[a] The superscript italic letter in parentheses on the catalyst indicates the relation used to evaluate the selectivity data (see Table 1).

promoters, phosphines and ruthenium cocatalysts can improve the catalysts' performance significantly.

Other promoters. No promoter was found to replace efficiently the corrosive iodine; the other halide derivatives are less effective in the order HI > HBr ≫ HCl ≃ HF; chloride and fluoride derivatives are virtually inactive [23].

Activating properties have also been reported in methanol homologation carried out in diols as solvents for cesium benzoate [57] and diammonium phosphate [28], but without any explanation of their role.

An interesting effect on the selectivity to ethanol, due to the nature of the cation added as iodide salt and to the presence of borates, was reported by Dumas [58]. Hard cations decreased both the activity and selectivity to ethanol; in contrast, the addition of borates increases activity and selectivity. The first effect is indicative of the importance in the catalysis of ionic species whose nature and concentration can be affected by the formation of ion-pairs promoted by the added salts. The second effect is related to the interaction of the borate ion with the acyl intermediate which increases its carbenoid character and also its reactivity towards hydrogen [58, 59].

Ligands. Modification of cobalt catalysts by ligands has a remarkable effect on the course of the reaction even if not so drastic as in the hydroformylation of olefins.

Electron donor ligands such as phosphines, phosphites, arsines and stibines stabilize the catalytic system and prevent its decomposition to metallic cobalt. This can be related to the higher σ-donor and lower π-acceptor character of these ligands with respect to carbon monoxide, resulting in an increase in electron density at the central atom and stabilization of the metal—CO bond via a stronger electron back donation [60]. As a consequence, a decrease in the catalytic activity is observed because of the more difficult dissociation of CO, necessary for the activation of the substrate. The higher electron density at the central atom, however, confers a stronger hydride character upon the catalyst which can improve the selectivity to hydrogenation products significantly. The addition of triphenylphosphine, for example, reduces the catalytic activity of $Co_2(CO)_8$ eight times and tributylphosphine inhibits the reaction [29].

The data reported in the literature on the effect of the nature and concentration of ligand on activity and selectivity of Co/I systems are often unclear and conflicting. However the following points seem to have been ascertained:

(1) *Type of ligand* – the activity and the selectivity of the system depends on the relative σ-donor and π-acceptor properties of the ligand: activity decreases with increasing the σ-donor and decreasing the π-acceptor power of the ligand; in an opposite way, the selectivity to ethanol and hydrogenated products increases. Thus $(n\text{-}C_4H_9)_3N$ completely inhibits the reaction probably due to the formation of a quaternary ammonium cation which leads to the precipitation of the salt $[(n\text{-}C_4H_9)_3CH_3N][Co(CO)_4]$ [31].

Triphenylphosphine is more active but less selective for ethanol with respect to tri-*n*-butylphosphine and tricyclohexylphosphine [30] (see Table 8). Moreover PPh_3, with π-acceptor properties like $AsPh_3$ and $SbPh_3$, but with a stronger σ-donor power is less active but more selective to ethanol than As and Sb compounds (Table 8) which on the contrary favour the formation of acetaldehyde and acetals [18]. $BiPh_3$, a very weak basic ligand, does not change the behaviour of the original catalyst significantly.

The systematic replacement of the phenyl group with alkyl groups in triarylphosphines causing an enhancement of the σ-donor and reduction of the π-acceptor properties of the ligands, results in an increase of ethanol selectivity at the expense of that of acetaldehyde [31]. Less basic phosphites reduce both the activity of the system and the selectivity to ethanol [31]. With diphosphines of type $(C_6H_5)_2P$-$(CH_2)_n$-$P(C_6H_5)_2$ (n = 1–6) the selectivity to ethanol increases with the increasing number of spacer -CH_2-groups (maximum selectivity 80% [19, 50]), whereas maximum selectivity to acetaldehyde and its derivatives is obtained using bis-diphenylphosphine-ethane [22].

(2) *P/Co ratio* – Apart from the expected negative effect on the catalytic activity of increasing the concentration of phosphine, no well-defined effect on the selectivity has been observed with a varying P/Co ratio in the range 0.5–3 [23, 30] (Table 9). In any case by increasing the amount of phosphine, the formation of acetic derivatives decreases with a simultaneous improvement of the selectivity to ethanol and especially to acetals [23, 30]. The optimum P/Co ratio recommended in the patents is in the range 0.5–2 [18, 19, 22, 29, 30].

In conclusion, phosphine ligands, especially diphosphines, improve the selectivity to ethanol (up to 80%), but depress activity and cause problems in recovery and recycling the catalyst. Moreover these ligands are not completely inert under reaction conditions and may undergo secondary reactions such as quaternization by iodide promoters, oxidation by aldehydes, phosphidation, etc. [61].

Solvents. The homologation reaction has been carried out mostly using pure methanol. However, several solvents have been used and claimed as media able to improve the catalytic activity and selectivity of the process. They can be classified as: (i) inert solvents and (ii) by-products of the reaction or their related products.

Hydrocarbons [29], chlorobenzene [19, 30], silicone oil [19, 30], and other inert compounds capable of forming a separate phase with methanol with up to 20% wt. of water, aliphatic, alicyclic or aromatic ethers such as di-*n*-propyl ether [46], tetrahydrofurane [19, 45, 46], dioxane [16], diphenylether [30], and acetone [18, 19], miscible in methanol belong to the class (i).

These inert solvents generally improve the selectivity to ethanol, reducing the amount of methyl acetate, without a significant decrease of activity (Table 10). However, these advantages are outweighed by serious disadvantages due to the difficulties in the products and catalyst recovery, in the separation of phases and

Table 8

EFFECT OF THE TYPE OF LIGAND ON METHANOL HOMOLOGATION TO ETHANOL WITH COBALT/IODINE CATALYSTS

Ligand (L)	Catalytic system	Co/L	MeOH conv. (%)	SELECTIVITY (%) [a] EtOH	AcH + acetals	AcOH + esters	Others	Ref.
$N(n\text{-}C_4H_9)_3$	$Co(acac)_2/I_2$ 1/0.25	1	–	–	–	–	–	31
$N(C_6H_5)_3$ (*c*) b	$Co(OAc)_2/I_2$ 1/0.5	0.6	53.9	38.7	7.2	23.5	30.6	30
$P(n\text{-}C_4H_9)_3$ (*c*) b	$Co(OAc)_2/I_2$ 1/0.5	0.6	60.3	47.8	12	11.4	28.8	30
$P(C_6H_5)_3$ (*c*) b	$Co(OAc)_2/I_2$ 1/0.5	0.6	64.7	44.8	17.4	15.6	22.2	30
$P(C_6H_{11})_3$ (*c*) c	$Co(OAc)_2/I_2$ 1/0.5	0.6	38.6	53.1	19.4	9.3	18.2	30
$P(C_6H_5)_3$ (*c*) c	$Co(OAc)_2/I_2$ 1/0.5	0.6	63.5	44.4	22.5	12.8	20.3	30
$P(C_6H_5)_3$ (*b*)	$Co(acac)_2/I_2$ 1/0.25	1.0	43.7	56.2	19.9	15.5	8.4	31
$As(C_6H_5)_3$ (*b*)	$Co(acac)_2/I_2$ 1/0.25	1.0	60.5	16.7	56.5	13.1	13.7	31

Table 8 (continued)

Ligand (L)	Catalytic system	Co/L	MeOH conv. (%)	SELECTIVITY (%) [a] EtOH	AcH + acetals	AcOH + esters	Others	Ref.
$Sb(C_6H_5)_3$ (*b*)	$Co(acac)_2/I_2$ 1/0.25	1.0	47.4	10.8	40.6	10.8	37.8	31
$Bi(C_6H_5)_3$ (*b*)	$Co(acac)_2/I_2$ 1/0.25	1.0	70.0	44.1	14.1	15.3	26.5	31
none (*b*)	$Co(acac)_2/I_2$ 1/0.25	–	70.9	45.2	21.8	17.3	15.7	31
$P(OC_6H_5)_3$ (*b*)	$Co(acac)_2/I_2$ 1/0.5	1.0	52.4	56.6	16.7	17.9	8.8	31
$(C_6H_5)_2P(CH_2)_6P(C_6H_5)_2$ (*c*)	$Co(OAc)_2$	0.57	50.6	65	12	5.0	18	19
$(C_6H_5)_2P(CH_2)_5P(C_6H_5)_2$ (*b*)	CoI_2	0.33	30.8	84.0	4.5	0.4	11.1	50

a The superscript italic letter in parentheses on the ligand indicates the relation used to evaluate the selectivity data (see Table 1).
b Methanol/chlorobenzene: 25.
c Methanol/chlorobenzene: 1.27.

Table 9

EFFECT OF P/Co RATIO ON THE METHANOL HYDROCARBONYLATION WITH COBALT-IODINE CATALYTIC SYSTEMS

Catalytic system and reaction conditions	P/Co	Methanol conversion (%)	SELECTIVITY (%)[a]				Ref.
			Ethanol	Acetaldehyde and acetals	Acetic acid and esters	Others	
$Co(OAc)_2/HI/P(C_6H_5)_3$ *(a)*	1	76	3.9	47	18	31.1	23
I/Co: 2	2	73	4.4	66	15	14.6	
Temperature: 200°C Pressure: 300 atm	3	71	3.1	68	11	17.9	
H_2/CO: 1/1	4	11	–	25	1.2	73.8	
$Co(OAc)_2/I_2/P(n\text{-}C_4H_9)_3$ *(c)*	0.5	60.1	40.9	15	9.6	34.5	30
I/Co: 0.5	0.9	47.3	54.8	12.4	8.4	24.4	
Temperature: 205 °C Pressure: 200 atm Solvent: chlorobenzene	1.75	51.9	50.7	10.2	6.5	32.6	

[a] The superscript italic letter in parentheses on the catalyst indicates the relation used to evaluate the selectivity data (see Table 1).

Table 10

EFFECT OF THE SOLVENT IN THE COBALT-CATALYZED METHANOL HOMOLOGATION

Catalytic system (mol/liter)	Solvent	MeOH/solvent (molar ratio)	Methanol conversion (%)	SELECTIVITY (%) [a]					Ref.
				Ethanol	Acetaldehyde and acetals	Acetic acid	Ethers	Others	
$Co(OAc)_2$ (0.31)	none	–	36.8	39.1	11.7	16	8	25.2	46
$Co(OAc)_2$ (0.23)	1,4-dioxane	6.66	31.5	54.9	16.0	20.9	7.6	0.6	46
$Co(OAc)_2$ (0.24)	tetrahydrofurane	6.66	33.2	55.1	12	23.8	7.5	1.6	46
$Co(OAc)_2$ (0.27)	acetone	6.66	40.4	39.9	17.4	17.3	6.9	28.5	46
$Co(OAc)_2/I_2/P(C_6H_5)_3$ (0.31)	none	–	45.1	41.9	24.1	7.3	n.d.	26.7	19
$Co(OAc)_2/I_2/P(C_6H_5)_3$ (0.28)	chlorobenzene	25	64.8	44.8	7.4	15.6	n.d.	32.2	19
$Co(OAc)_2/I_2/P(n\text{-}C_4H_9)_3$ (0.31)	none	–	52.8	41.7	13.25	13.8	3.6	27.6	30

$Co(OAc)_2/I_2/P(n\text{-}C_4H_9)_3$ (0.28)	silicone oil	9.8 [b]	47.9	53.9	12.1	16.3	2.3	15.4	30
$Co(OAc)_2$ (0.25)	methyl acetate	3	30.6	52.6	n.d.	16	n.d.	n.d.	47
$Co(OAc)_2$ (0.21)	acetic acid	3	21.6	66.2	n.d.	n.d.	n.d.	n.d.	47
$Co(OAc)_2$ (0.24)	*n*-propanol	6.66	41.3	40.2	7.2	15.7	4.6	32.3	46
$Co(OAc)_2$ (0.94)	*n*-butanol	1	32.6	52.8	12.6	11	3	20.6	46

a The selectivity data were evaluated according to the relation (*c*) (see Table 1).
b v/v ratio.

in a higher cost. Moreover, the real inertia of ethers and acetone is in doubt especially when they are used with a cobalt-iodine catalyst in acid medium.

The addition of solvents of the class (ii) to methanol (such as propionic and acetic acid, their esters or higher alcohols, by-products in the hydrocarbonylation reaction), has a positive effect on the course of the reaction. In fact, using an appropriate methanol/by-products ratio, it is possible to maintain the concentration of these latter practically constant so increasing the ethanol selectivity (Table 10) [19, 30, 46–48]. Moreover, these by-products do not complicate the separation of the products and the recycling of the catalyst which remains dissolved in the bottom products.

Other even more unusual solvents have been proposed for particular aims: α or β diols [57] to increase the reaction rate; methyl benzoate to prevent the formation of dimethyl ether [55]. However no practical use can be foreseen for these compounds.

The accelerating effect of some solvents is not well elucidated and probably is not the same for different classes of compounds. Polar solvents, in fact, may play a positive effect on the formation of the catalytically active species if this is an ionic derivative; the by-products may reduce the formation of themselves if they are involved in equilibrium steps of the catalytic cycle. Finally, inert unpolar solvents may affect the course of the reaction by changing the solubility of gases and of the liquid reaction products.

Metallic cocatalysts. Since the majority of the side-products arising from methanol homologation consist in acetaldehyde, acetals and aldol condensation products or in acetic acid and esters, an attempt to increase the selectivity to ethanol was made by adding compounds, active in the hydrogenation of carbonyl and carboxylic products, to cobalt catalysts.

Thus ruthenium compounds, known as the most active catalysts in homogeneous phase for this type of hydrogenation, were tested by several research groups and enabled an improvement of up to 80% in the selectivity to ethanol and ethyl derivatives (Table 11).

Ruthenium, generally supplied in a Ru/Co ratio of 0.1–0.3 as soluble organic or inorganic salt or as bimetallic cluster [38, 40, 44] strongly reduces the formation of acetaldehyde [55] and its derivatives but probably increases the hydrogenation to hydrocarbons (the patents however are reticent on this point).

The use of a Co–Ru system with Ru as the main component ($Ru/Co > 2$) and with a large excess of iodide promoters both as ionic iodide ($I^-/Co > 5$) and alkyl iodide ($CH_3I/Co > 2$) seems to improve the selectivity and the reaction rate [37].

Other VIII group metals added as soluble salts or as bimetallic Co-clusters do not give generally as satisfactory results in the ethanol selectivity as ruthenium derivatives (Table 12). Only Co–Fe clusters [44], $PtCl_4$ [42] and $OsCl_3$ [36] are sufficiently active and selective but not very practical.

The activity of different salts of $[FeCo_3(CO)_{12}]^-$ and of mixtures of $Fe(CO)_5$

Table 11

COBALT-RUTHENIUM SYSTEMS AS CATALYSTS FOR METHANOL HOMOLOGATION

Catalytic system	Temp. (°C)	Pressure (atm)	MeOH conv. (%)	SELECTIVITY (%) [a] EtOH (total ethyl derivatives)	AcH and acetals	AcOH and esters	Other alcohols, ethers and esters	Hydrocarbons (unidentified products)	TN × 10^3 (sec^{-1}) [b]	Ref.
$[Co(CO)_3PPh_3]_2/Ru(acac)_3/I_2$ (*b*) 1/0.12/0.5	200	215	44	72.1	–	9.6	6.5	n.d. (11.8)	72	34 35
$Co(OAc)_2/RuCl_3/I_2/PR_3$ (*c*) 1/0.12/0.5/1.8	190	200	53	68.4	3.5	7.2	2.5	n.d. (18.4)	5.3	36
$CoCp(CO)_2/RuCl_3/I_2/PR_3$ (*c*) 1/0.25/0.5/1	190	200	60	54	0.5	7	2.3	n.d. (36.2)	5.3	20
$CpRu(PPh_3)_2Co(CO)_4/CH_3I$ (*b*) 1/2	220	270	54	80	traces	5	13	1	180	38
$Co(OAc)_2/RuCl_3/I_2/PR_3$ (*b*) 1/0.03–0.3/0.5–25/1.5–10	180	210	31	80	n.d.	n.d.	n.d.	n.d. (20)	15	39
$[Et_4N][RuCo_3(CO)_{12}]$ (*d*)	180	120	41	51 (54)	5.4	1	n.d.	n.d. (42.6)	95	40

a The superscript italic letter in parentheses on the catalyst indicates the relation used to evaluate the selectivity data (see Table 1).
b Moles of converted methanol/[(moles of Co + Ru) × sec.].

Table 12

VIII GROUP METALS COMPOUNDS AS CATALYSTS FOR THE METHANOL HOMOLOGATION

Catalytic system	Temp. (°C)	Pressure (atm)	MeOH conv. (%)	SELECTIVITY (%)[a] EtOH (total ethyl derivatives)	AcH and acetals	AcOH and esters	Other alcohols, ethers and esters	Hydrocarbons (unidentified products)	TN × 10^3 (sec^{-1})[b]	Ref.
$Co(OAc)_2/Ni(OAc)_2/HI$ (*b*)	200	290	57.8	–	88.2	5.8	–	– (6)	–	41
$Co_2(CO)_8/Rh(acac)_3$ (*b*)			55	56.8	5.1	14.7	10.4	– (13)		43
$Co_2(CO)_8/Pd(acac)_2$ (*b*)			58.6	26.6	33.1	17	10.5	– (12.8)		43
$Co_2(CO)_8/Ir(acac)_2(CO)_2$ (*b*)			73.7	37	24	14.9	6.9	– (17.2)		43
$Co_2(CO)_8/Pt(acac)_2$ (*b*)			69.1	30.8	36.8	16	8.8	– (7.6)		43
$CoCO_3/PtCl_4/NaI/Ph_2PC_3H_6PPh_2$ (*b*) 1/0.03/0.4/1.3	185	500	55	72	1.6	1.3	11	11 (31)	9	42
$Co(OAc)_2/OsCl_3/I_2/P(C_6H_5)_3$ (*c*) 1/0.17/0.5/1.75	190	200	46.0	58.5	19.5	6	2.6	– (13.4)		36
$[(C_4H_9)_4N][FeCo_3(CO)_{12}]$ (*d*)	220	270	75	70	6	6	15	3	130	44
$PdCo_2(CO)_7(Ph_2PCH_2CH_2PPh_2)$ (*b*)	180	120	61	2.4	61.2	8.9	27.5	–	220	40
$PtCo_2(CO)_7(Ph_2PCH_2CH_2PPh_2)$ (*b*)	180	120	49	5.8	51.9	6.9	35.4	–	170	40
$RhCo_3(CO)_{12}$ (*b*)	180	120	52	3	53.6	14.9	28.5	–	180	40
$Cs[RuCo_3(CO)_{12}]$ (*b*)	180	120	43	58.9	5.1	9.3	26.7	–	150	40

a The superscript italic letter in parentheses on the catalyst indicates the relation used to evaluate the selectivity data (see Table 1).
b Moles of methanol converted/(moles of metals × sec).

and $Co_2(CO)_8$ in the presence or absence of $(C_4H_9)NI$ was tested and compared. All salts show about the same activity, higher than that shown by $Fe(CO)_5/Co_2(CO)_8$ mixtures. The addition of the ammonium salt to the mixtures of the carbonyls improves the activity which nearly approaches that of $[FeCo_3(CO)_{12}]^-$ salts [44]. However, the I.R. study indicates that the major species present in the solution were $Fe(CO)_5$, $HCo(CO)_4$, and $[Co(CO)_4]^-$, both when starting from the clusters and from cobalt and iron carbonyl mixtures. In any case, it remains unclear what the role is of the iron, which during the run and at the end of the reaction could only be detected as $Fe(CO)_5$.

Heterogeneous hydrogenation catalysts – such as rhenium on carbon, copper and zinc chromite, palladium and rhodium on carbon in combination with cobalt catalysts – have also been proposed [62, 63], but the performances of these systems are not superior to the homogeneous ones.

In conclusion, the use of mixed Co–Ru catalysts has the following advantages: (i) increase in catalytic activity with respect to that shown by the individual metal components; (ii) increase in the selectivity to ethanol and ethyl derivatives with reduction of the formation of acetaldehyde, acetals and products of aldol condensation; but it also has these disadvantages: (i) difficulties in catalyst recycling due to the different characteristics of the cobalt and ruthenium carbonyl and iodocarbonyl species; (ii) increase in the formation of hydrocarbons by complete hydrogenation of the substrate.

The strong synergic effect of the ruthenium, which increases not only the selectivity but also the reaction rate in comparison with individual cobalt and ruthenium systems, must be related to an intervention of the two metals at the level of the catalytic intermediates rather than to a successive action of ruthenium on the primary products produced by cobalt (i.e., acetaldehyde and acetals).

Reaction temperature. The reaction temperature has an important effect on catalytic activity, on catalyst stability and on reaction selectivity.

The catalytic activity, as is to be expected, increases with increasing temperature from 120°C, which is the minimum reaction temperature with cobalt catalysts, up to 200–220°C; then, depending on the catalytic system, on CO and H_2 partial pressures, a rapid decrease of activity is observed probably due to a decomposition of the catalyst.

For unpromoted cobalt catalysts, the decomposition of $Co_2(CO)_8$ and $HCo(CO)_4$ to metallic cobalt occurs at temperatures higher than 200°C for p_{CO} lower than 200 atm [64]. The cobalt-iodine or the cobalt–iodine–phosphine systems are more stable and resist up to 220–240°C under a p_{CO} of 100–150 atm [31].

Lower temperatures (120–130°C) favour aldehyde and acetal synthesis or methyl acetate formation depending on CO/H_2 ratio [17]. Practically no ethanol is produced at temperatures lower than 150°C [23, 44]. Higher temperatures favour the selectivity to ethanol and ethyl derivatives; however, temperatures $>$ 220°C must be avoided because the amount of hydrocarbon (methane and ethane) and heavy oxygenated products rapidly increases [44, 48]. Thus the optimum temperature range for ethanol formation is 180–210°C.

Pressure. Methanol homologation to ethanol must be carried out at high pressure, generally 250–400 atm, to avoid the decomposition of the cobalt carbonyl catalyst at the high temperature ($\simeq$200°C) required for the homologation to take place with a satisfactory selectivity.

The increase of the total CO + H_2 pressure generally leads to an increase of the methanol conversion but it has little effect on the selectivity above the threshold of catalyst stability [31, 44, 48].

With unpromoted cobalt catalysts, at pressures $>$ 900 atm, no further improvements in activity are observed [8] whereas during the initial stages of the reaction (methanol conversion $<$ 20%), the reaction rate is approximately first order with respect to total CO + H_2 pressure (CO/H_2 = 1) [31].

Moreover an increase of pressure favours the formation of higher alcohols and glycol derivatives [65, 66] and at the same time that of ethanol and acetaldehyde at the cost of acetals [44].

CO/H_2 ratio. The composition of syngas has a relevant effect on the rate and selectivity of the reaction.

A maximum of catalytic activity has been found at a CO/H_2 ratio of 1 using a cobalt-iodine catalytic system [48] and about this value has been claimed as the optimum ratio for all other cobalt-phosphine and cobalt-ruthenium systems [23, 34, 35].

Increasing percentages of CO favour the formation of products of simple carbonylation (acetic acid and its esters) whereas increasing percentages of hydrogen favour the formation of ethanol but also of acetaldehyde and acetals [16, 23, 48].

As a result, the highest ethanol selectivity is afforded using a CO/H_2 ratio of about 1 even if the stoichiometry of the homologation reaction requires a CO/H_2 ratio of 0.5. The selection of the optimum CO/H_2 ratio inevitably involves a compromise, since p_{CO} and p_{H_2} effects are interrelated and can often affect conversion and selectivity in an opposite way.

Reaction time. Considering the complicated nature of the reaction and the fact that successive and side reactions may take place, it is to be expected that reaction time will affect the product selectivity considerably.

Actually this effect has been clearly shown for a cobalt–iodine–phosphine system [23] and for cobalt-ruthenium and cobalt-iron systems [38, 44]. In all cases, both at low and high temperatures it has been seen that acetaldehyde and acetals are the dominant products at the early stage of the reaction [23, 38]. Their concentration reaches a maximum and then decreases, whereas the concentration of ethanol and products of condensation such as crotonaldehyde and butanal increases with time, decreasing only at very high methanol conversion due to the formation of products of further carbonylation.

On the other hand, methyl acetate is present from the start of the reaction and its selectivity remains practically constant in time, only decreasing after very long times and at high temperatures [44].

2.1.2. IRON CATALYSTS

A very poor catalytic activity of $Fe(CO)_5$ in the methanol homologation under typical hydrocarbonylation conditions (methanol convn.: 2% in 6 h at 220°C and 270 atm of H_2/CO = 1.5) was reported by Doyle [44] and compared with the surprisingly high activity of iron-cobalt clusters.

A different catalytic activity in methanol homologation is shown by iron carbonyls in the presence of a tertiary amine [67]. $[HFe(CO)_4]^-$ formed *in situ* from $Fe(CO)_5$ (Equation 19) catalyzes the homologation of methanol to ethanol with formation of CO_2 rather than H_2O (Equation 20).

$$Fe(CO)_5 + H_2 + NMe_3 \rightleftharpoons [HFe(CO)_4]^- + [NMe_3H]^+ + CO \tag{19}$$

$$CH_3OH + 2CO + H_2 \xrightarrow{[HFe(CO)_4]^-/NMe_3} C_2H_5OH + CO_2 \tag{20}$$

The reaction pathway is really more complicated since the methylating agent for $[HFe(CO)_4]^-$ is the tetramethylammonium cation, formed from methanol via methyl formate (Equations 21, 22, and 23)

$$CH_3OH + CO \rightleftharpoons HCOOCH_3 \tag{21}$$

$$HCOOCH_3 + NMe_3 \rightleftharpoons HCOO^- + Me_4N^+ \tag{22}$$

$$HFe(CO)_4^- + Me_4N^+ \longrightarrow MeFe(CO)_4 + Me_3N \tag{23}$$

Carbon dioxide arises from the decomposition of the formate anion (Equation 24)

$$HCOO^- + [Me_3NH]^+ \longrightarrow H_2 + CO_2 + Me_3N \tag{24}$$

and ethanol is formed through the carbonylation and hydrogenation of the methyl-iron derivative.

On the other hand this system is catalytically inactive for the simple carbonylation reaction to acetic derivatives. The data reported in Table 13 are also indicative of some drawbacks of this catalyst: (i) possibility of deactivation of $[HFe(CO)_4]^-$ due to the precipitation of $FeCO_3$; (ii) high consumption of CO necessary for the formation of methyl formate by base catalyzed methanol carbonylation; (iii) low reaction rate; (iv) formation of a considerable amount of methane and dimethyl ether; (v) the necessity of a large amount of trimethyl amine which is the reactive compound involved in the homologation reaction.

Table 13

METHANOL HOMOLOGATION WITH $Fe(CO)_5/Me_3N$ SYSTEM [67]

REAGENTS (moles/liter)			PRODUCTS (moles/liter)				
Methanol	Methyl formate	Trimethylamine	Methanol	Ethanol	Trimethylamine	TN × 10^3 (sec^{-1}) [a]	Ethanol yield (%) [b]
11.7	3.0	3.0	7.0	3.0	1.5	1.1	23.8
12.3	3.3	3.3	12.1	2.3	2.4	1.0	16.8
2.5 [c]	3.3	3.3	2.9	0.64	0.3	0.17	9.5
0	11.7	3.3	7.7	2.64	2.1	0.72	18.6

Reaction conditions: Temperature, 200° C; Pressure, 306 atm; CO/H_2, 3/1.

[a] Moles of methanol converted/(Moles of Fe × sec).

[b] Calculated as: Ethanol (moles/liter)/(methanol + methyl formate + trimethylamine) $(moles/liter)_{initial}$.

[c] In the presence of dyglime: 2.7 moles/liter.

2.1.3. NICKEL CATALYSTS

Little is known about nickel catalysts: only one patent deals with nickel-catalyzed methanol homologation [68], and the data reported are not sufficient to evaluate the methanol conversion and the ethanol selectivity.

2.1.4. RUTHENIUM CATALYSTS

Contrary to cobalt and iron, ruthenium carbonyls or their precursors without iodide promoters are not catalytically active for the methanol homologation [38, 56].

Solutions of $Ru_3(CO)_{12}$ or $Ru(CO)_5$ under high pressure of CO/H_2 (1000 atm) and temperature $> 200°C$ in alcohols catalyze the formation of methanol and alkyl formate from syngas [56, 69–71] and also of small amounts of ethylene glycol and its esters when carboxylic acids or molten quaternary phosphonium or ammonium salts are used as solvents [72, 73].

An iodide promoter supplied as CH_3I or NaI and the presence in the reaction medium of an acid component (HI directly added or formed by hydrolysis or hydrogenolysis of CH_3I) or a carboxylic acid are the essential requirements for the homologation reaction to take place with ruthenium catalysts [56, 74].

The catalytic activity for the ethanol formation is comparable to that of cobalt-iodine catalysts only if a mixture of covalent (CH_3I) and ionic (NaI) iodide promoters is used together with ruthenium compounds [74]; with methyl iodide alone, the most rapid reaction is the formation of dimethyl ether whereas with sodium iodide, the reaction proceeds very slowly [56] (Table 14).

Although the ruthenium systems are highly selective to C_2 products and particularly to ethyl ($\simeq$69%) and acetyl derivatives, considerable amounts of ethers (dimethyl, methylethyl and diethyl ether) due to the acidity of the medium, and methane and ethane, due to the high hydrogenation activity of the catalyst are formed [56, 75] (Table 14).

Apparently the same catalytic system, $Ru_3(CO)_{12}/CH_3I$, but really an anionic hydrido-Ru_3-cluster produced *in situ*, under a pressure of 400–800 atm in 1-methyl-2-pyrrolidone or sulfolane solution catalyzes the direct synthesis of ethylene glycol from CO and H_2 [76].

2.1.5. RHODIUM CATALYSTS

The papers dealing with rhodium-catalyzed methanol carbonylation to acetic acid, in emphasizing the extraordinary selectivity of this catalyst, exclude the formation of hydrogenation products (acetaldehyde and ethanol) even when H_2/CO mixtures (1/1) [77] or p_{H_2} of 40 atm [78] are used under the typical reaction conditions ($RhCl_3/CH_3I$ catalyst; temperature: 140–160°C; p_{CO}: 70–100 atm).

Deluzarche [26] working at higher temperatures (180°C) and with a 2/1

Table 14

RUTHENIUM CATALYZED METHANOL HOMOLOGATION

Catalytic system	Temp. (°C)	Pressure (atm)	Methanol conversion (%)	SELECTIVITY (%) [a]					TN × 10^3 (sec^{-1}) [b]	Ref.
				Ethanol (total ethyl derivatives)	Acetic acid and esters	Dimethyl ether, methylethylether and diethylether	Other alcohols, ethers and esters	$CH_4 + C_2H_6$		
$Ru(acac)_3$	220	270	very low	–	–	–	–	–	–	38
$Ru(acac)_3/CH_3I$ 1/3.3	200	380	56.1	33.7 (41)	–	65.1	1.2	–	39	75
$Ru(CO)_4I_2/NaI$ 1/10	200	150	34	9.5 (20)	2.5	59	–	29	0.8	56
$Ru(CO)_4I_2/CH_3I/NaI$ 1/1.2/10	200	380	95.5	28.4 (43)	27.9	28.4	12	3.3	60	75
$Ru(CO)_4I_2/NaI$ [c] 1/10	200	250	53	34 (55)	34.5	28	–	3.5	1.2	56
$Ru_3(CO)_{12}/CH_3I$ 1/38	180	120	18	8.1 (8.1)	2	89.9	–	–	64	41

[a] The selectivity data were evaluated according to the relation (*a*) (see Table 1).
[b] Moles of methanol converted/(moles of Ru × sec).
[c] In methyl acetate solution.

H_2/CO mixture in the presence of $Rh_4(CO)_{12}$ and $Rh_6(CO)_{16}$ as catalysts without iodide promoters, observed a considerable formation of ethanol and ethyl acetate. Only using a large excess of hydrogen in the H_2/CO mixture, for example 40–60/1, is it possible to obtain an ethanol yield of 20% [25]. These data however are not comparable with those of other catalytic systems being related to the carbon monoxide consumed and not to methanol. Moreover, even operating at high hydrogen content, acetic derivatives and ethers largely prevail over ethanol.

2.2. Homologation of Higher Alcohols

Several aliphatic and aromatic alcohols were used as substrates to be homologated; however, because most of the results are reported in papers of 1950–1960, no definite conclusions on conversion and product selectivity can be drawn.

2.2.1. ALIPHATIC ALCOHOLS

The homologation of higher aliphatic alcohols strongly differs from that of methanol in two respects: reaction rate and product distribution (Table 15). Concerning the reaction rate, a dramatic decrease of catalytic activity was observed passing from methanol to ethanol and higher linear homologs [9] (Table 16). Moreover the branched alcohols are more reactive than the straight chain ones [3, 9]. The easily dehydrated tertiary alcohols show the highest reactivity due to a different reaction pathway including the formation of an alkene (Equation 25), its hydroformylation to aldehydes and hydrogenation to the corresponding alcohols (Equation 26) [15].

$$R{-}\overset{\displaystyle R'}{\underset{\displaystyle CH_3}{C}}{-}OH \xrightarrow[-H_2O]{H^+} R{-}\overset{\displaystyle R'}{C}{=}CH_2 \tag{25}$$

$$R{-}\overset{\displaystyle R'}{C}{=}CH_2 \xrightarrow{CO + H_2} \begin{cases} R{-}\overset{\displaystyle R'}{C}HCH_2CHO \xrightarrow{H_2} R{-}\overset{\displaystyle R'}{C}HCH_2CH_2OH \\ R{-}\overset{\displaystyle R'}{\underset{\displaystyle CHO}{C}}{-}CH_3 \xrightarrow{H_2} R{-}\overset{\displaystyle R'}{\underset{\displaystyle CH_2OH}{C}}{-}CH_3 \end{cases} \tag{26}$$

It is interesting to note that the presence of water in the reaction system (20–40 mol%) strongly increases the reaction rate [79] suggesting that not only the activation mechanism of the substrate but also the nature and concentration of the catalytic intermediates, influenced by the composition of the reaction medium, determine the kinetics of the process.

As regards product distribution, in the homologation of higher alcohols containing β-hydrogen, branched products, arising from a dehydration-hydroformylation

Table 15

HOMOLOGATION OF ALIPHATIC ALCOHOLS

Alcohol	Catalyst	P (atm)	T (°C)	Conv. (%)	Products	Ref.
ethanol	Co	900	225	60	*n*-propanol 20%, *n*-butanol 4%, 2-methyl butanol 5%, diethylether 16%, various 17% [a]	8
ethanol + H_2O	$Co(OAc)_2/I_2$ Co/I = 2	400	220	66	*n*-propanol 41.3%, *n*-butanol 4.5%, main by-products: diethylether, ethyl acetate, isobutanol [b]	79
n-propanol	Co	1000	225	50	*n*-butanol 11%, *i*-butanol 4%, *n*-pentanol 3%, di-*n*-propyl ether 9%, ethylene glycol monopropyl ether 8% [a]	8
n-propanol + H_2O	$Co(OAc)_2/I_2$ Co/I = 3.8	240	200	10.1	*n*-butanol 32.2% [b]	79
2-propanol	Co	1000	225	66	ethers 25%, *i*-butanol 16%, *n*-butanol 18% [a]	8
n-butanol	Co	1000	225	38	ethylene glycol monopropyl ether 4%, *n*-C_5-alcohols 11% [a]	8
n-butanol	$Co(OAc)_2/I_2$ Co/I = 3.8	240	200	17.2	*n*-amyl alcohol 31.4% [b]	79
2-methyl propanol	Co	1000	225		*n*-butanol 1.1%, 2-methyl-1-butanol 0.8%, *n*-pentanol 0.5% [a]	8
2-butanol	Co	1000	225	70	first runnings 7%, 2-methyl-1-butanol 18%, *n*-pentanol 33%, C_6-alcohols 4% [a]	8
t-butanol	$Co(OAc)_2$	200	130	n.d.	3-methyl-1-butanol 51%, 2,2-dimethyl-1-propanol 10% [b]	80
t-butanol	$Co_2(CO)_8$	265	200	88.8	3-methyl-1-butanol 60%, 2,2-dimethyl-1-propanol 4%, isobutane 3%, isobutene 3% [b]	81
cyclohexanol	Co/Fe	1000	225	68	cyclohexylcarbinol 44% [a]	8
pinacol	$Co_2(CO)_8$	225	185	n.d.	pinacolone 17%, pinacolyl alcohol 4%, 3,4-dimethyl-1-pentanol 26% [b]	82

[a] Vol% of dry products.
[b] Molar yield (%).

Table 16

HOMOLOGATION REACTION RATE OF VARIOUS ALIPHATIC ALCOHOLS (CATALYST: $Co(acac)_2$: 0.025 mol.; I_2: 0.006 mol; T: 195–205°C; P: 200–250 atm) [9]

Alcohol	Maximum reaction rate (mmol gas/min)
Methanol	58
Ethanol	1.4
Isopropanol	1.2
t-butanol	140
Ethylene glycol	5

pathway (Equations 25, 26), are also formed besides the products of the linear increase of the chain.

The carbonylation at the β-carbon atom of an alcohol was demonstrated by Burns [83] who found that, among the products of ^{14}C-methanol homologation, *n*-propyl alcohol had an approximately equal distribution of the label between the 2- and 3-position carbon atoms. This was explained assuming that *n*-propanol was formed through the carbonylation of a symmetrical intermediate, probably ethylene derived from ethyl alcohol.

As a consequence of this behaviour, low selectivity in the single products is generally observed. However, any change in the activation mechanism of the substrate caused by a modification of the catalytic system or of the reaction medium may result in an increase of the selectivity up to an acceptable value. Thus, for instance, an initial addition of water to ethanol and *n*-propanol increases remarkably the selectivity to linear homologs [79].

2.2.2. AROMATIC ALCOHOLS

Only benzyl alcohol and a series of nuclear-substituted benzyl alcohols have been used as substrates in the homologation reactions (Table 17).

Interest in these compounds is related to the possibility of obtaining at low cost β-phenylethyl derivatives, valuable intermediates in the preparation of fragrances and perhaps useful for making styrene.

Benzyl alcohol, in the presence of unpromoted cobalt catalysts, undergoes homologation at a distinctly fast rate (about a half that of methanol); however, the reaction is accompanied by a remarkable hydrogenation to toluene [3, 5]. Improvements in selectivity to β-phenylethanol of up to 80% could be attained with cobalt–ruthenium iodide systems [86, 87] adding some water initially and also maintaining the conversion at a low level ($<40\%$). On the contrary Ru/I catalytic systems resulted

Table 17

HOMOLOGATION OF AROMATIC ALCOHOLS [a]

Alcohol	Conv. (%)	Homologated alcohol (%)	Hydrocarbon (%)	Relative reaction rate [b]	Ref.
benzyl alcohol	100	32	63	1	5
p-methoxybenzyl alcohol	100	44	16	1×10^4	84
p-hydroxymethyl benzyl alcohol	100	39	27	n.d.	84
p-methyl-benzyl alcohol	100	24	58	2×10^2	84
m-methyl-benzyl alcohol	100	36	52	50	84
p-t-butyl-benzyl alcohol	100	28	54	50	84
2,4,6-trimethyl-benzyl alcohol	100	18	58	n.d.	84
p-chloro-benzyl alcohol	69	16	41	0.8	84
m-methoxy-benzyl alcohol	44	2	23	0.3	84
m-trifluoromethyl-benzyl alcohol	22	–	5	0.01	84
methyl-phenyl carbinol	n.d.	30	70	n.d.	85
benzhydrol	n.d.	0	95	n.d.	85
triphenyl carbinol	n.d.	–	95	n.d.	85
1-naphthalene methanol	n.d.	–	72	n.d.	5

[a] Reaction conditions: catalyst: $Co_2(CO)_8$, T: 185°C, P: 238 atm, CO/H_2: 1/2.
[b] On the assumption that the reaction rate doubles for each 10° rise, this value indicates the ratio of time necessary for one mole of gas to be absorbed for mole of benzyl alcohol at 190°C to the time required for the other aromatic alcohols.

in exclusive production of hydrocarbons and traces of phenylacetic acid derivatives [88].

The presence of electron releasing groups in *meta* and *para* positions of the aromatic ring markedly increased the rate of both homologation and reduction to hydrocarbons. Selectivity to homologation products as compared to reduction ones increased according to the increase in electron release capability (Table 17).

On the contrary, electron-attracting groups strongly deactivated the substrate: *p*-nitrobenzyl alcohol practically did not undergo homologation.

Secondary and tertiary aromatic alcohols were mainly reduced rather than homologated and the same occurred for 1-naphthalene ethanol [5, 85, 89].

3. CARBONYLATION AND HOMOLOGATION OF ETHERS

The reaction of ethers with CO and H_2 leads to higher homologous alcohols and esters rather than to their homologous derivatives.

Only a few examples are reported in the literature of simple carbonylation of non-cyclic ethers to the corresponding esters and carboxylic acids, if sufficient water is present (Equations 27 and 28) [13, 90].

$$R{-}O{-}R + CO \longrightarrow R{-}COOR \tag{27}$$

$$R{-}O{-}R + CO + H_2O \longrightarrow 2R{-}COOH \tag{28}$$

No advantage, however, may be foreseen in this reaction with respect to the carbonylation carried out on the corresponding alcohol which leads to the same products (Equations 29 and 30).

$$R{-}OH + CO \longrightarrow R{-}COOH \tag{29}$$

$$2R{-}OH + CO \longrightarrow R{-}COOR + H_2O \tag{30}$$

The picture, however, is quite different when the homologation of the alkyl moiety of the ether takes place: in fact from one mole of an ether two moles of the homologous alcohol may be obtained with production of only one mole of water instead of two formed when starting from the corresponding alcohol (Equations 31 and 32).

$$R{-}O{-}R + 2CO + 3H_2 \longrightarrow 2R{-}CH_2OH + H_2O \tag{31}$$

$$2R{-}OH + 2CO + 4H_2 \longrightarrow 2R{-}CH_2OH + 2H_2O \tag{32}$$

Thus, since the separation of water from the homologation products is generally an expensive and not easy process, the possibility of utilizing industrially available ethers instead of alcohols seems to be an interesting point. In particular, dimethyl ether, a by-product of methanol synthesis, easily prepared either from methanol or directly from carbon monoxide and hydrogen [91], is a very interesting substrate.

No explicit reference on the direct homologation of ethers with cobalt catalysts is reported in the literature. However, from the data on methanol homologation with cobalt-iodine systems, where dimethyl ether is formed, it is possible to infer that this substrate is only very slightly active for successive carbonylation reactions [48].

Thus the ability of ruthenium systems to catalyze the homologation of dimethyl ether and higher ethers to higher homologous alcohols and their esters seems to be peculiar to this metal [74].

The carbonylation and homologation of dimethyl ether took place in inert solvents (toluene, dioxane, etc.) [56] and more rapidly in acetic acid–methyl acetate solutions using a ruthenium catalyst with an iodide promoter (Table 18) [74, 90, 93]. Together with the products of simple carbonylation, which are methyl acetate and acetic acid (Equations 33 and 34),

$$CH_3{-}O{-}CH_3 + CO \longrightarrow CH_3COOCH_3 \tag{33}$$

$$CH_3{-}O{-}CH_3 + 2CO + H_2O \longrightarrow 2CH_3COOH \tag{34}$$

a significantly high conversion into ethyl acetate and ethanol, arising from the reductive carbonylation of the substrate, was observed (Equations 35 and 36).

$$CH_3{-}O{-}CH_3 + 2CO + 2H_2 \longrightarrow CH_3COOC_2H_5 + H_2O \tag{35}$$

$$CH_3{-}O{-}CH_3 + 2CO + 3H_2 \longrightarrow 2CH_3CH_2OH + H_2O \tag{36}$$

The presence of hydrogen (3–5 atm) was required to start the reaction; with CO alone not even the simple carbonylation occurred, the presence of H_2 and/or an alcohol being necessary to cleave the acyl intermediate involved in the catalytic cycle (see Scheme 1) [93].

Working in acetic acid–methyl acetate solution with an appropriate ether/acid/ester molar ratio and with CO and H_2 partial pressures which allowed a balanced production of carboxylate and ethyl groups, it was possible to convert the ether mainly into ethyl acetate (selectivity up to 60–74%) [74, 92, 93].

The effect of the different reaction parameters on reaction rate and selectivity of the process has been studied in acetic acid–methyl acetate solution and the most significant results are summarized in Table 19.

In contrast to methanol carbonylation to acetic acid with rhodium iodide catalysts [94], the carbonylation and homologation of dimethyl ether with ruthenium iodide catalysts were not first-order dependent on methyl iodide concentration and the course of the reaction was different using CH_3I or NaI as promoter, the former being more specific for the homologation steps of the reaction and the latter for the carbonylation ones [56, 93].

Non-cyclic ethers of higher molecular weight alcohols were more easily carbonylated than homologated with respect to dimethyl ether under comparable reaction conditions (Table 20). Accordingly, using for instance methyl ethyl ether, the production of more propionic than acetic derivatives emphasized the higher reactivity of ethyl than the methyl group for the simple carbonylation, the opposite occurring for the homologation reaction.

Table 18

CARBONYLATION AND HOMOLOGATION OF DIMETHYL ETHER WITH RUTHENIUM CATALYSTS[a] [56]

Substrate (mmol)	P_{CO} initial P_{H_2} initial (atm)	Conv. (%)		SELECTIVITY (%)[b]						
		Me_2O	AcOH	C_1	C_2	AcOMe	AcOH	AcOEt	Heavy prod.	$CH_4 + C_2H_6$
Me_2O in toluene (125) (25)	125 25	69	–	0.9	0.8	52	15	19.3	3.8	8.2
Me_2O and AcOH (100) (430)	135 30	74	20	0.7	0.7	45.2	–	45	2.4	6.0
Me_2O, AcOH and AcOMe (100) (250) (194)	123 27	58	8	5.4	3.6	18	64	–	1.5	7.5

a Reaction conditions: T = 200°C; $Ru(acac)_3 = 1.5 \times 10^{-2}$ M; I/Ru = 10.
b The selectivity data were evaluated according to the relation (*a*) (see Table 1).

Table 19

EFFECT OF DIFFERENT REACTION VARIABLES ON CARBONYLATION-HOMOLOGATION OF DIMETHYL ETHER [93] [a]

VARIABLE	PROMOTER	REACTION RATE			SELECTIVITY		
		Homologation (ethyl groups)	Carbonylation (acyl groups)	Hydrogenation (hydrocarbons)	Ethyl Acetate	Methyl Acetate	Hydrocarbons
P_{H_2} increase	CH_3I	+	–	+	+	–	+
	NaI	+	+	+	+	disappears	–
P_{CO} increase	CH_3I	Maximum at 20–40 atm	+	Maximum at 40 atm	±	Minimum at 20 atm	+
Concentration of Me_2O, increase [b]	CH_3I	±	+	±	–	+	–
I/Ru increase	CH_3I	Maximum at I/Ru: 50	+	+	–	+	+

+: increase; –: decrease

a Reaction conditions: $Ru(acac)_3$: 1.5×10^{-2} M; $Me_2O/AcOH/AcOMe$: 1/1/1; T: 200°C; I/Ru: 10.

b In acetic acid solution.

Table 20

CARBONYLATION AND HOMOLOGATION OF HIGHER MOLECULAR WEIGHT NON-CYCLIC ETHERS IN TOLUENE [56] [a]

Groups (mmol.) produced by:	Carbonylation		Homologation		Hydrogenation	
CH_3OCH_3	acetyl	84	ethyl	27	$CH_4 + C_2H_6$	44
	propionyl	1				
$CH_3OC_2H_5$	acetyl	33	ethyl	9	$CH_4 + C_2H_6$	42
	propionyl	79	propyl	–		
$CH_3OC_4H_9$ [b]	acetyl	16	ethyl	3	$CH_4 + C_2H_6 + C_4H_{10}$	13
	pentanoyl	56	pentyl	–		
$C_2H_5OC_2H_5$	propionyl	143	propyl	13	$C_2H_6 + C_3H_8$	64

[a] T: 200°C; P: 250 atm; CO/H_2 : 2/1 ; $Ru(acac)_3/CH_3I$: 1/10; ether: 120 mmoles; reaction time: 28 h.
[b] 12.5 h.

4. HOMOLOGATION OF ALDEHYDES

Aldehydes, mainly formaldehyde and its acetals, were used as substrates for the hydrocarbonylation reaction to obtain the more valuable aldehydo-alcohols (Equation 37) rather than the higher homologous aldehyde (Equation 38).

$$HCHO + CO + H_2 \longrightarrow \begin{array}{l} CH_2OH \\ | \\ CHO \end{array} \qquad (37)$$

$$HCHO + CO + 2H_2 \longrightarrow CH_3CHO + H_2O \qquad (38)$$

The difficulty of performing the reaction with a satisfactory rate and selectivity is mainly due to the easy hydrogenation of the aldehyde to alcohol and to the formation of aldol condensation products. Thus the characteristics required of the catalyst are: high carbonylation and low hydrogenation activity for the carbonyl compounds.

4.1. Formaldehyde

Cobalt and mainly rhodium catalysts with the addition of suitable promoters and ligands, under well-controlled temperature and pressure conditions in amide solvents, have the required requisites for the hydrocarbonylation of formaldehyde with a satisfactory selectivity to glycolaldehyde (Equation 37).

The hydrocarbonylation with cobalt carbonyls as catalysts of formaldehyde to a mixture of polyhydroxy compounds including ethylene glycol, glycerol and higher polyols was reported by Gresham in 1948 [95]. Attempts to drive the reaction towards the production of glycolaldehyde gave unsatisfactory results when using cobalt catalysts (Table 21). Only the stoichiometric hydroformylation with $HCo(CO)_4$ at 0°C gave glycol aldehyde yields in the range 60–90% based on $HCo(CO)_4$ consumed [98] whereas better yields were obtained with rhodium derivatives.

An optimization of the performance of the rhodium catalytic systems has been obtained through a detailed study of the effect of the reaction parameters and of the role played by promoters, ligands and solvents on the rate and selectivity of the reaction (Table 22) [99–101].

Thus cationic Rh(I) phosphine complexes were recognized as the most active catalysts [101]. *N,N*-disubstituted amides were the most useful solvents to attain high formaldehyde conversions [100]; optimum temperatures, and pressures (CO/H_2 = 1) were in the range 110–125°C and 130–160 atm, respectively [100]. The addition of phosphine ligands was essential for a high efficiency of the catalyst but the phosphine to rhodium ratio must be maintained at < 3 since an excess of the ligand suppressed the hydroformylation reaction [100, 101]. In contrast, an organic base, i.e., triethylamine, markedly increased the hydroformylation rate and permitted operation with catalytic systems stabilized by an excess of phosphine [101]. Amines also promote the reaction in a variety of organic solvents with a striking effect

Table 21

HYDROCARBONYLATION OF FORMALDEHYDE WITH Co-CATALYSTS

Catalyst	Conditions (T, P, solv.)	Products (yield %)[a]	Ref.
Co oxide	150°C, 500–700 atm, in HCOOMe	Ethylene glycol, glycerol, methanol and poly-hydroxy derivatives	95
$Co_2(CO)_8$	110°C, 200 atm, in acetone and water	glycol aldehyde (20)	96
$Co_2(CO)_8$	120°C, 200 atm, in dioxan, dimethyl acetamide and water	glycol aldehyde (50)	96
$Co_2(CO)_8 + Rh_2O_3$	150°C, 210 atm, in ethanol	2-ethoxy ethanol + ethylene glycol (35)	97
$HCo(CO)_4$	0°C, 1 atm CO in CH_2Cl_2	glycol aldehyde (60–90)	98

[a] With respect to the charged formaldehyde.

Table 22

RHODIUM-CATALYZED HYDROFORMYLATION OF FORMALDEHYDE

Rhodium complex	Ligand (P/Rh)	Base (N/Rh)	Solvent (conc. of CH_2O)	P (atm) (CO/H_2)	T (°C)	conv. (%)	SELECTIVITY[a] Glycol aldehyde	MeOH	Ethylene glycol (by-products)	Ref.
$Rh(CO)_2(acac)$	–	–	*N*-methyl pyrrolidone (2 M)	178 (4/1)	130	92.8	51.4	26.8	–	99
$RhCl(CO)(PPh_3)_2$	– (2)	–	*N*-methyl pyrrolidone (1.9 M)	180 (4/1)	140	98.7	65.4	3.9	3.9	99
$RhCl(CO)(PPh_3)_2$	– (2)	–	Dimethyl formamide (1 M)	80 (1/1)	110	n.d.	23.5	0.4	n.d.	100
$RhCl(CO)(PPh_3)_2$	PPh_3 (3)	–	*N-N*-Dimethyl acetamide (2 M)	180 (1/1)	110	82	83	4	n.d.	101
$RhCl(CO)(PPh_3)_2$	PPh_3 (30)	–	*N-N*-Dimethyl acetamide (2 M)	180 (1/1)	110	44	43	11	n.d.	101
$[Rh(COD)(PPh_3)_2]BF_4$	– (2)	–	*N-N*-Dimethyl acetamide (1 M)	130 (1/1)	110	83	73	7	n.d.	101
$[Rh(NBD)(PPh_3)_2]BF_4$	PPh_3 (32)	NEt_3 (4)	*N-N*-Dimethyl acetamide (2 M)	180 (1/1)	110	91	67	3	n.d.	101
$[Rh(COD)(PPh_3)_2]PF_6$	– (2)	NEt_3 (4.4)	Acetone (2 M)	285 (1/1)	110	94	80	10	n.d.	101
$RhCl(CO)(PPh_3)_2$	– (2)	NEt_3 (90)	*N-N*-Dimethyl acetamide (2 M)	180 (1/1)	110	95	17	6	– (77)	101

[a] The selectivity data were evaluated according to the relation (*d*) (see Table 1).

[101]. The base concentration however was very critical for selectivity to glycolaldehyde since aldol condensation might prevail in a more basic medium [101].

In conclusion, with an appropriate choice of metallic precursor, solvent, and level of phosphine and amine concentration, it is possible to maintain the catalytic system's activity and obtain high formaldehyde conversions (80–90%) with selectivity to glycolaldehyde of 70–90%.

Since the formation of glycoaldehyde is usually completed in relatively short reaction times, to complete the synthesis of ethylene glycol, it was proposed that the hydrogenation of the aldehyde be carried out with the same rhodium catalyst in a second stage under more drastic conditions [99].

In conclusion, this process presents heavy drawbacks among which are the formation of high molecular weight by-products arising from the aldol condensation of glycolaldehyde with formaldehyde, which are difficult to separate, the operation taking place in diluted solution, using high boiling, expensive and not completely inert solvents, and an easy catalyst deactivation requiring a difficult recycle.

Ruthenium-carbonyl catalysts with chloride or bromide promoters in aprotic organic amides at 120–150°C and 300 atm drive the hydrocarbonylation of formaldehyde towards acetaldehyde (Equation 38) (maximum selectivity 25%) and ethanol (maximum selectivity 20%), the main side-product being methanol formed by hydrogenation of formaldehyde [102]. However, no advantages are seen at present to using this route if it is compared with methanol homologation to the same products.

4.2. Acetals

By carrying out hydrocarbonylation on acetals and hemi-acetals of aldehydes (i.e., methylal, dimethyl acetal etc.) with cobalt-ruthenium systems used for methanol homologation, products of hydrogenation and homologation of the parent aldehyde (methanol, dimethyl ether, acetaldehyde, ethanol, ethylene glycol), together with products of alcohol carbonylation (methyl and ethyl acetate) were obtained [103].

The excessively wide range of products, however, discouraged the use of these substrates as useful starting materials for the preparation of valuable chemicals.

Aromatic acetals, under hydroformylation conditions, gave mainly aryl-alkyl ethers as products of a reductive cleavage (Equation 39) [104].

$$PhCH(OR)_2 \xrightarrow[CO/H_2:\ 200\ atm;\ 160^\circ C]{Co_2(CO)_8} PhCH_2OR + ROH \qquad (39)$$

5. HOMOLOGATION OF CARBOXYLIC ACIDS

If the general Scheme 1 is considered, the homologation of a carboxylic acid to the higher homologous one (Equation 40) occurs through a primary hydrogenation of the starting acid to an alcohol followed by its carbonylation to acid.

$$C_nH_{2n+1}COOH + (m-n)CO + 2(m-n)H_2 \longrightarrow C_mH_{2m+1}COOH + (m-n)H_2O \qquad (40)$$

Thus suitable catalysts for this type of process must have the rather uncommon capability of reducing, in homogeneous phase, carboxylic acids and their derivatives to alcohols. Up to now only ruthenium derivatives are reported to be active for this reaction [105–109] and actually ruthenium-iodide systems have been claimed as the more suitable catalysts for the homologation of aliphatic carboxylic acids [72, 110–112] (Tables 23 and 24). Also rhodium-iodide systems seem to be active and more selective for this reaction even if they require higher $CO + H_2$ pressures and higher I/Rh ratios [112] (Table 23). Palladium and nickel catalysts were generally less active and selective [112].

From a systematic study of the reaction parameters in the ruthenium catalyzed process the following statements have been drawn [12, 111]:

(i) a broad range of ruthenium oxides, salts, carbonyl and hydrocarbonyl compounds are effective catalyst precursors for acid homologation, whereas ruthenium complexes with bulky stabilizing ligands (i.e., triphenylphosphine) are generally less active;
(ii) alkyl iodides or hydrogen iodide are the best iodide promoters;
(iii) a first order dependence of the rate on the initial ruthenium concentration, for a I/Ru of about 10, is valid up to a concentration of 70 mM;
(iv) a maximum of yield of homologous acid is achieved with a CO/H_2 ratio of 1, even though the stoichiometry of the reaction calls for 2 moles of H_2 per mol of CO (Equation 40);
(v) the homologation reaction may even be carried out at low pressure (50 atm) but the selectivity to propionic acid reaches a maximum at about 300 atm, whereas the selectivities of higher acids (such as butyric and valeric) continue to increase at least up to 500 atm;
(vi) the homologation of higher straight-chain acids, i.e., propionic, valeric, mainly yields the acid containing one additional carbon atom per molecule, the linear-chain isomer being predominant over the branched-chain;
(vii) the homologation of branched-chain aliphatic acids is often accompanied by a substantial skeleton rearrangement with a tendency to produce tertiary acids.

The main side reactions, which lower the acid's selectivity, are the reduction of the substrate to hydrocarbons and the formation of esterification products. The quantity of these last by-products can be reduced maintaining a low acid conversion and a high water concentration (7–30% wt.) in the reaction mixture. As a consequence a remarkable formation of CO_2 occurs due to the water gas shift reaction catalyzed by the same ruthenium and rhodium systems [56, 113] and expensive fractionation processes are necessary to break the azeotropes with water.

A possible flow-sheet of the process for making C_3—C_5 aliphatic carboxylic acids from acetic acid and syngas has been proposed and an estimate of the cost demonstrated that it could be of practical interest [112].

Table 23

HOMOLOGATION OF ACETIC ACID [112]

Catalytic system (molar ratio)	Reaction conditions Pressure (atm)	Temperature (°C)	Conv. (%)	TN (sec^{-1}) [a]	SELECTIVITY (%) [b] Propionic acid	Butyric acid	Valeric acid	Butyric acids: (n/iso ratio)	Ref.
RuO_2/CH_3I (1/1)	272	220	52	1.6×10^{-3}	71.1	13.2	1.9	1.6	12
$Rh(acac)_3/CH_3I$ (1/20)	410	220	46	2.9×10^{-3}	75	21	1.8	1.3	112
$Pd(OAc)_2/PPh_3/CH_3I$ (1/8/31)	410	220	46	n.d.	26	1.9	–	0.6	112
$Ni(OAc)_2/PPh_3/CH_3I$ (1/3/28)	410	220	4	n.d.	75	13	–	2.0	112

a Moles of acetic acid converted/(moles of metal × sec).

b The selectivity data were evaluated according to relation (*d*) (see Table 1); hydrocarbons not accounted.

Table 24

HOMOLOGATION OF ALIPHATIC CARBOXYLIC ACIDS IN THE PRESENCE OF $RuO_2 \cdot H_2O/MeI$ AT 220°C and 270 atm [12]

Acid	Conv. (%)	T.N. sec^{-1} [a]	Major acid homologues formed Composition	mmoles	n/iso ratio
C_2H_5COOH	69	1.3×10^{-3}	C_3H_7COOH	35	8.5
$(CH_3)_2CHCOOH$	45	1.1×10^{-3}	$(CH_3)_3CCOOH$	26	–
n-C_4H_9COOH	67	1.3×10^{-3}	$C_5H_{11}COOH$	35	4.2
$(CH_3)_2CHCH_2COOH$	43	n.d.	$(CH_3)_2CH_2CH_2COOH$	8	–
			$CH_3CH_2C(CH_3)_2COOH$	7	–
$C_2H_5CH(CH_3)COOH$	48	n.d.	$C_2H_5CH(CH_3)CH_2COOH$	12	–
			$C_2H_5C(CH_3)_2COOH$	16	–
$(CH_3)_3CCOOH$	43	n.d.	$CH_3CH_2C(CH_3)_2COOH$	26	–

[a] Moles of acid converted/(moles of Ru × sec).

6. HOMOLOGATION OF CARBOXYLIC ACID ESTERS

One of the few new developments in industrial organic chemistry in the last five years is the Halcon process of carbonylation of methyl acetate to acetic anhydride (Equation 41): the first commercial plant (250 000 t/yr) is being built by Tennessee Eastman at Kingsport and it is to start up in 1983 [114].

$$CH_3COOCH_3 + CO \xrightarrow{[Rh]} (CH_3CO)_2O \tag{41}$$

A methyl acetate/dimethyl ether reductive carbonylation process to ethylidene diacetate (Equation 42), intermediate for the production of vinyl acetate, is technologically ready for commercialization (Equation 43) [114].

$$2CH_3COOCH_3 + 2CO + H_2 \longrightarrow CH_3CH(OCOCH_3)_2 + CH_3COOH \tag{42}$$

$$CH_3CH(OCOCH_3)_2 \longrightarrow CH_2{=}CHCOOCH_3 + CH_3COOH \tag{43}$$

The homologation of methyl acetate to ethyl acetate carried out in the presence of ruthenium catalysts with selectivity of 50–70% (Equation 44) [92, 115] shows interesting practical prospectives.

$$CH_3COOCH_3 + CO + 2H_2 \longrightarrow CH_3COOC_2H_5 + H_2O \tag{44}$$

All the three examples are indicative of the great interest in the carbonylation and hydrocarbonylation reactions on esters which were for a long time disregarded as not being useful reactive substrates. In reality the homologation of an ester, as shown in Scheme 1, is not a simple reaction as side and/or successive carbonylation and hydrogenation steps both on the alkyl and acyl group of the substrate are possible. Accordingly the homologation of *n*-butyl formate and *n*-butyl acetate gives products of hydrogenation, carbonylation and homologation of the acyl group such as methyl, acetyl, ethyl and propionyl derivatives together with products of carbonylation and homologation of the *n*-butyl group such pentanoyl and *n*-pentyl derivatives [116, 117].

6.1. Homologation of Methyl Acetate

Most of the work done in the field of ester homologation has been devoted to the homologation of methyl acetate or methyl acetate/acetic acid mixtures which are easily available products from the methanol carbonylation process.

The product distribution is dependent on the hydrogenation/carbonylation properties of the catalytic system used (Table 25). Thus cobalt-iodine systems exhibit the same drawbacks as in methanol homologation such as low activity and formation of large amounts of homologation and condensation products with high molecular weight [118]. Rhodium-iodine systems, in contrast, due to their high

Table 25

PROCESS CONDITIONS FOR DIMETHYL ETHER AND METHYL ACETATE HOMOLOGATION

Company	MONTEDISON [92]	CNR [124]	IMHAUSEN [118]	RHONE-POULENC [125]	HALCON [119]
Substrate	Me_2O/AcOMe	AcOMe	AcOMe	AcOMe	Me_2O/AcOMe
Catalyst (wt %)	$Ru(acac)_3$ (0.5)	$Ru(acac)_3$ (0.5)	$Co(OAc)_2$	$Ru_3(CO)_{12}$ $Co_2(CO)_8$ (0.2–10)	$RhCl_3 \cdot 3H_2O$ (1.2)
Promoter	HI, CH_3I, NaI	CH_3I/pyridine	I_2	CH_3I/LiI/PPh_3/MeI	CH_3I/picoline
I/Metal	10	10	1	2–50	38
Solvent	acetic acid	acetic acid	–	acetic acid	–
Pressure (atm)	150–200	180	440	150–350	70–100
Temperature °C	200	220	180	215	150
Conversion (%)	50–80	84	71	40–50	n.d.
Selectivity [a]	(*a*)	(*a*)	(*b*) [b]	(*a*) [b]	c
AcOEt	50–75	30	78	60	–
AcOH	3–10	45	–	25	1–75
AcH	–	–	–	1	0–40
Ac_2O	–	–	–	–	0–80
$MeCH(OAc)_2$	–	–	–	–	1–60
T.N. ($sec.^{-1}$) [d] (time of the run, h)	0.007–0.01 (8)	0.004 (8)	n.d.	0.07–0.05 (0.6)	n.d.

[a] The italic letter in parentheses indicates the relation used to evaluate the selectivity (see Table 1).
[b] Hydrocarbons not accounted.
[c] % Wt., hydrocarbons not accounted.
[d] Number of moles of AcOEt produced per second per 1 at-g of metal catalyst.

carbonylation and low hydrogenation activity, give useful acetaldehyde/acetic acid/acetic anhydride/ethylidene diacetate mixtures [119]. This last product probably arises from the reaction of acetic anhydride and acetaldehyde (Equation 45) which is the primary product of hydrogenation of the acyl-rhodium intermediate formed in the reaction of dimethyl ether or methyl acetate with rhodium catalysts [120].

$$(CH_3CO)_2O + CH_3CHO \xrightarrow{H^+} CH_3CH(OCOCH_3)_2 \qquad (45)$$

Only $Rh(acac)_2/LiI$ catalysts lead to an appreciable formation of ethyl acetate and acetic acid [121].

Ruthenium-iodine and cobalt-ruthenium-iodine systems, owing to the strong hydrogenation properties of ruthenium on acyl derivatives, leads mainly to the formation of ethyl acetate [116, 121]. Detailed studies on the effect of the reaction parameters [115] and of the role and effect of the nature of the iodide promoters (covalent derivatives and ionic salts) [122], of phosphine ligands and nitrogen bases [123] on the ruthenium catalyzed methyl acetate homologation demonstrates interesting possibilities of balancing the carbonylation and reduction properties of the catalytic system to make the process selective towards the desired product.

The data reported in Table 26 show the important effect of the nature of the iodide promoter and of the presence of nitrogen bases on the course of the reaction.

With CH_3I, the main reaction is the homologation of methyl acetate to ethyl acetate with formation of water (Equation 44). Products initially present in low concentration are alcohols (methanol and ethanol), arising from the hydrolysis of the corresponding esters, and small amounts of ethers. The concentration of methyl derivatives (MeOH, Me_2O, and MeOEt) reaches a maximum and then decreases due to their transformation into carbonylation and homologation products. At methyl acetate conversions above 50%, products of further carbonylation and homologation of ethyl derivatives begin to be produced in appreciable amounts. However, the main by-products of the reaction are hydrocarbons (methane and ethane) which account for the consumption of 15–25% of the starting methyl groups. In the end reduction of the acetyl prevails over carbonylation of the methyl groups [115].

With NaI, the catalytic system is less active and the simple carbonylation of methyl acetate to acetic acid (Equation 46) is the predominant reaction and consumes great part of the water produced in the homologation and hydrogenation reactions [122].

$$CH_3COOCH_3 + CO + H_2O \longrightarrow 2CH_3COOH \qquad (46)$$

With NH_4I or Me_4NI, the simple carbonylation prevails still more over the reduction reactions [122].

Table 26

HOMOLOGATION OF METHYL ACETATE WITH RUTHENIUM CATALYSTS IN PRESENCE OF DIFFERENT IODIDE PROMOTERS AND NITROGEN BASES[a] [115, 122, 123]

Reaction products	SELECTIVITY (%)							
	CH_3I		NaI		$(CH_3)_4NI$ [b]		pyridine [c]	
	After 8 h	After 26 h	After 8 h	After 56 h	After 8 h	After 32 h	After 8 h	After 16 h
$Me_2O + MeOH + MeOEt + CH_3I$	4.2	2.6	1.6	0.3	3.7	–	0.1	0.1
$EtOH + Et_2O$	4.6	6.1	0.3	0.1	0.1	0.1	0.1	0.1
AcOH (produced)	6.7	3.1	51.9	46.7	41.9	50.7	35.5	44.9
AcOEt	61.9	51.8	19.5	24.3	31.2	24.5	38.5	29.2
n-PrOH + AcOPr + EtCOOPr	3.0	6.6	–	–	–	–	–	0.5
EtCOOEt + EtCOOH	1.0	5.4	–	–	0.2	1.3	–	–
$CH_4 + C_2H_6$	18.6	24.4	26.6	28.6	22.8	23.3	25.6	25.9
H_2O (% wt.)	5.0	5.6	1.0	0.78	0.69	0.48	0.2	0.35
AcOMe (conv. %)	53.5	84.6	34.8	81.1	41.8	88.2	44.3	83.8
Acyl groups formed (mmol)	–	–	136	236	280	376	77.6	80.6
Acyl groups disappeared (mmol)	184	461	–	–	–	–	–	–
Ethyl groups (mmol)	753	1229	124	350	279	364	312	445

Reaction conditions: $Ru(acac)_3$: 3.6×10^{-3} moles; I/Ru: 10; AcOMe: 1.8 mol; AcOH: 1.8 mol; P_{CO}: 40 atm; P_{H_2}: 80 atm; T: 200°C.
[a] The selectivity data were calculated according to relation *(a)* (see Table 1).
[b] AcOMe: 1.65 mol; AcOH: 1.68 mol.
[c] p: 180°C; H_2/CO: 2; T: 220°C.

With phosphines, the whole activity of the system is lowered but the selectivity to ethyl derivatives and unfortunately to hydrocarbons is enhanced. This effect becomes more evident when the ligand-to-metal ratio and the phosphine basicity increase [123].

With pyridine, the reaction proceeds at a rate comparable to that with CH_3I alone, but with a better balance between carbonylation and hydrogenation activity. As result very low formation of hydrolysis and etherification products and smaller quantities of propionates and *n*-propyl derivatives are observed. Moreover the concentration of water in the reaction mixture is very low ($< 0.5\%$ wt.) thus facilitating the separation process of the products and the recycling of the catalyst which remains highly stable and active [125].

An interesting co-promoting and optimizing effect which strongly enhances the hourly productivity and selectivity to ethyl acetate is obtained by adding a cobalt catalyst to the ruthenium system (Co/Ru = 0.01–0.5) and using a mixture of covalent, ionic, and ammonium or phosphonium iodides [121, 125]; ruthenium/rhodium or cobalt/rhodium catalysts seem to be less active [125].

However, an optimized Ru/Rh ratio (2), the addition of a suitable iodide promoter (ZnI_2) and of a ligand (α-picoline), and a proper choice of reaction conditions enables one to direct the reaction towards the formation of ethyl acetate and acetic acid in a ratio 1/2 (Equation 47) [126].

$$2CH_3COOCH_3 + 2CO + 2H_2 \longrightarrow CH_3COOC_2H_5 + 2CH_3COOH \qquad (47)$$

This is of special interest when this reaction is part of an integrated process wherein the acetic acid produced is to be recycled.

6.2. Homologation of Formic Esters

To date, literature data seem to exclude the possibility of homologating formic to acetic esters. In fact, rhodium catalysts are only able to decarbonylate formic esters and carbonylate the alcoholic moiety to acid [127].

Ruthenium-phosphine [109] and ruthenium-iodide-phosphine systems [128] are unable to reduce the formyl group but only catalyze the carbonylation and homologation of the alcoholic part of the ester with pronounced decomposition of the ester (Equations 47, 48).

$$HCOOR \longrightarrow CO + ROH \qquad (47)$$

$$HCOOR \longrightarrow CO_2 + RH \qquad (48)$$

In contrast, recent results obtained with $Ru(CO)_4I_2/CH_3I$ systems working under appropriate carbon monoxide and hydrogen pressure in an acid medium (acetic or formic acid) demonstrate the possibility of reducing the formyl to methyl group and subsequently of homologating the formyl to acetyl derivatives (Table 27). Thus methyl formate, directly formed from CO and H_2 [69, 129, 130] or easily prepared by carbonylation of methanol with sodium methoxide as catalyst [131], may become a convenient source of C_2 chemicals (acetic and ethyl derivatives) [116].

7. HOMOLOGATION REACTIONS WITH HETEROGENEOUS CATALYSTS

The possibility of carrying out the homologation reaction in the gas or liquid phase with heterogeneous catalysts was immediately considered when methanol homologation was discovered.

In the most cases it is not easy to ascertain whether the complex mixture of the oxygenated products obtained with heterogeneous catalysts arises from CO and H_2 by a direct Fischer-Tropsch synthesis or from a carbonylation and hydroformylation reaction on ethylene produced from methanol [132].

In general the performances of the catalysts based on cobalt and on iron are very poor, with respect to either the activity and/or the selectivity of the process [10, 133, 134], and no information is available on the possible losses of metal transformed into soluble or volatile carbonyls.

Of major interest are some homologation processes of low molecular weight alcohols such as methanol, ethanol, *n*-propanol and ethylene glycol to a mixture of higher linear primary alcohols. The process, first described in two old Du Pont patents [6, 7] and later developed by Celanese [135] is catalyzed by ruthenium, rhodium and palladium oxides, hydroxides or basic salts, pure or impregnated on silica or alumina supports in the presence of basic salts of alkaline metals as promoters. Depending on reaction conditions and catalyst composition, the process may be directed either towards the preferential production of high molecular weight homologous alcohols (C_6—C_{18}) (liquid phase, ruthenium catalyst, pressure > 700 atm, temperature 220–250°C) [6, 7], or towards lower molecular weight alcohols (C_4—C_6) (gas phase, rhodium or palladium catalyst, pressure 70–120 atm, temperature 350–380°C) [135].

8. HOMOLOGATION REACTIONS WITH CO AND H_2O

The possibility of using CO/H_2O instead of CO/H_2 mixtures for homologation reactions was considered for a long time, since this might represent a relevant economic improvement.

The first attempt was the combination of a soluble homologation catalyst i.e., $Co_2(CO)_8$ with a heterogeneous catalyst active for the water gas shift reaction, i.e., copper or iron on kieselguhr [133]. A second suggestion was the modification of the typical homologation catalysts based on cobalt or ruthenium derivatives with the addition of a basic inorganic compound (i.e., an alkali or alkaline–earth metal carbonate: 0.1–0.3 parts per part of methanol [136]) or a nitrogen base (i.e., pyridine or piperidine [124]) which promote the water gas shift reaction.

Table 27

HOMOLOGATION OF METHYL FORMATE

Catalytic system (molar ratio)	HCOOMe / RCOOH	P (atm)	T (°C)	HCOOMe conv. (%)	Carboxylic acid conv. (%)	SELECTIVITY (%)[a] MeOH	EtOH	HCOOEt	AcOMe (total acetyl derivatives)	$CH_4 + C_2H_6$ (others)	Decomposition of reacted formyl groups (%)	Ref.
$Ru(acac)_3/CH_3I/PPh_3$ [b] (1/10/4)	–	270	220	48.9	–	35.7	12	42.2	5	n.d. (5)	100	128
$Ru(acac)_3/CH_3I/PPh_3$ (1/10/1)	5.3 ($R=CH_3$)	148	200	69	93.1	10.4	4.3	14.5	33.8 (39.5)	29.9 (1.1)	91.0	116
$Ru(acac)_3/CH_3I$ (1/10)	8.3 ($R=H$)	150	200	25	85.7	59.3	5.0	7.4	15.6 (17.4)	6.9 (4)	17.0	116 117
$Ru(acac)_3/CH_3I$ (1/10)	7.9 ($R=CH_3$)	130	200	43	87.8	32	2.8	3.1	35.9	15.5 (7.1)	0	116 117

a The selectivity data were calculated according to relation (*a*) (see Table 1).
b Extensive formation of methane observed but not accounted in the evaluation of the selectivity.

In all cases, however, catalytic activity and selectivity were unsatisfactory and a partial decomposition of methanol to carbon monoxide and hydrogen sometimes took place [56].

An extreme case was the suggestion of using methanol alone for homologation, since the CO and H_2 necessary for the reaction are supplied by the decomposiiton of methanol itself in a catalytic preconverter [137].

9. REACTION MECHANISM

It is not possible to deduce a general mechanism for homologation reactions comprising all types of reaction and catalysts since the single steps of catalysis differ according to the nature of the catalyst, promoter, ligand, and substrate.

We shall simply attempt here to discuss each step of the reaction separately, first considering the various metals, then the substrates, and finally the effect of promoters.

9.1. Cobalt Catalysts

A six-step mechanism of cobalt-catalyzed methanol homologation was first proposed by Wender [138, 139] and has recently been critically reviewed by Slocum [15]. Moreover mechanisms for the cobalt-iodine catalytic systems were reported by Bahrmann and Cornils [13] and by Roeper [24].

9.1.1. FORMATION OF CATALYTICALLY ACTIVE SPECIES FROM PRECURSORS

Non-promoted cobalt catalysts

Various cobalt compounds, soluble or insoluble in the reaction medium, were used as catalyst precursors for the homologation reaction; it is generally assumed that under the drastic reaction conditions (150–200°C; 200–300 atm of CO/H_2 mixtures), they are completely transformed into $Co_2(CO)_8$ or $HCo(CO)_4$. In fact, the synthesis of these carbonyls occurs at temperatures and pressures of carbon monoxide and hydrogen lower or similar to those used for homologation [140].

$HCo(CO)_4$, rapidly formed from $Co_2(CO)_8$ in hydrocarbon solutions under the reaction conditions (Equation 49) was proposed as the key intermediate in the homologation reaction [138, 139].

$$Co_2(CO)_8 + H_2 \rightleftharpoons 2HCo(CO)_4 \qquad (49)$$

In fact, in methanol under CO/H_2 high pressure, only small amounts of $Co_2(CO)_8$ and $HCo(CO)_4$ are present [31], as an extensive disproportion of $Co_2(CO)_8$ is present according to the equilibrium represented in Equation 50 [141].

$$12CH_3OH + 3Co_2(CO)_8 \rightleftharpoons 2[Co(CH_3OH)_6]^{2+} + 4[Co(CO)_4]^- + 8CO \qquad (50)$$

Even if the reversibility of this reaction in an alcoholic medium has been demonstrated by Orchin [142], the anion $[Co(CO)_4]^-$ is the dominant cobalt species in the homologation reaction medium under the reaction conditions [31]. However, other species of oxidized cobalt, containing both methanol and carbon monoxide in the coordination sphere, seem to be present when cobalt acetate or acetylacetonate are used as precursors (I.R. evidence) and are probably directly involved in the catalytic cycle as for the promoted cobalt catalysts [31].

In any case, the direct correlation between the $[Co(CO)_4]^-$ concentration, as indicated by I.R., and the reaction rate, as indicated by the syngas consumption, clearly associates the catalytic activity for methanol homologation with the presence of the anion.

When a phosphine ligand is added to the cobalt-carbonyl catalyst a disproportion of $Co_2(CO)_8$ initially occurs, forming the ionic inactive product $[Co(CO)_3(PR_3)_2][Co(CO)_4]$; successively some $[Co(CO)_3PR_3]_2$ as well as $HCo(CO)_3PR_3$ can be produced. However, due to the weaker acidity of this last hydrido species with respect to $HCo(CO)_4$, a very low activity for the methanol activation is ultimately observed [31].

Iodine-promoted cobalt catalysts

The addition of iodide promoters, CH_3I, HI, and I_2 to cobalt catalysts or the use of CoI_2 strongly modifies the nature of the cobalt-carbonyl species involved in the catalysis.

Mizoroki and Nakayama [52, 53] demonstrated the presence, under homologation conditions, of Co(II) ions of the type $[Co(AcO)_{4-n}I_n]^{2-}$ (n = 0–4) in methanol–acetic acid solution. Moreover evidence based on I.R. data on the formation of $[Co(CH_3OH)_x(CO)_yI_z]^{n+}$ species in methanol solution (I/Co = 0.5) have also been reported by Pretzer [31] who suggested that species of the type $[Co(CH_3OH)_x(CO)_yI_z]^{n+}[Co(CO)_4]_n^-$ (z = 1, 2) are the dominant forms of cobalt throughout the reaction when iodide promoters are added.

Using higher I/Co ratios (I/Co > 2) when the iodide promoter is CH_3I, I_2 or HI, a complete elimination of $[Co(CO)_4]^-$ was observed together with a decrease of catalytic activity in the carbonylation of methanol to acetic acid [53] and in the homologation to ethanol [31]; in contrast, the addition of an ionic iodide such as NaI increased the reaction rate [53]. This fact was related to the formation of an inactive cobalt complex, i.e., $[Co(OAc)_2I_2]^{2-}$ by action of the strong HI [31], whereas the iodide ions may positively influence the disproportion of $Co_2(CO)_8$ favouring the formation of $[Co(CO)_4]^-$ (Equation 51) [21, 143].

$$2Co_2(CO)_8 + 4A^+I^- + 2H_2 \rightleftarrows 4A^+[Co(CO)_4]^- + 4HI \qquad (51)$$

(A = Na, K, etc.)

Also, hydrido-iodocarbonyl-cobalt derivatives such as $HCo(CO)_3I$ [13] or $[HCoCOX_4]^{2-}$ (X = I or AcO) [48] have been proposed as intermediates involved in the catalysis, but no direct experimental evidence has yet been adduced.

When a phsophine ligand is added to a cobalt-iodide system, new unidentified iodo-phosphine complexes are formed and are probably involved in the catalytic cycle [31]. According to Roeper [23, 24] the products might be Co(I) or Co(III) derivatives of type $CoCO(PR_3)_2I$ or $HCoCO(PR_3)_2I_2$.

9.1.2. ACTIVATION OF THE SUBSTRATE AND FORMATION OF ALKYL-COBALT INTERMEDIATES

Alcohols

The first mechanism suggested fo the carbonylation reaction of alcohols explained the formation of alkyl-cobalt intermediates via an esterification of the alcohol by the strong acid $HCo(CO)_4$ (Equation 52) [144].

$$ROH + HCo(CO)_4 \longrightarrow RCo(CO)_4 + H_2O \tag{52}$$

This over-simplified view of the problem was successively developed and treated by most authors' distinguishing two successive steps: activation of the substrate and formation of the alkyl-cobalt derivative, and proposed various and sometimes conflicting hypotheses.

Dehydration of the alcohol to alkene and its addition to $HCo(CO)_4$. Ziesecke [8], on the basis of the nature of the products obtained in the homologation of secondary and tertiary alcohols (i.e., 3-methylbutanol was obtained from *t*-butanol and not neopentyl alcohol as would be expected), proposed that the reaction could proceed via an acid-catalyzed dehydration of the alcohol to alkene followed by a hydroformylation-type process in which the alkene is added to $HCo(CO)_4$ (Equation 53).

$$R{-}\underset{\displaystyle OH}{\underset{|}{C}}H{-}CH_3 \xrightarrow[-H_2O]{H^+} RCH{=}CH_2$$

$$\xrightarrow{HCo(CO)_4} R\underset{\displaystyle CH_3}{\underset{|}{C}}HCo(CO)_4 + RCH_2CH_2Co(CO)_4 \tag{53}$$

This proposal has been confirmed as effective for alcohols having hydrogen in a β-position in a study on ^{14}C-labelled methanol homologation previously discussed [83], but fails to rationalize the much higher reactivity of methanol and benzyl alcohol with respect to other alcohols.

Formation of a carbene intermediate and addition to $HCo(CO)_4$. Ziesecke himself [8] proposed a carbene or phenylcarbene intermediate to explain the high reactivity of methanol and benzyl alcohol without however supplying any experimental support (Equation 54).

$$:CH_2 + HCo(CO)_4 \longrightarrow CH_3Co(CO)_4 \tag{54}$$

Recently Roeper [23], using a perdeuterated methanol (CD_3OH) as substrate, has shown beyond all doubt that the CD_3 group of the alcohol remains intact during the homologation reaction. It is integrally transferred to the reaction products (acetaldehyde, ethanol, methyl acetate). These findings make carbene intermediates highly improbable in homologation reactions.

Formation of carbonium ions or ion pairs by reaction with $HCo(CO)_4$. In studying homologation reactions on a series of *p*-substituted benzyl alcohols, Wender [84] found a correlation between the reactivity of the alcohol and its tendency to form the corresponding carbonium ion. He put forward the suggestion that a benzyl cation produced according to Equations 55–57 was involved in the reaction, leading to a benzyl-cobalt intermediate [139].

$$C_6H_5CH_2OH + HCo(CO)_4 \rightleftharpoons C_6H_5CH_2OH_2^+ + [Co(CO)_4]^- \quad (55)$$

$$C_6H_5CH_2OH_2^+ \rightleftharpoons C_6H_5CH_2^+ + H_2O \quad (56)$$

$$C_6H_5CH_2^+ + [Co(CO)_4]^- \rightleftharpoons C_6H_5CH_2Co(CO)_4 \quad (57)$$

The formation of pinacolone and pinacolyl alcohol, observed in the reaction of pinacol with CO + H_2 [82], was indicative of an acid-catalyzed carbonium ion rearrangement and the identified carbonylation products, 3,4-di-methyl-1-pentanol and 2,2,3-trimethyl tetrahydrofuran, most likely arise via a dehydration of the alcohol and hydroformylation or hydrogenation of the corresponding alkene according to Equation 53.

This mechanism, probably operative for benzyl alcohols, must be ruled out for methanol as it is unable even in the presence of strong Lewis acids to dissociate giving a methyl cation [145].

Protonation of methanol and nucleophilic attack to $[Co(CO)_4]^-$. The most widely accepted hypothesis on methanol activation with non-promoted cobalt catalysts suggests an initial protonation of the alcohol by $HCo(CO)_4$ with formation of an ion pairing (Equations 58, 59) and a successive nucleophilic attack of $[Co(CO)_4]^-$ in a Sn2-type process with displacement of the water and formation of a methyl-cobalt intermediate [139].

$$CH_3OH + HCo(CO)_4 \longrightarrow \left[CH_3\underset{\displaystyle H}{\underset{|}{O}}H\right]^+ [Co(CO)_4]^- \quad (58)$$

$$\overset{\delta-}{Co(CO)_4}\text{---}\overset{H\ H}{\overset{\backslash\,/}{\underset{\underset{\displaystyle H}{|}}{C}}}\text{---}\overset{+\delta}{OH_2} \longrightarrow CH_3Co(CO)_4 + H_2O \quad (59)$$

From this point of view methanol is particularly susceptible to nucleophilic attack because of the lack of steric hindrance.

Since the turnover rates for methanol carbonylation to acetic acid with iodide promoters are comparable to those reported for homologation in the absence of iodide [146], the formation of methyl-cobalt intermediates may not involve an iodide even if it is present. Moreover the positive effect on the reaction rate claimed for the use of ionizing solvents [46, 47] might be indicative of the formation of intermediates possessing some charge separation.

Slocum [15] proposes as an alternative pathway for the formation of a methyl–cobalt intermediate, an oxidative addition of protonated methanol to a tricarbonyl species (Equation 60),

$$CH_3OH_2^+ + HCo(CO)_3 \xrightarrow{CO} \left[(CO)_3Co(CH_3)(OH_2)(CO)_2 \right]^+ \xrightarrow[-H_2O]{} CH_3Co(CO)_4 \qquad (60)$$

or a displacement of water from a hydrogen-bounded pseudo-four-centred intermediate (Equation 61):

$$CH_3-\overset{\delta+}{O}(H)\cdots H\cdots\overset{\delta-}{Co}(CO)_n \longrightarrow CH_3Co(CO)_n + H_2O \qquad (61)$$

The occurrence of methanol protonation as the initial activation step has also been proposed by Mizoroki [53] for the cobalt-catalyzed carbonylation of methanol to acetic acid in the presence of iodide promoters.

However the successive attack of protonated methanol on a Co(II) carbonyl derivative (Equation 62) as suggested by the same author is more doubtful and needs careful investigation.

$$CH_3OH_2^+ + Co(II)CO \rightleftharpoons CH_3Co(III)CO^+ + H_2O. \qquad (62)$$

Nucleophilic attack of $[Co(CO)_4]^-$ on methyl group of coordinated methanol. Another possibility, based on the well-known disproportion of $Co_2(CO)_8$ in methanol (Equation 50), suggested by Wender [139] considers a nucleophilic attack of $[Co(CO)_4]^-$ on the methyl group of a methanol molecule in the coordination sphere of Co(II) cation (Equation 63).

$$[Co(CH_3OH)_6]^{2+} + [Co(CO)_4]^- \longrightarrow (CH_3OH)_5\overset{2+}{Co}\cdots O(H)-CH_3\cdots Co(CO)_4^- \longrightarrow CH_3Co(CO)_4 + [Co(CH_3OH)_5OH]^+ \qquad (63)$$

Direct reaction of methanol with $Co_2(CO)_8$. An anomalous activation path taking place in the absence of or under low hydrogen pressure was proposed by Albanesi [17] to explain the formation of small amounts of methylale: $Co_2(CO)_8$ is cleaved by methanol giving $HCo(CO)_4$ and a methoxy-cobalt intermediate which releases formaldehyde according to Equations 64 and 65.

$$Co_2(CO)_8 + CH_3OH \longrightarrow HCo(CO)_4 + CH_3OCo(CO)_4 \tag{64}$$

$$CH_3OCo(CO)_4 \longrightarrow H(CH_2O)Co(CO)_4 \longrightarrow HCo(CO)_4 + CH_2O \tag{65}$$

Intermediate transformation of methanol into methyl iodide and reaction of methyl iodide with $[Co(CO)_4]^-$. Controversal hypotheses exist in the literature on the activation step of methanol in the presence of iodide promoters (HI, I_2, CH_3I).

Berty [9] excluded methyl iodide, formed by esterification of methanol (Equation 17), as the intermediate reacting with $HCo(CO)_4$ since it could not be homologated in anhydrous benzene.

Bahrmann and Cornils [13] and Falbe [147] proposed on the basis of Berty's results a direct oxidative addition of methanol to a hydrido-iodocarbonyl co-ordinatively-unsaturated derivative, favoured by the labilizing effect of the iodine ligands of the cobalt carbonyl (Equation 66).

$$CH_3OH + I{-}\underset{H}{\overset{CO}{Co}}{-}CO \longrightarrow I{-}\underset{CO\ \ CO}{\overset{H\ \ CH_3}{Co}}{-}OH \xrightarrow{-H_2O} I{-}\underset{CH_3}{\overset{CO}{Co}}{-}CO \tag{66}$$

However some evidence casts doubt on these conclusions: in fact methyl iodide undergoes carbonylation in methanol solution in the presence of $[Co(CO)_4]^-$ to give methyl acetate (Equation 67) [148].

$$CH_3I + CO + ROH \xrightarrow[{[Co(CO)_4]^-;\ B}]{} CH_3COOR + H^+ + I^- \tag{67}$$

(R = alkyl; B = base)

Moreover, via a stoichiometric hydrocarbonylation at room temperature with $HCo(CO)_4$ it gives acetaldehyde in high yield (Equation 68) [149].

$$3CH_3I + 2[Co(CO)_4]^- + 3H^+ \longrightarrow 3CH_3CHO + 5CO + 2Co^{2+} + 3I^- \tag{68}$$

A convincing test in favour of methyl iodide as an intermediate in methanol hydrocarbonylation has been recently given by Roeper [23]. Working with deuterium labelled reagents (CH_3OH/CH_3I: 4/1), he found a selective reaction of CH_3I in a short time.

In conclusion, most authors, Wender [139], Pretzer [31], Mizoroki [55], Roeper [23, 24], Gauthier [21], consider methyl iodide as the active methylating species which produces the methyl–cobalt intermediate either by an oxidative addition to a coordinatively unsaturated cobalt species (Equations 69, 70) [139], present under oxo conditions in a concentration of about 0.3%,

$$HCo(CO)_4 \rightleftharpoons HCo(CO)_3 + CO \tag{69}$$

$$CH_3I + HCo(CO)_3 \rightleftharpoons HCo(CH_3)(CO)_3I \tag{70}$$

or by a nucleophilic Sn2 displacement of $[Co(CO)_4]^-$ according to Equation 71 [55]

$$CH_3I + [Co(CO)_4]^- \longrightarrow CH_3Co(CO)_4 + I^- \tag{71}$$

The successive steps (Equations 72, 73) parallel those proposed for the rhodium or iridium-catalyzed methanol carbonylation [55, 94] but they are not supported by any experimental data on Co(I) ⟶ Co(III) transformations on cobalt carbonyl derivatives and on reductive elimination of HI from a hydrido-carbonyl derivative.

$$HCo(CH_3)(CO)_3I \longrightarrow CH_3Co(CO)_3 + HI \tag{72}$$

$$CH_3Co(CO)_3 + CO \longrightarrow CH_3Co(CO)_4 \tag{73}$$

On the other hand, according to Roeper [23], the oxidative addition route seems to be more likely for cobalt-iodide-phosphine systems in which no hexacoordinated too labile hydrido-carbonyls are involved (Equation 74).

$$Co(CO)(PR_3)_2I + CH_3I \longrightarrow CH_3Co(CO)(PR_3)_2I_2 \tag{74}$$

The nucleophilic displacement of $[Co(CO)_4]^-$ (Equation 71) seems to be more probable in the absence of phosphine and may account for the deactivation effect of CH_3I when it is used in a large excess with respect to cobalt. In fact it is important to emphasize that covalent iodide promoters (HI, I_2, CH_3I) can play two opposite roles: activators of methanol via methyl-iodide formation and inactivators of cobalt via oxidation of $[Co(CO)_4]^-$ to non-carbonyl complexes [31, 53]. The 0.6 order with respect to CH_3I over a I/Co range 0.5–2 and the decay of activity operating at higher I/Co ratios is indicative of this behaviour.

On the other hand, the ionic iodides (NaI, KI, etc.) act only as promoters facilitating catalyst formation (Equation 51) and perhaps also regeneration of the $[Co(CO)_4]^-$ (Equation 75) [21].

$$HCo(CO)_4 + I^- \longrightarrow [Co(CO)_4]^- + HI \tag{75}$$

Aldehydes

Controversial hypotheses have been suggested concerning the type of interaction between an aldehyde, particularly formaldehyde, and $HCo(CO)_4$. In fact the addition of the cobalt hydrocarbonyl can occur in two directions to give either a hydroxymethyl intermediate (Equation 76) or a methoxy-cobalt intermediate (Equation 77).

$$CH_2O + HCo(CO)_4 \longrightarrow HO{-}CH_2Co(CO)_4 \quad (76)$$

$$CH_2O + HCo(CO)_4 \longrightarrow CH_3OCo(CO)_4 \quad (77)$$

Although no hydroxy or alkoxy-methyl cobalt carbonyl has been isolated, the catalytic or stoichiometric formation of glycolaldehyde form $HCo(CO)_4$, CO and H_2 (Equation 78) clearly substantiates this pathway [98].

$$HOCH_2Co(CO)_4 + CO + HCo(CO)_4 \longrightarrow HOCH_2CHO + Co_2(CO)_8 \quad (78)$$

Moreover Dombeck recently reported [150] a number of model compounds of type $ROCH_2Mn(CO)_5$ which reacted at elevated temperatures with CO and H_2 leading to carbonylation and hydrogenolysis products. Even hydroxymethyl complexes of iridium have been isolated recently; these compounds however are preferentially deprotonated rather than hydrogenolyzed by another molecule of a hydrido-iridium derivative (Equation 79) [151].

$$\left[\begin{array}{l}L_4Ir{-}CH_2OH\\ \;\;\;|\\ \;\;\;H\end{array}\right]PF_6 + HIr(PMe_3)_4 \longrightarrow$$

$$\left[\begin{array}{l}L_4Ir(CH_2O)\\ \;\;\;|\\ \;\;\;H\end{array}\right] + [H_2Ir(PMe_3)_4]PF_6 \quad (79)$$

In contrast, the formation of methanol and methyl formate (products which accompany glycolaldehyde particularly at high temperatures) presupposes the intermediate formation of a methoxy derivative (Equation 80) [152, 153].

$$\begin{array}{lcl} CH_3OCo(CO)_4 & \xrightarrow{H_2} & CH_3OH + HCo(CO)_4 \\ \big\downarrow CO & & \\ CH_3OCOCo(CO)_4 & \xrightarrow{H_2} & HCOOCH_3 + HCo(CO)_4 \end{array} \quad (80)$$

The competition between these two pathways is affected by temperature and carbon monoxide pressure and accordingly the distribution of the products varies.

9.1.3. CARBONYLATION OF ALKYL TO ACYL INTERMEDIATE

Abundant review literature exists on the formation of acyl from alkyl-metal derivatives, i.e., on methyl migration–CO insertion reactions. Practically all authors interested in the alkene hydroformylation reaction have dealt with this reaction step [64]. For the homologation reaction, Slocum reported a lengthy discussion on this point [15].

Here, only those aspects important for the kinetics and the selectivity of the reaction will be reviewed.

Direct proof of CO group exchange in the acyl intermediate (Equation 81) [154] and of the equilibrium transformation of the alkyl tetracarbonyl into acetyltricarbonyl cobalt intermediate has been provided for unmodified cobalt catalysts (Equation 82) [155, 156] and for phosphine cobalt catalysts (Equation 83) [157].

$$RCOCo(CO)_3 + CO^* \rightleftharpoons RCO^*Co(CO)_3 + CO \quad (81)$$

$$RCo(CO)_4 \rightleftharpoons RCOCo(CO)_3 \quad (83)$$

$$R'Co(CO)_3PPh_3 \rightleftharpoons R'COCo(CO)_2PPh_3 \quad (83)$$

$$(R = C_3H_7;\ R' = PhCH_2)$$

The carbonylation step which occurs through alkyl migration on a coordinated CO [60, 157] is followed by an addition of gaseous CO to the coordinatively unsaturated species (Equation 84) [60].

$$RCOCo(CO)_3 + CO \rightleftharpoons RCOCo(CO)_4 \quad (84)$$

The energetic barrier to methyl migration estimated by a molecular orbital approach [158] and by an *ab initio* approach [159] is 8–13 kcal/mol.

Analogous evolutions of the alkyl intermediate have been also proposed for the iodide promoted cobalt systems (Scheme 3) [147]. The alkyl intermediate, formed through an oxidative addition of methanol, either by splitting of water or by CO insertion, depending on reaction conditions, evolves towards an acyl derivative giving successively acetic acid or acetaldehyde and then ethanol.

However, from a practical point of view, the most important point in the homologation process is the acceleration of the alkyl migration-CO insertion reaction to avoid the hydrogenation of the alkyl intermediate to hydrocarbons (Equation 85) under the drastic reaction conditions.

$$CH_3Co(CO)_4 + H_2 \longrightarrow CH_4 + HCo(CO)_4 \quad (85)$$

All promoters and ligands able to accelerate the carbonylation steps might turn out to be very useful in improving the selectivity to carbonylation and homologation products with respect to hydrocarbons.

Unfortunately these effects are frequently masked by successive and side-reaction steps and sometimes it is impossible to distinguish and attribute them with sufficient certainty. However knowledge of these trends can be of help in leading research towards better performance of the catalytic systems.

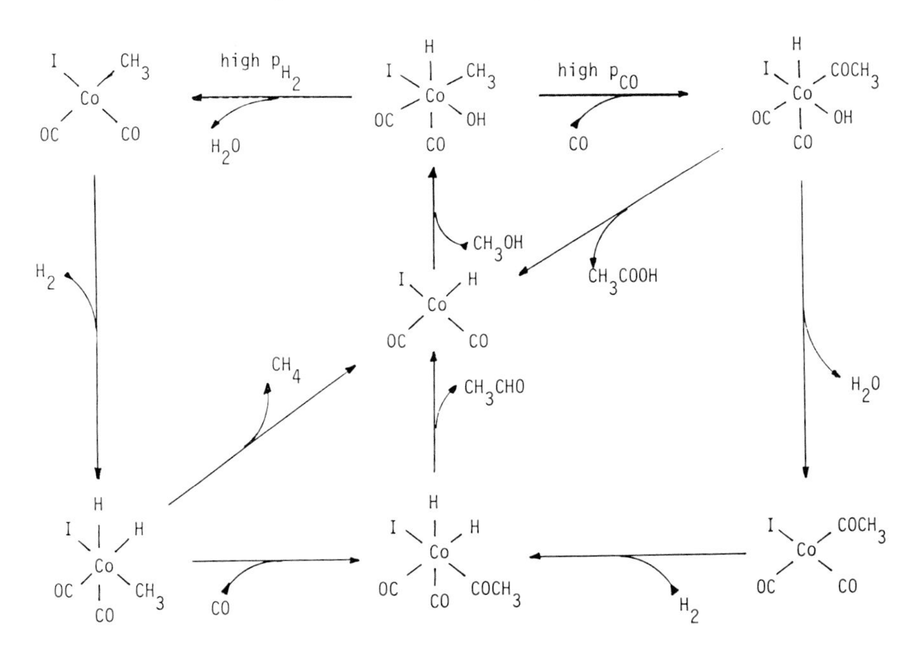

Scheme 3. Evolution of alkyl iodocarbonyl cobalt intermediates.

Effect of hard or soft bases. In principle all labilizing ligands (hard bases) as halogen ions or amines can accelerate the carbonylation whereas non-labilizing ligands (soft bases) such as phosphines depress the same reaction [53]. For instance the rate constant for the formation of acetyl tetracarbonyl from methyl pentacarbonyl manganese is halved in the presence of triphenyl phosphine (Equation 86) [60].

$$CH_3Mn(CO)_5 + PPh_3 \longrightarrow CH_3COMn(CO)_4PPh_3 \quad (86)$$

Experimental results actually agree with this picture: iodide promoters increase and phosphines decrease the rate and in some cases increase the selectivity to hydrocarbons.

Effect of ion pairing (Lewis acids). An important effect of the nature of the counter-ion of the anionic alkyl-metal carbonyl in the kinetic of the alkyl-acyl migratory insertion has been reported and thoroughly discussed by Collman [160, 161] in the case of alkyl-acyl tetracarbonyl iron systems.

Apparently migration of the alkyl to an adjacent terminal carbonyl is facilitated by an electrostatic interaction of small polarizing cations (Li^+, Na^+) with electron-rich carbonyl oxygen which probably promotes the formation of highly reactive ion-paired coordinatively unsaturated acyl species (Equation 87).

$$\left[(CO)_2(CO)\overset{R}{\underset{\delta^-}{Fe}}-CO\cdots Na^+\right] \rightleftharpoons \left[R-\overset{O^-\,Na^+}{C}-Fe(CO)_3\right] \xrightarrow{PR_3} R-\overset{O^-Na^+}{C}-Fe(CO)_3PR_3 \tag{87}$$

Larger cations or solvent separated cations strongly reduce the reaction rate [161].

From this point of view Mizoroki reported a strong promoting effect of NaI on the reaction rate in methanol carbonylation to acetic acid using as catalytic precursor $Co(OAc)_2$ in methanol-acetic acid solution [53].

Effect of strong Lewis acids and protonic acids. Strong Lewis acids such as $AlBr_3$, $AlCl_3$ etc. can also strongly accelerate the methyl migration of alkyl-metal carbonyl derivatives. Evidence on this effect has been reported by Shriver for methyl pentacarbonyl manganese in toluene solution and correlated with the initial formation of an adduct of type Mn–CO–$AlBr_3$ which rapidly evolves to a more stable acyl derivative (Equation 88) [162].

$$CH_3Mn(CO)_5 + AlBr_3 \longrightarrow (OC)_4Mn(Br{-}AlBr_2{-}O{=}C{-}CH_3) \xrightarrow{CO} (OC)_5Mn{-}C(CH_3){=}O{-}AlBr_3 \tag{88}$$

Protonic acids too may play an important role in accelerating the methyl migration. Accordingly, a ten-fold increase of the reaction rate was found by Shriver for the methyl pentacarbonyl manganese in the presence of suitable amounts of trifluoroacetic acid [163]. However, the enhancement of the rate by protonic acids is significantly less than that caused by strong Lewis acids, and the use of mineral acids is detrimental because of the possibility of a complete cleavage of the carbonyl metal derivative.

The accelerating effect of acetic acid and other carboxylic acids in the homologation reaction reported by some authors [19, 30, 46–48] and the inhibiting effect of mineral acids [53] may be related to this type of interaction.

Isomerization during alkyl–acyl interconversion. The possibility of isomerization of alkyl intermediates clearly concerns the homologation of higher alcohols. The data reported in the literature are indicative only for the formation of isomer alcohols coming from dehydration of the starting alcohol and hydroformylation in the double position of the intermediate alkene, without any subsequent isomerization of the alkene.

This agrees with the data reported by Bianchi [164] who found that alkyl cobalt carbonyls do not isomerize signifiantly under Oxo conditions ($p_{CO} > 50$ atm; $p_{H_2} > 50$ atm; temperature > 80°C) even in the presence of free alkene. However, further investigation seems to be necessary for clarifying this point because of the questionable accuracy of the analyses carried out in old papers on the cobalt catalyzed homologation of higher alcohols.

In this context it is of interest to recall that skeletal isomerization has been found in the ruthenium catalyzed homologation of higher carboxylic acids [12].

9.1.4. EVOLUTION OF THE ACYL INTERMEDIATES

The acylcobalt intermediates C-bounded to the metal ("end on") may undergo either an electrophilic attack by hydrogen to give aldehydes and then alcohols or a nucleophilic one to give acids and esters according to Scheme 4.

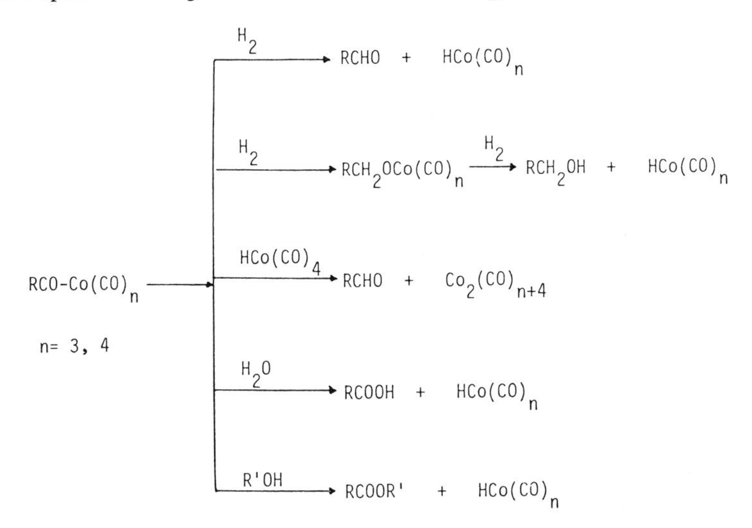

Scheme 4. Evolution of acylcobalt carbonyl derivatives.

Product distribution is determined by reaction conditions and by the presence of suitable promoters and ligands which preferentially direct the reaction towards a particular product, probably by changing the character of the acyl intermediate.

Hydrogenation by molecular hydrogen or by $HCo(CO)_4$ to aldehyde. An electrophilic oxidative addition of H_2 (Equation 89), followed by a reductive elimination of aldehyde (Equation 90) should preferentially take place with unpromoted cobalt catalysts in view of the low oxidation state of the metal (differently from Rh(III) and Ir(III) corresponding systems) [55].

$$RCOCo(CO)_3 + H_2 \rightleftharpoons RCO-\underset{H}{\overset{H}{|}}\!\!Co(CO)_3 \qquad (89)$$

$$RCO-\underset{H}{\overset{H}{|}}\!\!Co(CO)_3 \longrightarrow RCHO + HCo(CO)_3 \qquad (90)$$

The addition of H_2 would probably involve a dissociation of CO from a acyl-tetracarbonyl-cobalt intermediate (Equation 91).

$$RCOCo(CO)_4 \rightleftharpoons RCOCo(CO)_3 + CO \qquad (91)$$

Attack by $HCo(CO)_4$ is equally possible [165] but the reaction with H_2 seems to be more important under hydroformylation conditions [166].

Hydrogenation to alcohol. On the basis of studies on aldehydes' side-formation and on their stoichiometric [167] and catalytic reduction to alcohols in the presence of $HCo(CO)_4$ [168, 169], most authors propose reaction schemes involving the formation of a π-complex of the aldehyde (Equation 92), followed by insertion in a metal-hydrogen bond with formation of an alkoxy intermediate and finally cleavage by H_2 to give the alcohol (Equation 93).

$$RCHO + HCo(CO)_4 \rightleftharpoons R-\underset{HCo(CO)_3}{\overset{H}{C}}\!\!=O + CO \qquad (92)$$

$$R-\underset{HCo(CO)_3}{\overset{H}{C}}\!\!=O \longrightarrow RCH_2OCo(CO)_3 \xrightarrow{H_2} RCH_2O\underset{H}{\overset{H}{|}}\!\!Co(CO)_3 \longrightarrow RCH_2OH + HCo(CO)_3 \qquad (93)$$

This pathway also accounts for the inverse dependence of the hydrogenation of the aldehyde on carbon monoxide pressure [168]. Another scheme involving a hydroxy-alkyl intermediate has been proposed (Equations 94 and 95) but no clear evidence for this has been supplied [169].

$$RCHO + HCo(CO)_4 \longrightarrow R\underset{OH}{\underset{|}{C}}HCo(CO)_3 + CO \qquad (94)$$

$$R\underset{OH}{\underset{|}{C}}HCo(CO)_3 + HCo(CO)_4 \longrightarrow RCH_2OH + Co_2(CO)_7 \qquad (95)$$

Another possibility which deserves consideration is the direct transformation of the acyl into an alkoxy intermediate without passing through the free aldehyde (Equation 96).

$$CH_3\overset{O}{C}\!\!=\!Co(CO)_3 + H_2 \longrightarrow CH_3\overset{O}{\overset{\|}{C}}\!=\!\underset{H}{\overset{H}{Co}}(CO)_3 \longrightarrow CH_3\underset{H}{\overset{H}{C}}-OCo(CO)_3$$

$$\xrightarrow{H_2,CO} C_2H_5OH + HCo(CO)_4 \qquad (96)$$

This may preferentially take place when the acyl intermediate has a partial character of "side on" coordination, i.e., is C- and O-bonded to the metal [60]. The addition of a phosphine ligand to the catalytic system favours this coordinative situation as indicated by the lowering of ν(CO) of the acyl group ($CH_3COCo(CO)_4$: ν(CO) 1725 cm^{-1} [170]; $CH_3COCo(CO)_3PBu_3$: ν(CO) 1684 cm^{-1} [171]) and accordingly increases the selectivity to alcohols.

Analogously, according to Dumas [58], the addition of borate ions by the complexation of boron to the oxygen of the acyl intermediate which confers a certain carbenoid character to the group, favours the electrophilic attack of hydrogen thus increasing selectivity to ethanol (Equation 97):

$$CH_3-\overset{O}{\overset{\|}{C}}-\overset{B(OH)_2\text{—}O}{[Co]} \rightleftharpoons CH_3-\overset{O\cdots B(OH)_2\text{—}O}{C}\cdots[Co] \qquad (97)$$

Hydrolysis or alcoholysis to acids and esters. The formation of acids and esters by nucleophilic attack by water and alcohols is always a possibility since water and alcohols are present in the homologation reaction medium.

As a consequence the main aim in the homologation process is to depress these side reactions. Accordingly the use of nitrogen bases, such as pyridine, which have an accelerating effect on the hydrolysis or alcoholysis of the acyl intermediate through the formation of *N*-acetyl pyridinium salt of the tetracarbonyl cobalt anion, is definitely to be avoided (Equation 98) [172, 173].

$$CH_3COCo(CO)_4 \xrightarrow{C_5H_5N} \left[\begin{array}{l} CH_3CO \\ \quad | \\ \quad N^+C_5H_5 \end{array}\right][Co(CO)_4]^-$$

$$\xrightarrow[\text{fast}]{} [C_5H_5NH]^+[Co(CO)_4]^- + CH_3COOCH_3 \qquad (98)$$

The same effect may be produced by iodide which is at least as efficient a nucleophile as pyridine and may give an acetyl iodide easily transformed into acetic acid and esters (Equations 99 and 100) [146].

$$CH_3COCo(CO)_4 + I^- \longrightarrow CH_3COI + Co(CO)_4^- \tag{99}$$

$$CH_3COI \begin{cases} \xrightarrow{H_2O} CH_3COOH + HI \\ \xrightarrow{R'OH} CH_3COOR' + HI \end{cases} \tag{100}$$

Accordingly the selectivity to homologation products appears to be decreased by increasing the I/Co ratio.

9.2. Iron Catalysts

Only one reaction mechanism for methanol homologation to ethanol catalyzed by $Fe(CO)_5/NMe_3$ has been reported and discussed [67, 174]. The course and the steps proposed for the reaction are quite different from those for cobalt catalysts and have been already described in Section 2.1.1 (Equations 19–24). In fact methanol is activated, via methyl formate, through transformation into a methyl quaternary ammonium salt which functions as methylating agent for $[HFe(CO)_4]^-$ [175, 176] (Scheme 5). For the methyl migration-carbonyl insertion step the considerations previously discussed for cobalt systems are probably equally effective. However, the intermediate formation of a hydrido-alkyl derivative which can give easily methane via a reductive elimination, just as the hydrido-acyl gives acetaldehyde, is cause of the large hydrocarbon formation.

On the other hand the high selectivity for methanol over ethanol homologation with this system is related to the much slower transfer of ethyl group from the quaternary ammonium salt to $[HFe(CO)_4]^-$ in a Sn2 type reaction due to the larger steric effect in a back-side attack [67, 174].

9.3. Ruthenium Catalysts

The essential requirements for the ruthenium catalysts to be active in homologation reaction is the presence of an iodide promoter which may be I_2, an alkyl iodide, a metal iodide or a quaternary ammonium or phosphonium iodide.

9.3.1. FORMATION OF CATALYTICALLY ACTIVE SPECIES FROM PRECURSORS

Different ruthenium compounds, soluble in the reaction medium ($Ru_3(CO)_{12}$, $Ru(acac)_3$, $RuCl_3 \cdot 3H_2O$, etc.) or insoluble (RuO_2, etc.) have been used as catalytic precursors and in all cases neutral or anionic iodocarbonyl ruthenium derivatives were detected in solution under reaction conditions or at the end of the runs (Scheme 6) [12, 76, 93].

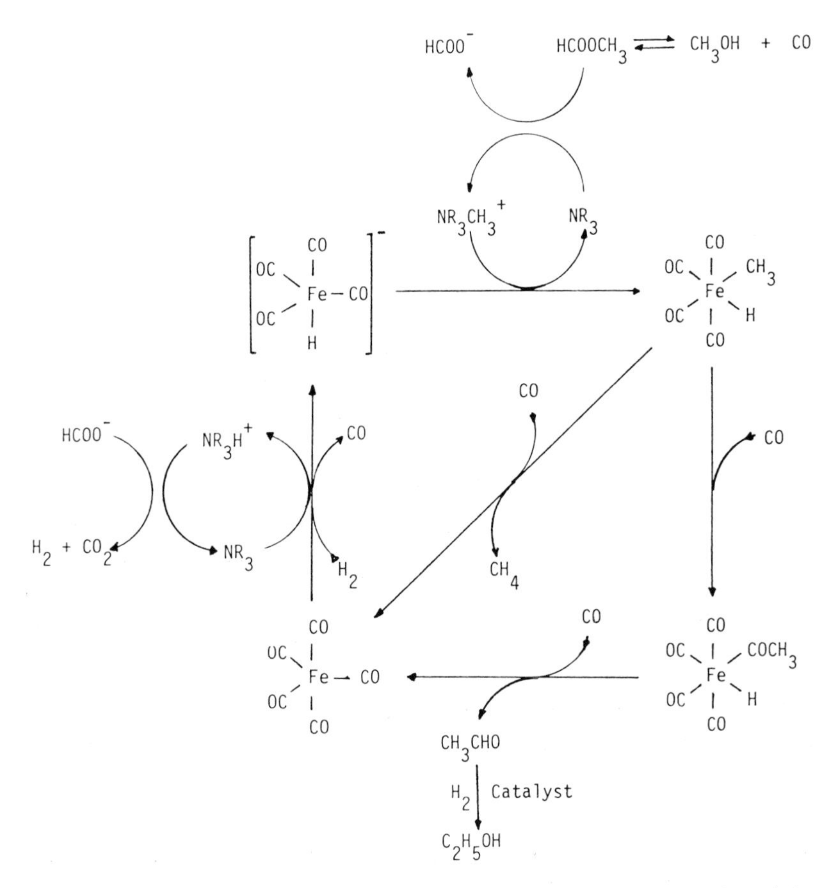

Scheme 5. Mechanism suggested for iron/amine-catalyzed methanol homologation.

The anionic hydridocarbonyl derivative $[HRu_3(CO)_{11}]^-$ has also been found at low I^-/Ru ratio [76]. Preformed neutral $Ru(CO)_4I_2$ is catalytically inactive and further addition of iodides such as CH_3I or NaI is necessary to activate the system [93]. On the other hand a large excess of iodide inhibits the homologation reaction due to the formation of $[Ru(CO)_2I_4]^{2-}$, catalytically inactive [177]. This suggests that the active species must have a $I/Ru > 2$, probably 3. Moreover, the observed necessity of a proton supplier for the reaction to take place, together with the higher catalytic activity in polar solvents and the strong salt effect, suggests an ionic nature for the catalytically active species [93, 122].

In the proposed reaction mechanism schemes [12, 93], however, the coordination sphere of ruthenium is not precisely defined and the species involved in catalysis is described as $[Ru(CO)_xI_y]^-$, it could be for example $[Ru(CO)_3I_3]^-$.

$$\left.\begin{array}{l} Ru(acac)_3 \\ Ru_3(CO)_{12} \\ RuCl_3 \end{array}\right\} \xrightarrow[CH_3I\ (HI)]{CO\ +\ H_2} \left\{ \begin{array}{c} cis\ \ Ru(CO)_4I_2 \xrightarrow{NaI,\ -\ 2\ CO} Na_2Ru(CO)_2I_4 \\ +I^- \ -CO \downarrow\uparrow + CO\ -\ I^- \\ [Ru(CO)_3I_3]^- \end{array}\right.$$

$$[Ru(CO)_3I_3]^- \xrightarrow{Na^+} fac\ \ Na^+[Ru(CO)_3I_3]^-$$

$$Ru_3(CO)_{12} \xrightarrow[CO\ (15\ atm),\ 100\ °C]{NaI,\ AcOH} fac\ \ Na^+[Ru(CO)_3I_3]^- \xleftarrow[NaI,\ AcOH]{CO\ +\ H_2,\ 200\ °C} Ru(acac)_3$$

Scheme 6. Ruthenium species present in solution in the homologation reaction.

Phosphine ligands, though affecting catalytic activity analogously to the rhodium catalyzed methanol carbonylation reaction, do not coordinate to the metal in the presence of CH_3I [178]. In fact free phosphine and/or phosphine originally attached to ruthenium is quaternarized by CH_3I and displaced from the complexes forming phosphonium salt of $[Ru(CO)_3I_3]^-$ [123, 128]. It is therefore likely that phosphonium salts may affect the kinetics of the reaction only in the successive reaction steps. On the other hand the formation of species of the type $Ru(CO)_2L_2I_2$ with ligand coordinated to the metal has been demonstrated using $SbPh_3$ not easily quaternarized with CH_3I, or PPh_3 and LiI as iodide promoter [128].

9.3.2. ACTIVATION OF THE SUBSTRATES AND FORMATION OF ALKY- AND ACYL-RUTHENIUM INTERMEDIATES

Ruthenium systems have the particular capacity to activate not only alcohols and ethers but also carboxylic acids and their esters in homologation reactions.

Two different hypotheses have been proposed for the activation of the substrates.

Protonation of the substrate and its nucleophilic attack to $[Ru(CO)_xI_y]^-$.

The ascertained necessity of a proton donor and the higher activity found at low pH in the methyl acetate homologation and in the carbonylation-homologation of dimethyl ether, are indicative of an activation of the substrate through a protonation step [93]. Protonating agents can be: HI, initially added or formed by hydrogenolysis or hydrolysis of CH_3I, weak mineral acids such as H_3BO_3 or carboxylic acids as CH_3COOH, HCOOH, and perhaps hydrido-iodocarbonyl ruthenium derivatives [93, 177].

The protonated substrate, probably the counterion of an anionic ruthenium iodocarbonyl derivative, could produce alkyl or acyl-ruthenium intermediates by a nucleophilic attack on $[Ru(CO)_x I_y]^-$ when the substrate is an alcohol and an ether in the former case or an acid in the latter (Scheme 7). Both alkyl or acyl derivatives can be formed from protonated esters as indicated by the formation of homologation products arising from both the alcoholic and acyl moiety.

$$\left.\begin{array}{l} \left[CH_3C\begin{smallmatrix}OH\\OCH_3\end{smallmatrix}\right]^+ \\ \left[CH_3C\begin{smallmatrix}OH\\OH\end{smallmatrix}\right]^+ \\ \left[CH_3O\begin{smallmatrix}H\\H\end{smallmatrix}\right]^+ \\ \left[CH_3O\begin{smallmatrix}H\\OCH_3\end{smallmatrix}\right]^+ \end{array}\right\} [Ru(CO)_xI_y]^- \begin{array}{l} \xrightarrow{-CO} CH_3CORu(CO)_{x-1}I_y \\ \quad -CO \downarrow \uparrow +CO \\ \xrightarrow{-CO} CH_3Ru(CO)_{x-1}I_y \end{array}$$

Scheme. 7. Activation of the substrates and formation of alkyl and acyl ruthenium intermediates.

Ion-pairing might also result from an interaction of a neutral iodocarbonyl ruthenium derivative (i.e., $Ru(CO)_4I_2$) with a protonated substrate containing I^- as a counterion (Equation 101).

$$\left[CH_3\overset{H}{\overset{|}{O}}CH_3\right]^+ I^- + Ru(CO)_{x+1}I_{y-1} \longrightarrow$$

$$\left[CH_3\overset{H}{\overset{|}{O}}CH_3\right]^+ [Ru(CO)_xI_y]^- + CO \qquad (101)$$

It must however be pointed out that a large excess of strong acids (HI, HCl) as in cobalt-iodine systems [53], due probably to the displacement of CO ligands from the metal, has an inhibiting effect on the reaction rate [177].

Intermediate transformation of the substrate into an alkyl or acyl iodide and their oxidative addition to an iodocarbonyl ruthenium derivative.

An alternative hypothesis on the activation of the substrate suggests the intermediate transformation of methyl esters into methyl iodide (Equation 102) [128] and of acids into acyl iodides (Equation 103) [12] followed by an oxidative addition to Ru(O) or Ru(II) intermediates (Equations 104 and 105).

$$RCOOCH_3 + HI \rightleftharpoons CH_3I + RCOOH \tag{102}$$

$$RCOOH + HI \rightleftharpoons RCOI + H_2O \tag{103}$$

$$CH_3I + [Ru] \longrightarrow CH_3[Ru]I \tag{104}$$

$$RCOI + [Ru] \longrightarrow RCO[Ru]I \tag{105}$$

These hypotheses are based on the positive dependence of the reaction rate on methyl iodide concentration but do not agree with the inertia of CH_3I in carbonylation and homologation reactions [74] and do not explain the actual homologation of formyl derivatives which definitely cannot presuppose the intermediate formation of a non-existent formyl iodide.

Moreover the oxidative addition of CH_3I seems improbable as the reaction of $Ru(CO)_3(PPh_3)_2$ with an excess of CH_3I does not occur with prolonged heating at 80°C [179].

9.3.3. CARBONYLATION OF ALKYL TO ACYL INTERMEDIATE

No experimental data exist on the alkyl migration–CO insertion step to transform alkyl into acyl ruthenium intermediate under homologation conditions.

An equilibrium step (Scheme 7) has been proposed [93] analogous to that reported for the $CH_3Ru(CO)_2(PR_3)_2Cl$ systems for which an initial intramolecular methyl migration to a CO group followed by a *trans* attack of gaseous CO to the acetyl ligand has been demonstrated (Equation 106) [180].

$$Ru(CO)_2(CH_3)(PR_3)_2Cl \rightleftharpoons Ru(CO)(COCH_3)(PR_3)_2Cl \underset{-CO}{\overset{+CO}{\rightleftharpoons}} Ru(CO)_2(COCH_3)(PR_3)_2Cl \tag{106}$$

Influences on this interconversion step of ion-pairing, proton and Lewis acids, analogous to those discussed for cobalt systems, may be foreseen and some experimental data obtained in the homologation of methyl acetate might be related to these effects [122, 123]; however, up to now no direct proofs exist to clearly define this point.

In any case, since hydrocarbon formation arises from hydrogenation of alkyl intermediates, all ligands and promoters able to accelerate the alkyl migration–CO insertion step may succeed in improving selectivity to valuable carbonylation and homologation products, thus limiting the formation of undesirable hydrocarbons. This is the key point of the process and certainly needs further inquiry to solve the main problem of low selectivity in homologation reactions.

9.3.4. EVOLUTION OF THE ACYL INTERMEDIATE

Acyl ruthenium intermediates, unlike the corresponding cobalt derivatives, are not hydrogenolyzed to aldehydes under homologation conditions but they can be hydrogenated directly to alcohols or hydrolyzed by water or alcohols to acids and esters (Scheme 8).

$$CH_3CORu(CO)_{x-1}I_y \xrightarrow{H_2} CH_3CH_2ORu(CO)_{x-1}I_y \xrightarrow{H_2} HRu(CO)_{x-1}I_y + C_2H_5OH$$

$$CH_3CORu(CO)_{x-1}I_y \xrightarrow{HA} HRu(CO)_{x-1}I_y + CH_3C(=O)A$$

(A = OH, OR)

Scheme 8. Evolution of the acyl intermediate.

The peculiar capacity of ruthenium of catalyzing homologation reactions without intermediate formation of aldehydes and the marked decrease in catalytic activity when passing from methyl and ethyl to higher alkyl derivatives may be related to the particular structure, C- and O-bonded, of the ruthenium acyl derivatives. This rather unusual structure has actually been found for the acetyl-iodocarbonyl ruthenium complex $Ru[\eta^2\text{-}C(O)CH_3]CO(PPh_3)_2I$ [181] and accordingly a particularly high selectivity to homologation products has been attained by addition of phosphine ligands to the iodocarbonyl ruthenium catalysts [123].

This type of acyl intermediate may react more easily with molecular hydrogen giving an alkoxy intermediate successively hydrogenolyzed to alcohol and restoring the starting hydrido-ruthenium derivative. Ethanol without traces of acetaldehyde has actually been obtained by stoichiometric hydrogenation with molecular hydrogen at 60 atm and 150°C of C- and O-bonded acyl derivative of $Ru[\eta^2\text{-}C(O)CH_3](CO)(PMe_2Ph)_2I$ [177]. This behaviour is analogous to that demonstrated in the reduction of CO promoted by hydrido-derivatives of permethylzircocene [182].

On the other hand the acyl intermediates with higher alkyl groups ($> C_2$) and a stronger carbenoid character of the Ru—C bond in the acyl undergo the nucleophilic attack of water and alcohols more easily, resulting in an increase on the selectivity to simple carbonylation products [93].

An analogous effect observed with alkaline iodides as promoters instead of methyl iodide can be attributed to the carbonyl group polarization by the alkaline cation of the acyl intermediate, formally represented as

$$Na^{+}\cdots O{=}\overset{\displaystyle R}{\overset{|}{C}}Ru(CO)_xI_y$$

which favours the nucleophilic attack on the carbon atom by water and alcohols [122]. The effect of the various cations (Li $\simeq$ Na $\gg$ Cs), analogous to that observed in the hydrogenation of ketones and esters with hydrido-ruthenium catalysts [109], accounts well for the change in activity and selectivity of the process. Thus the slightly lower activity of LiI with respect to NaI and KI may be related to the higher solvation of Li^{+} which hinders the interaction with the acyl group [122]. On the other hand ammonium or tetramethyl ammonium cations, unable to interact with the oxygen of the acyl as alkaline cations do, produce effects on activity and selectivity analogous to those observed by adding *N*-bases such as pyridine or piperidine. These bases, according to what is observed in the reaction of acetyl tetracarbonyl cobalt with methanol [172] and in the cobalt catalyzed hydrocarboxylation of alkenes [173], lead to the formation of *N*-acetyl ammonium or pyridinium salts of the iodocarbonyl ruthenium derivative (Equations 107–109) which undergo the nucleophilic attack of water or alcohols more quickly than the covalent acyl ruthenium intermediate [123].

$$CH_3R_3NI \rightleftharpoons R_3N + CH_3I \qquad (107)$$

$$CH_3COR''(CO)_xI_y + R_3N \xrightarrow{CO} [CH_3CONR_3]^{+}[Ru(CO)_{x+1}I_y]^{-} \qquad (108)$$

$$[CH_3CONR_3]^{+}[Ru(CO)_{x+1}I_y]^{-} \xrightarrow[\text{fast}]{HA}$$

$$[HNR_3]^{+}[Ru(CO)_{x+1}I_y]^{-} + CH_3COA \qquad (109)$$

($R = CH_3$; $A = OH, OR$; R_3N = tertiaryamine, pyridine)

The presence in the reaction solution of *N*-acetyl ammonium or piperidinium iodocarbonyl ruthenium salts is supported by the appearance in the carbonyl stretching region of the I.R. spectrum of a new broad absorption in the range 1925–1945 cm^{-1}, frequencies which are considerably lower than those typical for $[Ru(CO)_3I_3]^{-}$ and for species with coordinated nitrogen ligands [183].

An alkoxy [93] or a hydroxymethyl iodocarbonyl ruthenium derivative [12] (Equations 110, 111) has been proposed as intermediate in the hydrogenation of acetyl ruthenium complexes to ethanol.

$$Ru(CH_3CO)(CO)_xI_y + H_2 \longrightarrow C_2H_5ORu(CO)_xI_y$$

$$\xrightarrow{H_2} HRu(CO)_xI_y + C_2H_5OH \qquad (110)$$

$$Ru(CH_3CO)(CO)_xI_y + H_2 \longrightarrow CH_3\underset{\displaystyle OH}{\underset{|}{C}}HRu(CO)_xI_y$$

$$\xrightarrow{H_2} HRu(CO)_xI_y + C_2H_5OH \qquad (111)$$

Even if at the present time no direct evidence exists for the presence of these derivatives, indirect proof of the formation of the alkoxy intermediate is given by the production of hydrocarbon products through a dehydrogenation-decarbonylation step. Thus for example, ethane has been found among the products coming from a homologation reaction on propionic derivatives [56] which presupposes the intermediate formation of a *n*-propoxy compound (Equation 112) [184].

$$CH_3CH_2CH_2ORu(CO)_xI_y \longrightarrow CH_3CH_3 + RuH(CO)_{x+1}I_y \qquad (112)$$

Moreover, reliable indications on the intermediate formation of alkoxy-ruthenium derivatives arise from studies on homogeneous dehydrogenation of primary and secondary alcohols catalyzed by carboxylato-carbonyl ruthenium complexes of the $Ru(RCOO)_2CO(PPh_3)_2$ type [185] and on hydrogen transfer from alcohols to aldehydes and ketones catalyzed by $H_2Ru(PPH_3)_4$ [186].

On the other hand α-hydroxyalkyl metal complexes are believed to be involved in Fischer-Tropsch syntheses [187] and in the hydrogenation of aldehydes with ruthenium catalysts [188] and stable α-hydroxymethyl rhenium and iridium complexes have been successfully isolated [151, 189].

9.3.5. REGENERATION OF THE STARTING CATALYTICALLY ACTIVE SPECIES

The reaction of the acyl intermediate with hydrogen, water or alcohols leads to the formation of a hydrido-iodocarbonyl ruthenium derivative of the $HRu(CO)_xI_y$ type. At this point two possibilities exist: the hydrido derivative is directly able to activate the substrate (i.e., by protonation) so initiating a new cycle; or the hydrido undergoes a reductive elimination of HI to give a coordinatively unsaturated species, $Ru(CO)_xI_{y-1}$, able to react with CO or with the activated substrate to start the catalytic cycle.

No supporting evidence for these steps exists at present. They may only be presumed by analogy with the corresponding cobalt or rhodium systems.

9.4. Rhodium Catalysts

Two differenti types of rhodium catalysts have been used in the homologation reactions: rhodium carbonyls with iodide promoters and carbonyl-phosphine rhodium complexes without iodide promoters.

The former system, certainly the best catalyst for carbonylation reactions on oxygenated substrates, is however only a poor catalyst for homologation reactions being active only under very high H_2/CO ratios. The reaction mechanism up to the step of formation of the acyl intermediate (Equations 113–116) is the same which operates in the methanol carbonylation to acetic acid [178].

$$CH_3OH + HI \rightleftharpoons CH_3I + H_2O \quad (113)$$

$$[Rh(CO)_2I_2]^- + CH_3I \longrightarrow [CH_3Rh(CO)_2I_3]^- \quad (114)$$

$$[CH_3Rh(CO)_2I_3]^- \longrightarrow [CH_3CORh(CO)I_3]^- \quad (115)$$

$$[CH_3CORh(CO)I_3]^- + CO \longrightarrow [CH_3CORh(CO)_2I_3]^- \quad (116)$$

Exhaustive discussions on these steps have been reported by Forster [94, 178] and by Hjortkjaer [190, 191] and further comments on these go beyond the limits of the homologation field. Concerning the hydrogenation of the acyl intermediate Dumas [25] proposed very schematically an oxidative addition of hydrogen to the acyl intermediate (Equation 117),

$$CH_3CO{-}[Rh] + H_2 \longrightarrow CH_3CO{-}\underset{\displaystyle H}{\underset{|}{[Rh]}}{-}H \quad (117)$$

followed either by a reductive elimination of aldehyde (Equation 118) or by formation, in the presence of a protolytic solvent, of a hydroxyethyl Rh(IV) derivative which successively produces ethanol or acetaldehyde by a reductive elimination (Equation 119).

$$CH_3CO{-}\overset{\displaystyle H}{\overset{|}{[Rh]}}{-}H \longrightarrow CH_3CHO + [Rh]{-}H \quad (118)$$

$$\underset{H-\overset{\uparrow}{O}-CH_3}{CH_3CO{-}\overset{\displaystyle H}{\overset{|}{[Rh]}}{-}H} \longrightarrow CH_3{-}\underset{OH}{\underset{|}{\overset{\displaystyle H}{\overset{|}{C}}}}{-}\underset{OCH_3}{\underset{|}{[Rh]}}{-}H \begin{cases} \nearrow C_2H_5OH + [Rh]{-}OCH_3 \\ \searrow CH_3CHO + [Rh]{-}H + CH_3OH \end{cases} \quad (119)$$

The methoxy derivative is then hydrogenolyzed (Equation 120) and the acyl intermediate is regenerated by reaction with CH_3I (Equation 121).

$$[Rh]{-}OCH_3 + H_2 \longrightarrow \underset{\displaystyle H}{\underset{|}{\overset{\displaystyle H}{\overset{|}{[Rh]}}}}{-}OCH_3 \longrightarrow [Rh]{-}H + CH_3OH \quad (120)$$

$$[Rh]{-}H + CH_3I + CO \longrightarrow CH_3CO{-}[Rh] + HI \quad (121)$$

Experimental support for this picture, however, is completely lacking.

Quite a different mechanism is thought to operate in the homologation of formaldehyde catalyzed by rhodium phosphine complexes without iodide promoters. Spencer [100] and Goetz [99], in consideration of the higher activity shown by phosphine chlororhodium derivatives and the recovery at the end of the runs of most of the rhodium as $RhClCO(PPh_3)_2$, suggested that the active species should be a di-hydrido-chlorocarbonylphosphine derivative (Equation 122).

$$RhClCO(PPh_3)_2 + H_2 \rightleftharpoons RhH_2ClCO(PPh_3)_2 \quad (122)$$

The dissociation of PPh_3 by intervention of the solvent would allow the activation of CH_2O which inserts into a Rh—H bond giving a hydroxymethyl intermediate and after carbonylation, the homologous aldehyde is formed by a reductive elimination step (Equations 123–125).

$$RhH_2ClCO(PPh_3)_2 \underset{+PPh_3 - S}{\overset{-PPh_3 + S}{\rightleftharpoons}} RhH_2ClCOPPh_3S$$

$$\underset{-CH_2O + S}{\overset{+CH_2O - S}{\rightleftharpoons}} RhH_2Cl(CH_2O)COPPh_3 \quad (123)$$

$$RhH_2Cl(CH_2O)COPPh_3 \longrightarrow HOCH_2RhHClCOPPh_3$$

$$\xrightarrow{+S} HOCH_2CORhHClPPh_3S \quad (124)$$

$$HOCH_2CORhHClPPh_3S + CO \longrightarrow$$
$$RhClCOPPh_3S + HOCH_2CHO \quad (125)$$

Chan's findings contradict this picture, since halide-containing species turn out to be less active than the corresponding cationic catalysts (i.e., $[Rh(diene)(PPh_3)_2]^+ BF_4^-$); for this reason $RhH(CO)_2(PPh_3)_2$ is suggested as the key species for the activation of formaldehyde. This occurs either by displacement of PPh_3 or, when an excess of ligand is used, by a Lewis base promoted deprotonation of the hydrido derivative (amides are the most useful bases) (Scheme 9). In the former case, the coordinatively unsaturated hydrido species may coordinate a CH_2O molecule which inserts into a Rh—H bond producing a hydroxymethyl intermediate. Examples of "side bound" formaldehyde complexes of osmium, $Os(\eta^2\text{-}CH_2O)(CO)_2(PPh_3)_2$ [192], and iron, $Fe(\eta^2\ CH_2O)(CO)_2[P(OMe)_3]_2$ [189], and of hydroxymethyl derivatives of rhenium, $(C_5H_5)ReCO(NO)(CH_2OH)$ [193], and iridium, $Ir(CH_2OCH_3)(C_2H_4)(PMe_3)_3$ [151], have been recently reported.

In the latter case, a direct nucleophilic attack of CH_2O to the anionic intermediate with formation of the hydroxymethyl derivative may be foreseen. Carbonylation and successive hydrogenolysis of the hydroxymethyl intermediate leads to the homologous glycol aldehyde whereas the direct hydrogenolysis gives methanol (Scheme 9).

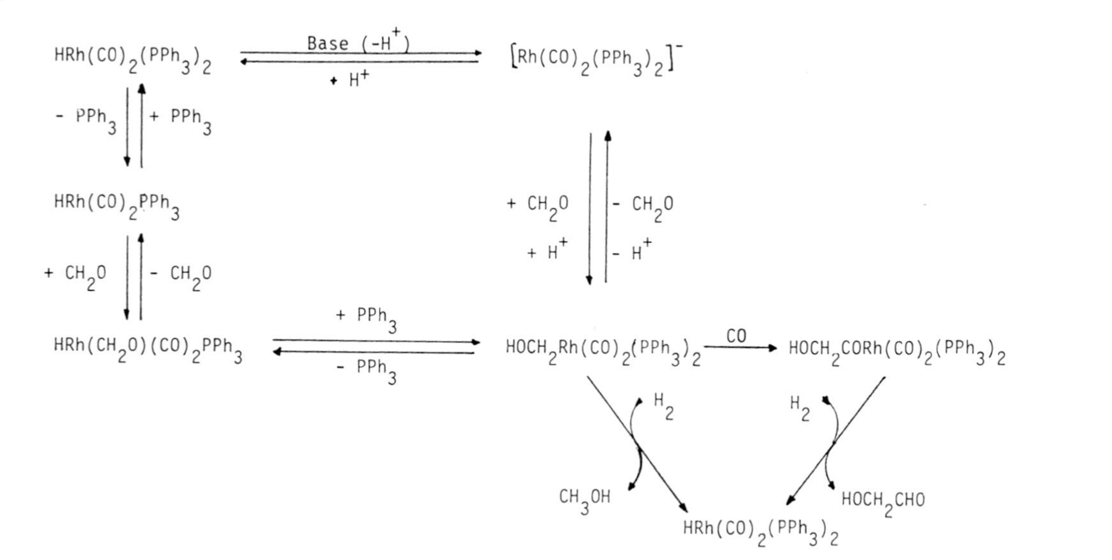

Scheme 9. Mechanism proposed for the activation of formaldehyde and its homologation to glycolaldehyde.

10. CONCLUDING REMARKS

At the end of this review on an argument of wide industrial interest two sorts of judgement must be made, one merely scientific and the other technical-economical.

The former is more easily made and in our case is certainly positive: the work done in the last five or six years in the field of homologation reaction has in fact been very productive and has achieved important goals. Thus homologation reactions, initially limited to alcoholic substrates, have been extended to aldehydes, ethers, carboxylic acids and carboxylic acids esters. New catalytic systems, based on different VIII group metals – Fe, Ru, Rh – with better performances with respect to the old cobalt catalysts have been developed and knowledge of the mechanism of their action has been in some cases reached. Of particular interest are the rationalization and the efforts to elucidate the role and the effects of the various promoters and ligands added to the metallic systems which had been considered too long simply as "the salt and pepper of the reaction" without any attempt to understand the mechanism by which they are able to modify strongly activity and selectivity of the systems. And it is from this knowledge which leads us to expect that in the near future better performances in the different homologation reactions will be obtained as the most recent results seem to indicate [12, 23, 31, 34, 67, 93, 101, 122, 123, 126, 178, 194].

Among the different substrates used for the homologation reaction those considered more recently, such as acetic acid, methyl acetate and formaldehyde seem to be the most interesting for the commerical value of the products obtained: short chain aliphatic acids, ethyl acetate, glycolaldehyde and ethylene glycol; these are also the most promising for industrial development due to the wide margin of profitability of the products. Moreover dimethyl ether and methyl formate, produced directly from syngas or easily formed from methanol with simple catalytic processes, can become economic sources of C_2 and $>C_2$ chemicals if their homologation reactions can be improved and optimized.

As far as methanol homologation to ethanol and/or to acetaldehyde is concerned, probably only an improvement in reaction kinetics may be expected but this will only be of practical interest if the catalytic system is not too expensive and is easy to recycle.

It is more difficult to discuss the technical-economical prospectives of the homologation processes because of the uncertain politics governing the supply of raw materials and the lack of information on pilot plant initiatives.

However, after a period (which coincided with the worst crude oil crisis) when economic analyses foresaw a medium term wide use of syngas, made from alternative non-conventional raw materials, for producing base chemicals and fuels [11a], the situation is now being reexamined and a more conservative forecast for the development of the syngas processes is being considered [11b, 195].

Thus for example the industrial development for the route: methanol (from syngas) $\longrightarrow$ ethanol (by homologation) $\longrightarrow$ ethylene (by fluid bed dehydration) which according to a 1981 estimation [11a] was to become economically attractive in the 1990s, seems, at the time of writing at least, to have been put off to the turn of the century due to the stop in the growing of the cost of crude oil and the recent difficult times in the industry [195].

More optimistic forecast for a short term development may be foreseen for the homologation reactions on acetic acid, methyl acetate and formaldehyde, however, in view of their integration in a comprehensive process of C_2 derivatives from syngas as indicated in Scheme 10. Some of the pathways reported in this scheme have obtained an industrial development and may constitute a support for developing the others [195]. In this situation there are in particular the processes which involve acetic acid, whereas those involving acetaldehyde and formaldehyde still need a pilot plant test. In this context the hydrocarbonylation of methanol to acetaldehyde rather than its homologation to ethanol may show better prospectives in view of the versatility of this intermediate for the production of a large number of chemicals, including C_4 derivatives, 2-ethylhexanol, penta-erythrol, pyridine and alkyl pyridines.

11. ADDENDUM

Since writing and printing this review, a number of relevant articles and patents have been published. We have listed, for the convenience of the reader the following additions in the order in which they have been reported in the text with the appropriate section number.

2.1.1. COBALT CATALYSTS

Historic development of the catalytic system

A well-detailed description of the correct way of carrying out the experiments and analyses and of evaluating the yields and selectivities of the methanol homologation has at last been reported [196]. Experimental procedure, particularly during product recovery and work up must be designed to quantitatively determine all product fractions, especially the volatile derivatives such as methane, ethane, dimethylether, carbon dioxide etc., which are frequently not considered in the literature data. This is most important with respect to accountability of the methanol charged and to the correct evaluation of the reaction selectivity.

The accurate results reported [196] evidence the large formation of dimethylether and methane, especially when the I/Co ratio is > 2, and the change of the relative distribution of ethanol and acetaldehyde as a function of the I/Co ratio. Accordingly, almost equal amounts of ethanol and acetaldehyde are produced at I/Co = 0.5, whereas at I/Co = 1, the amount of acetaldehyde doubles and ethanol remains constant and at I/Co = 4.5 the formation of ethanol is practically suppressed.

A two-step process of production of ethanol from methanol and synthesis gas has been proposed as an alternative route to the direct homologation reaction [220]. The first step is the methanol carbonylation with rhodium catalysts to an

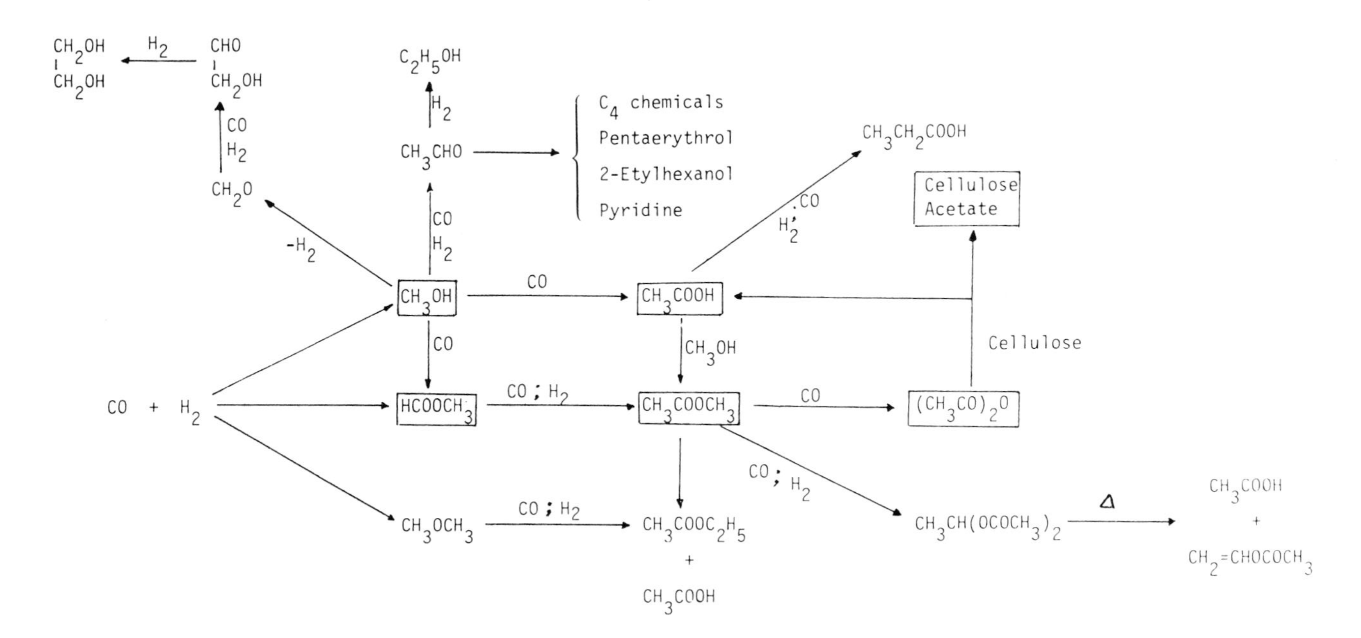

Scheme 10. Alternative paths to C_2 chemicals, starting from synthesis gas. Commercial processes exist for the products shown in rectangles.

acetic acid–methyl acetate mixture which is separated from the unconverted reagent, dimethyl ether and methyl iodide and hydrogenated in the second step with a heterogeneous catalyst ($CuO/Cr_2O_3/ZnO$ on silica and MgO and BaO as promoters) to a methanol-ethanol mixture.

Nature and concentration of the cobalt compound. Catalysts based on cobalt sulphides coupled with a phosphine ligand in a nitrogen-containing solvent such as an amide or *N*-methyl pyrrolidone, operating at high temperature (230–250°C) and pressure (250 atm) allow one to obtain very high selectivities to ethanol (maximum 92%) but at low methanol conversions (10–20%) [221]. Analogous results under comparable conditions are obtained using cobalt phosphate or cobalt phosphate coupled with heterogeneous ruthenium, palladium or rhodium catalysts [222].

Other promoters. The use of an inert particulate solid comprising a high surface area form of silica, alumina, silica-alumina or carbon black together with a cobalt-based catalyst or a cyclopentadienyl VIII group metal complex in the presence of iodide promoters and phosphine ligands increases the reaction rate and the ethanol selectivity in the methanol homologation reaction [223].

Ligands. The addition to the cobalt/iodide catalytic systems of a divalent *N*-, *P*- or As-ligand with the VA group atoms separated by a sterically constrained carbon—carbon bond of general formula **I** (see below) such as for example *cis*-bis(1,2-diphenyl-phosphino)ethylene or bis(1,2-diphenylphosphino)benzene, improves the selectivity to acetaldehyde and acetals up to 73% wt. [197].

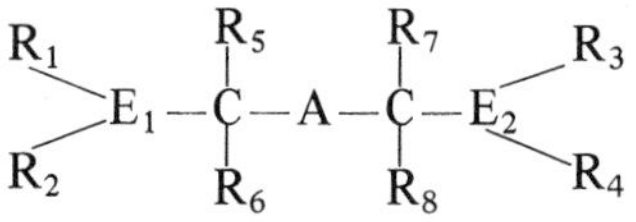

$$(R_1)(R_2)E_1-C(R_5)(R_6)-A-C(R_7)(R_8)-E_2(R_3)(R_4) \qquad \textbf{I}$$

E_1; E_2 = N, P, As; R_1; R_2; R_3; R_4 = alkyl or preferably aryl groups; R_5; R_6; R_7; R_8 = H, alkyl, aryl groups; A = olefinic, acetylenic or aromatic divalent organic group such as

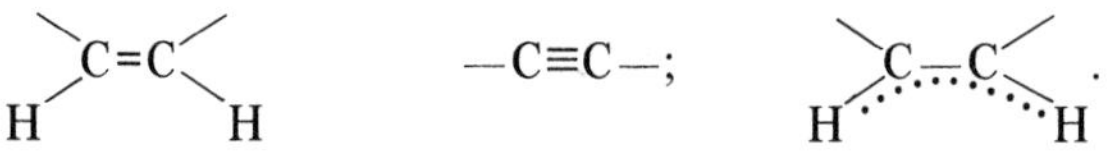

$$>C=C< \text{ (H, H)} \qquad -C\equiv C-; \qquad >C\overset{...}{-}C< \text{ (H, H)}.$$

An analogous effect has been obtained using diphosphines of general formula **II** [198].

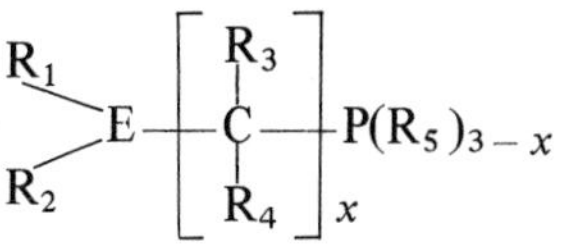

$$(R_1)(R_2)E-\left[C(R_3)(R_4)\right]_x-P(R_5)_{3-x} \qquad \textbf{II}$$

R_1; R_2; R_5 = alkyl, aryl group; R_3; R_4 = H, alkyl, aryl group; E = P; As.

The addition to the cobalt carbonyl catalysts of arsine and stibine ligands in conjunction with iodide promoters at an appropriate I/Co ratio (1/2), increases the selectivity to acetaldehyde up to 55% at a methanol conversion of 60% [199].

The same type of effect is produced by the use of preformed arsenic–cobalt tricarbonyl catalysts of type $R_3AsCo(CO)_3-Co(CO)_3AsR_3$ in conjunction with iodide promoters [200]. When triphenylphospine-cobalt-tricarbonyl dimer or triphenylstibine-cobalt-tricarbonyl dimer were used in place of the specified arsine–cobalt complexes, the selectivity to acetaldehyde was greatly reduced and correspondingly the selectivity to ethanol increased.

The addition of a tertiary phosphine oxide to a cobalt catalyst in absence of iodide promoters stabilizes the catalytic system and allows the reaction to proceed without hindrance; thus the products can be distilled without precipitation or loss of the cobalt catalyst which can be readily recycled [201].

Solvents. The use of inert solvents such as dioxane, tetrahydrofurane, toluene etc. at low temperatures (120–160°C) and at I/Co ratios of 4/1–8/1 is claimed to increase the selectivity to acetaldehyde and dimethyl acetal up to 80–85% [202].

Metallic cocatalysts. An increasing attention has been payed to the development of the Co—Ru mixed catalysts which undoubtedly offer the best performances for the methanol homologation to ethanol. Addition to a Co/Ru/I system (1/0.1/2) of a ligand (Co/Ligand: 1/0.5) containing atoms from Group VA of the periodic table separated by a sterically constrained carbon—carbon bond of general formula **I** improves the selectivity to realizable ethanol up to 70–80% (free ethanol 40–55%) at a high methanol conversion (70–80%) [203].

The addition also of a quaternary phosphonium salt (P/Co: 0.5–2/1) to the same Co/Ru/I systems (Co/Ru preferably 1/1) causes improvements in the ethanol selectivity of the same order (55–70%) at a methanol conversion of 60–80% [204].

The use of a diphosphine of formula **III** or of a diphosphite of formula **IV** is claimed to improve the selectivity to ethanol of the Co/Ru/I systems up to 80% at a methanol conversion of 30–50% [205].

$$(R_1)(R_2)P-(CH_2)_n-P(R_3)(R_4) \quad \textbf{III} \qquad (R_1O)(R_2O)P-O-(CH_2)_n-O-P(OR_3)(OR_4) \quad \textbf{IV}$$

R_1; R_2; R_3; R_4 = H, alkyl, aryl; n = 1–10.

Comparative data on the catalytic properties of bimetallic preformed Co—Ru derivatives versus mixtures of individual cobalt or ruthenium complexes seem to demonstrate the non-participation of bimetallic species in the homologation; the two metals however do operate independently but in a synergetic manner [206]. The methanol conversions and the selectivities to ethanol with the mixed catalysts are, in fact, considerably higher than those expected by a simple additive effect of the same type of individual catalysts. The selectivity of the reaction varies greatly with residence time and temperature and depends on the nature of the two metallic precursors only when operating at low temperatures. Thus at very short times the major product is acetaldehyde (mainly in the form of dimethylacetal), but it is impossible to obtain good yields in this product at high methanol conversions. The situation is the opposite for ethanol whose selectivity can reach 80–85%.

The catalytic activity of some salts of the anionic cluster $[RuCo_3(CO)_{12}]^-$ in the methanol homologation in the presence of CH_3I as promoter has been studied in detail [207]. An important effect of the nature of the counter-ion of the cluster on the course of the reaction has been observed: the tetra-phenyl-phosphonium ion allows one to obtain the highest selectivity to ethanol (60–65%) and the lowest to acetaldehyde and dimethylacetal. The optimal range of the molar ratio CH_3I/Ru was 10–15 as in the case of monometallic ruthenium catalysts. From the recovered solution after catalysis, salts of the anion $[Ru(CO)_3I_3]^-$ were isolated indicating a considerable decomposition of the bimetallic cluster. In spite of this, the authors believe that ruthenium–cobalt mixed carbonyl clusters containing iodide groups may be present under reaction conditions and responsible for the better performances of the ruthenium–cobalt 1/3 mixed catalyst [207].

The use of bimetallic carbonyl catalysts Co/Cr, Co/Mo, Co/W, Co/Ni, Co/Ti, Co/U or Ru/Mo with iodide promoters and polydentate phosphines, arsines or stibines is claimed to produce acetaldehyde–ethanol mixtures with a selectivity of 65–86% at a 62–80% conversion of methanol [224].

Homologation of methanol to C_4 compounds. A catalytic system based on cobalt carbonyls in the presence of iodide promoters (Co/I = 2/1) and tiols [208] or organic sulphide derivatives (Co/S = 1/1) [209] is selective for the production of C_4 derivatives (*n*-butanal and *n*-butanol). The addition of the sulfur derivatives improves the selectivity to the C_4 products (maximum: 32%) decreasing that of ethanol.

2.2.1. ALIPHATIC ALCOHOLS

A comparative study of the homologation of several alcohols has been carried out using the catalytic system CoI_2/NaI/bis(1,2-diphenylphosphine) ethane and charging pure or aqueous C_1–C_4 alcohols [196]. The order of reactivity in the absence of added water is: methanol $\gg$ ethanol $\gg$ 2-propanol (the ratio of reactivity is 157/13/1); in the presence of 5% vol. of water the reactivity is strongly enhanced and the order is: methanol > ethanol > 1-propanol $\simeq$ 1-butanol > 2-propanol $\simeq$ 2-butanol $\gg$ *t*-butanol. This behaviour differs from that reported in prior works using cobalt carbonyls and cobalt carbonyls promoted with iodine catalysts [210] where the order was: tertiary alcohols > secondary > primary, with the exception of methanol and benzyl alcohol. This difference can be attributed to the different nature and composition of the catalytic species: in the prior systems the reaction medium, due to the high concentration of the cobalt catalyst which produces $HCo(CO)_4$, is quite acidic and the protonation of alcohols should be facile with subsequent dehydration and formation of olefins; the olefins are then subjected to the hydroformylation catalyzed by $HCo(CO)_4$. The system now used, characterized by a lower cobalt concentration, a higher I/Co ratio (8.5) and a mild promoter system (NaI and a diphosphine) does not produce high amounts of $HCo(CO)_4$. As a consequence, the activation of alcohols involves the formation of alkyl iodides by a Sn2 attack by I^- on the protonated alcohols which favours the primary against the secondary and tertiary alcohols. In contrast, the hydroformylation of the olefin is drastically disfavoured due both to the presence of a diphosphine and to the low concentration of the active $HCo(CO)_4$.

3. CARBONYLATION AND HOMOLOGATION OF ETHERS

Palladium complexes in conjunction with a large concentration of alkyl iodides and phosphine ligands at a temperature of 160°C and 30–40 atm of CO/H_2 6/1 are active in the reductive carbonylation of dimethyl ether to acetaldehyde in methyl acetate solution with a selectivity of about 55%. To attain this object the volatile reaction product must be continuously moved away from the reaction mixture [211].

6. HOMOLOGATION OF CARBOXYLIC ACID ESTERS

A copromoting effect on the reaction rate of methyl acetate homologation to ethyl acetate due to the concurrent use of manganese and ruthenium soluble compounds in the presence of iodide promoters has been claimed [212]. Actually the catalyst performances ($\overline{TN} = 6.2 \times 10^{-3}$ sec^{-1}) are analogous to those reported for ruthenium or cobalt–ruthenium catalysts when used under appropriate conditions (see Tables 25 and 26).

The reductive carbonylation of dimethyl ether and methyl acetate in various solvents has been studied with mixed cobalt–ruthenium catalysts using different Co/Ru ratios and different iodide promoters [213]. The use of mixed catalysts instead of simple ruthenium catalysts seems to improve the activity and selectivity of the reaction in the case of methyl acetate homologation but not when dimethyl ether is used as substrate. The reason of this behaviour is not clear and no experimental data are at present available to elucidate this point. Our opinion is that the strong acid, $HCo(CO)_4$, could play a promoting effect in the activation of the substrate and/or in the evolution of the alkyl or acyl ruthenium intermediates.

Another interesting effect on the course of the homologation reaction of methyl acetate is produced by the addition to the carbonyl iodide ruthenium catalysts, together with the alkaline iodides, of small amounts of crown ethers. A remarkable increase both of the reaction rate (higher conversions) and of the selectivity to valuable products (ethyl acetate and acetic acid) was observed. Particularly with cesium and potassium iodides and 18-crown-6 ether, the improvement is important so that the selectivity becomes higher than 90% [214]. It seems that crown ethers, but also linear high molecular weight ethers, probably by removing the solvated molecules of the solvents from the ions, allow them to interact and activate the carbonyl groups and the acyl group of the catalytic intermediates. As a consequence, there is less formation of hydrocarbons produced by hydrogenation of the alkyl intermediates, and higher selectivities to ethyl acetate and acetic acid are obtained.

In an analogous manner, the promoting effect of some Lewis acids (AlI_3, TiI_4 etc.,) takes place when they are added in catalytic amounts to the carbonyl iodide ruthenium systems [214]. Lewis acids, especially AlI_3, increase the conversion of methyl acetate and reduce the selectivity towards hydrocarbons, simultaneously increasing that of acetic acid and acetic acid + ethyl acetate.

Protonic non-complexing acids, such as hexafluorophosphoric and tetrafluoboric acids, have an effect analogous, even if less pronounced than the effect Lewis acids have [214]. The addition of the acids (acid/Ru: 10–50/1) in fact decreases the selectivity towards hydrocarbons and favours the homologation reaction.

Ruthenium catalysts coupled with quaternary phosphonium salts have been used in the synthesis of ethyl esters of aliphatic carboxylic acids starting from synthesis gas and the appropriate acid [215, 216]. When propionic acid is used as co-reactant, high yields of C_1-C_4 alkyl propionates (75–95%) are obtained; among these, ethyl propionate is the major product. The more likely path of formation of ethyl propionate is a sequential route passing through methyl propionate (Equation 1′):

$$C_2H_5COOH + CO + 2H_2 \longrightarrow C_2H_5COOCH_3 + H_2O \xrightarrow{CO + H_2} C_2H_5COOC_2H_5 \qquad (1')$$

Under the reaction conditions, in fact, the hydrogenation of the carboxylic acid takes place only to a very minor extent.

Ethyl propionate and other alkyl propionates can be pyrolyzed with high yields to ethylene, propylene etc. and propionic acid which can be recycled to the synthesis [215]. This two-step process for the manufacture of ethylene from synthesis gas gives a total yield of olefin of about 50% and may become economically competitive with traditional pyrolysis of paraffinic cuts [215].

The homologation of the alkyl moiety of an ester can be stopped at the aldehyde level (Equation 2′) by using palladium catalysts operating at low temperatures (100°C) and low pressures (2–40 atm) of CO/H_2 in a 5–7/1 ratio [225].

$$CH_3COOCH_3 + CO + H_2 \longrightarrow CH_3COOH + CH_3CHO \qquad (2')$$

The catalytic system is based on a soluble palladium salt with an alkyl iodide and a phosphine ligand; the selectivity to acetaldehyde is in the range 50–80%.

9.3. Ruthenium Catalysts

9.3.1. FORMATION OF CATALYTICALLY ACTIVE SPECIES FROM PRECURSORS

When an iodide salt, such as KI, is added to $Ru_3(CO)_{12}$ under high CO + H_2 pressure in a polar solvent, such as sulfolane, *N*-methylpyrrolidone etc., a catalytic system, active in the methanol and/or ethylene glycol synthesis from CO + H_2 and in the homologation of methanol to ethanol is produced [217]. The simultaneous presence under reaction conditions of anionic hydridocarbonyl and iodocarbonyl ruthenium species produced according to Equation 3′ has been demonstrated by I.R.

$$7/3\ Ru_3(CO)_{12} + 3I^- + H_2 \longrightarrow 2HRu_3(CO)_{11}^- + Ru(CO)_3I_3^- + 3CO \qquad (3')$$

It was seen that $Ru(CO)_3I_3^-$, which is involved in the homologation reaction of oxygenated substrates, was not active for CO hydrogenation either in the presence or absence of KI. In contrast, $HRu_3(CO)_{11}^-$ showed some activity alone in the presence of KI, displaying, however, maximum activity when $Ru(CO)_3I_3^-$ is also present in a ratio of $HRu_3(CO)_{11}^-$: $Ru(CO)_3I_3^-$ of about near 2:1. $HRu_3(CO)_{11}^-$ must be considered as a precursor of the species active in CO hydrogenation, which is probably the mononuclear carbonyl derivative $HRu(CO)_4^-$.

The same species is also produced from $Ru_3(CO)_{12}$ and a quaternary phosphonium salt in acetic acid solution; the catalytic system is active in the synthesis of ethylene glycol acetates from CO + H_2 [218, 219].

12. GLOSSARY OF ABBREVIATIONS

acac: acetylacetonate
Ac: CH_3CO
COD: 1,5-cyclooctadiene
Cp: cyclopentadienyl
Et: ethyl
NBD: nor-bornadiene
Me: methyl
Ph: phenyl
P--P: bis-diphenylphosphinethane
TN: turn-over number

13. REFERENCES

1. G. Wietzel, K. Eder, and A. Scheuerman (B.A.S.F.): Ger. Patent 867,849 (1941); *C.A.* 48, 1408a (1954).
2. G. Wietzel, G. Vorbach, and A. Scheuerman (B.A.S.F.): Ger. Patent 875,346 (1953).
3. I. Wender, R. Levine, and M. Orchin: *J. Am. Chem. Soc.* 71, 4160 (1949).

4. I. Wender, R. A. Friedel, and M. Orchin: *Science* **113**, 206 (1951).
5. I. Wender, H. Greenfield, and M. Orchin: *J. Am. Chem. Soc.* 73, 2656 (1951).
6. W. F. Gresham (E. I. Du Pont de Nemours): U.S. Patent 2,535,060 (1950).
7. B. W. Howk and G. L. Hager (E. I. Du Pont de Nemours): U.S. Patent 2,549,470 (1951).
8. K. H. Ziesecke: *Brennst. Chem.* **33**, 385 (1952).
9. J. Berty, L. Markò, and D. Kallò: *Chem. Tech.* (Leipzig) 8, 260 (1956).
10. G. Marullo and A. Baroni (Montecatini): Italian Patent 484,182 (1953).
11. J. Haggin: *Chemical Eng. News* **59**(20), 52 (1981); *ibidem*, **60**(20), 28 (1982).
12. J. F. Knifton: *J. Mol. Catal.* **11**, 91 (1981).
13. H. Bahrmann and B. Cornils: *Homologation of Alcohols* (New Syntheses with Carbon Monoxide, ed. J. Falbe), pp. 226–242. Springer Verlag (1980).
14. I. Fleming and H. I. Bolker: *Can. J. Chem.* **54**, 685 (1976).
15. D. W. Slocum: *Chemicals from Coal: The Cobalt Octacarbonyl Catalyzed Homologation of Methanol to Ethanol* (Catalysis in Organic Syntheses, ed. W. Jones), pp. 245–276. Academic Press (1980).
16. G. S. Koermer and W. E. Slinkard: *Ind. Eng. Chem., Prod. Res. Dev.* **17**, 231 (1978).
17. G. Albanesi: *Chim. Ind. (Milan)* **55**, 319 (1973).
18. B. R. Gane (British Petroleum Co.): Eur. Patent 1936 (1978).
19. B. R. Gane and D. G. Stewart (British Petroleum Co.): Eur. Patent 10373 (1979).
20. M. T. Barlow (British Petroleum Co.): Eur. Patent 29723 (1980).
21. (a) J. Gauthier-Lafaye, R. Perron, and Y. Collenille: *J. Mol. Catal.* **17**, 339 (1982);
    (b) J. Gauthier-Lafaye and R. Perron (Rhone-Poulenc Ind.): Eur. Patent 11042 (1979);
    (c) *idem*., Eur. Patent 22735 (1981).
22. K. H. Keim, J. Korff (Union Rheinische Braunkohlen Kraftstoff A.G.): Ger. Patent 3,031,558 (1980).
23. M. Röper, H. Loevenich, and J. Korff: *J. Mol Catal.* **17**, 315 (1982).
24. M. Röper and H. Loevenich: '*The Homologation of Methanol*' in *Catalysis in* C_1 *Chemistry*, ed. by Wilhelm Keim (Catalysis by Metal Complexes, vol. 4, ed. R. Ugo), pp. 105–134. Reidel (1983).
25. H. Dumas, J. Levisalles, and H. Rudler: *J. Organomet. Chem.* **177**, 239 (1979).
26. A. Deluzarche, G. Jenner, A. Kiennemann, and F. Abou Samra: *Erdoel, Kohle-Erdgas-Petrochem., Brennst. Chem.* **32**, 436 (1979).
27. W. E. Walker (Union Carbide Corp.): U.S. Patent 4,277,634 (1981).
28. A. D. Riley and W. O. Bell Jr. (Commercial Solvents Corp.): U.S. Patent 3,248,432 (1966).
29. L. H. Slaugh (Shell Int. Res. Maatsch.): Ger. Patent 2,625,627 (1976).
30. B. R. Gane and D. G. Stewart (British Petroleum Co.): Eur. Patent 1937 (1978).
31. W. R. Pretzer and T. P. Kobylinski: *Ann. N.Y. Acad. Sci.* **333**, 58 (1980).
32. M. Kuraishi and A. Takahashi: U.S. Patent 3,356,734 (1967).
33. G. N. Butter (Commercial Solvents Corp.): U.S. Patent 3,285,948 (1966).
34. (a) W. R. Pretzer, T. P. Kobylinski, and J. E. Bozik (Gulf Res. and Dev. Co.): U.S. Patent 4,133,966 (1979);
    (b) *idem*., U.S. Patent 4,239,924 (1980).
35. T. J. Devon (Eastman Kodak Co.): U.S. Patent 4,328,379 (1982).
36. B. R. Gane, D. G. Stewart (British Petroleum Co.): U.K. Patent 2,036,739 (1979).
37. J. Gauthier-Lafaye and R. Perron (Rhone-Poulenc Ind.): Eur. Patent 22038 (1980).
38. G. Doyle (Exxon Res. and Eng. Co.): Eur. Patent 30434 (1980).
39. (a) R. A. Fiato (Union Carbide Corp.): U.S. Patent 4,233,466 (1980);
    (b) *idem*., U.S. Patent 4,253,987 (1980).
40. M. Idahi, M. Orisaku, M. Ue, and Y. Uchida: *Chem. Lett.* **143** (1981).
41. J. Korff, M. Fremery, and J. Zimmermann (Union Rheinische Braunkohlem Kraftstoff A.G.): Ger. Patent 2,913,677 (1980).
42. C. D. Frohning, G. Diekhaus, E. Wiebus, and H. Bahrmann (Ruhrchemie A.G.): Eur. Patent 53792 (1981).

43. H. Beuther, T. P. Kobylinski, G. M. Singerman, and W. R. Pretzer: *Am. Chem. Soc. Div. Petr. Chem. Prepr.* **25**, 93 (1980).
44. G. Doyle, *J. Mol. Catal.* **13**, 237 (1981).
45. W. E. Slinkard and A. B. Baylis (Celanese Corp.): U.S. Patent 4,168,391 (1979).
46. B. R. Gane and D. G. Stewart (British Petroleum Co.): Eur. Patent 3876 (1979).
47. J. L. Barclay and B. R. Gane (British Petroleum Co.): U.K. Patent 1,583,084 (1977).
48. T. Mizoroki and M. Nakayama: *Bull. Chem. Soc. Jpn.* **37**, 236 (1964).
49. G. N. Butter (Commerical Solvents Corp.): Belg. Patent 618,413 (1962).
50. Y. Sugi, K. Bando, and Y. Takami: *Chem. Lett.* **63** (1981).
51. M. Novotny and I. L. Mador: *Synthesis of C_2-oxygenated Chemicals from Methanol* (Catalysis of Organic Reactions, vol. 5, ed. W. Moser), pp. 249–277. Dekker (1980).
52. T. Mizoroki and M. Nakayama: *Bull, Chem. Soc. Jpn.* **38**, 1876 (1965).
53. T. Mizoroki and M. Nakayama: *Bull. Chem. Soc. Jpn.* **41**, 1628 (1968).
54. W. Reppe, H. Kroper, N. von Kutepov, and H. J. Pistor: *Ann. Chem.* **582**, 72 (1953).
55. T. Mizoroki, T. Matsumoto, and A. Ozaki: *Bull. Chem. Soc. Jpn.* **52**, 479 (1979).
56. G. Braca, G. Sbrana, G. Valentini, G. Andrich, and G. Gregorio: *Carbonylation and Homologation of Methanol, Methyl Ethers and Esters in the Presence of Ruthenium Catalysts* (Fundamental Research in Homogeneous Catalysis, vol. 3, ed. M. Tsutsui), pp. 221–238. Plenum Publ. Co. (1979).
57. A. Deluzarche, G. Jenner, and A. Kiennemann: *Tetrahedron Lett.* **40**, 3797 (1978).
58. H. Dumas, J. Levisalles, and H. Rudler: *J. Organomet. Chem.* **187**, 405 (1980).
59. J. P. Collman and S. R. Winter: *J. Am. Chem. Soc.* **95**, 4089 (1972).
60. F. Calderazzo: *Angew. Chem.* **89**, 305 (1977).
61. G. Gregorio, G. Montrasi, M. Campieri, P. Cavalieri d'Oro, G. Pagani, and A. Andreetta: *Chim. Ind. (Milan)* **62**, 389 (1980).
62. P. D. Taylor (Celanese Corp.): U.S. Patent 4,111,837 (1978).
63. P. D. Taylor (Celanese Corp.): U.S. Patent 4,150,246 (1979).
64. B. Cornils: *Hydroformylation, Oxo Synthesis, Roelen Reaction* (New Syntheses with Carbon Monoxide, ed. J. Falbe), pp. 1–225. Springer Verlag (1980).
65. W. F. Gresham (E. I. Du Pont de Nemours): U.S. Patent 2,623,906 (1952).
66. D. Riley and W. O. Bell (Commercial Solvents Corp.): Ger. Patent 1,173,075 (1965).
67. M. J. Chen and H. M. Feder: *Mechanism of a New Process for Methanol Homologation* (Catalysis of Organic Reactions, vol. 5, ed. W. R. Moser), pp. 273–288. Dekker (1980).
68. M. Fremery, M. Elstner, and J. Zimmermann (Union Rheinische Braunkohlen Kraftstoff A.G.): Ger. Patent 2,726,978 (1977).
69. J. S. Bradley: *J. Am. Chem. Soc.* **101**, 7419 (1979).
70. J. S. Bradley: *Homogeneous Catalysis of Carbon Monoxide Hydrogenation* (Fundamental Research in Homogeneous Catalysis, vol. 3, ed. M. Tsutsui), pp. 165–179. Plenum Publ. Co. (1979).
71. G. Jenner, A. Kiennemann, E. Bagherzadah, and A. Deluzarche: *React. Kinet. Catal. Lett.* **15**, 103 (1980).
72. B. D. Dombek: *J. Am. Chem. Soc.* **102**, 6855 (1980).
73. J. F. Knifton: *J. Am. Chem. Soc.* **103**, 3959 (1981).
74. G. Braca, G. Sbrana, G. Valentini, G. Andrich, and G. Gregorio: *J. Am. Chem. Soc.* **100**, 6240 (1978).
75. G. Braca, G. Sbrana, and G. Andrich (Montedison SpA): Ital. Patent Appl. 24250/A (1977).
76. B. D. Dombek: *J. Am. Chem. Soc.* **103**, 6508 (1981).
77. J. F. Roth, J. H. Craddock, A. Hershman, and F. E. Paulik: *Chem. Techn.* **600** (1971).
78. T. Matsumoto, K. Mori, T. Mizoroki, and A. Ozaki: *Bull, Chem. Soc. Jpn.* **50**, 2337 (1977).
79. M. Kuraishi, S. Asano, and Y. Shinozaki (Commercial Solvents Corp.): U.S. Patent 3,387,043 (1968).

80. H. Kroeper, H. Hauber, and W. Hagen (B.A.S.F.): Ger. Patent 921,936 (1955); *C.A.* **53**, 222 (1959).
81. I. Wender, J. Feldman, S. Metlin, B. H. Gwynn, and M. Orchin: *J. Am. Chem. Soc.* 77, 5760 (1955).
82. I. Wender, S. Metlin, and M. Orchin: *J. Am. Chem. Soc.* 73, 5704 (1951).
83. G. R. Burns: *J. Am. Chem. Soc.* 77, 6615 (1955).
84. I Wender, H. Greenfield, S. Metlin, and M. Orchin: *J. Am. Chem. Soc.* 74, 4079 (1952).
85. M. Orchin: *Adv. Catal.* 5, 393 (1953).
86. M. B. Sherwin and A. M. Brownstein: U.S. Patent 4,158,100 (1979).
87. M. B. Sherwin, A. M. Brownstein, and J. Peress: Eur. Patent 4732 (1979).
88. L. Paladini, G. Braca, and G. Sbrana: unpublished results.
89. Y. C. Fu, H. Greenfield, S. Metlin, and I. Wender: *J. Org. Chem.* 32, 2837 (1967).
90. F. Piacenti and M. Bianchi: *Carbonylation of Saturated Oxygenated Compounds* (Organic Syntheses via Metal-Carbonyls vol. 2, eds. I. Wender and P. Pino), pp. 1–42. Wiley (1977).
91. T. Floris: *II Nat. Congr. on Catal. Soc. Chim. Ital.*, Prepr., Siena (1980) p. 27.
92. (a) G. Braca, G. Sbrana, and G. Gregorio (Montedison SpA): U.S. Patent 4,189,441 (1980);
(b) *idem.*, Ital. Pat. Appl. 25782/A (1976);
(c) *idem.*, Ital. Pat. Appl. 22223/A (1977).
93. G. Braca, L. Paladini, G. Sbrana, G. Valentini, G. Andrich, and G. Gregorio: *Ind. Eng. Chem., Prod. Res. Dev.* 20, 115 (1981).
94. D. Forster: *J. Am. Chem. Soc.* 98, 846 (1976).
95. W. F. Gresham and R. E. Brooks (E.I. Du Pont de Nemours): U.S. Patent 2,451,333 (1948); *C.A.* **43**, 673d (1949).
96. T. Yukawa and K. Kawasaki (Ajiinomoto Co.): Ger. Patent 2,427,954 (1975).
97. R. G. Wall (Chevron): U.S. Patent 4,079,085 (1978).
98. J. A. Roth and M. Orchin: *J. Organomet. Chem.* 172, C 27 (1979).
99. R. W. Goetz (National Distillers and Chem. Co.): U.S. Patent 4,200,765 (1980).
100. A. Spencer; *J. Organomet. Chem.* 194, 113 (1980).
101. A. S. C. Chan, W. E. Carroll, and D. E. Willis: *J. Mol. Catal.* **19**, 377 (1983).
102. D. W. Smith (National Distillers & Chem. Co.): U.S. Patent 4,267,384 (1981).
103. M. Dubeck and G. G. Knapp (Ethyl Co.): U.S. Patent 4,062,898 (1977).
104. B. I. Fleming and H. I. Bolker: *Can. J. Chem.* **54**, 685 (1976).
105. J. E. Lyons: *J. Chem. Soc., Chem. Commun.* **412** (1975).
106. P. Morand and M. Kayser: *J. Chem. Soc., Chem. Commun.* **314** (1976).
107. M. Bianchi, F. Piacenti, P. Frediani, U. Matteoli, C. Botteghi, S. Gladiali, and E. Benedetti: *J. Organomet. Chem.* 141, 107 (1977).
108. M. Bianchi, G. Menchi, F. Francalanci, F. Piacenti, U. Matteoli, P. Frediani, and C. Botteghi: *J. Organomet. Chem.* 188, 109 (1980).
109. R. A. Grey, G. P. Pez, and A. Wallo: *J. Am. Chem. Soc.* **103**, 7536 (1981).
110. J. F. Knifton: *J. Chem., Soc. Chem. Commun.* **41** (1981).
111. J. F. Knifton: *Chem. Techn.* **609** (1981).
112. J. F. Knifton: *Hydr. Proc.* **60** (12), 113 (1981).
113. P. C. Ford, C. Ungermann, V. Landis, S. A. Moya, R. C. Rinker, and R. M. Laine: *Homogeneous Catalysis of Water Gas Shift Reaction by Metal Carbonyls* (Advances in Chem. Series no. 173), pp. 81–93. Am. Chem. Soc. (1979).
114. J. L. Ehrler and B. Juran: *Hydr. Proc.* **61** (2), 109 (1982).
115. G. Braca, S. Busni, G. Sbrana, and G. Valentini: *Chim. Ind.* (*Milan*) **63**, 516 (1981).
116. G. Braca, G. Sbrana, and G. Valentini (Consiglio Nazionale delle Ricerche): Ital. Pat. Appl. 21462/A (1982).
117. G. Braca, G. Guainai, A. M. Raspolli, G. Sbrana, and G. Valentini: *Ind. Eng. Chem., Prod. Res. Dev.*, in press (1984).

118. Imhausen Chemie GmbH: Ger. Patent 2,731,962 (1979).
119. N. Rizkalla and Ch. N. Winnich (Halcon Int. Inc.): Belg. Patent 839,321 (1976).
120. E. H. Man, J. J. Sanderson, and C. R. Hauser: *J. Am. Chem. Soc.* **72**, 847 (1950).
121. H. Keradmand, J. Jenner, H. Kiennemann, and A. Deluzarche: *Chem. Lett.* **395** (1982).
122. G. Braca, G. Sbrana, G. Valentini, and M. Cini: *J. Mol. Catal.* **17**, 323 (1982).
123. G. Braca, G. Sbrana, G. Valentini, and C. Barberini: *C_1 Mol. Chem.* **1**, 9 (1984).
124. G. Braca, G. Sbrana, and G. Valentini (Consiglio Nazionale delle Ricerche): Ital. Pat. Appl. 20954/A (1981).
125. J. Gauthier-Lafaye and R. Perron (Rhone-Poulenc Ind.): Eur. Patent 31784 (1980).
126. E. Drent (Shell Int.): Eur. Patent 31606 (1981).
127. F. J. Bryant, W. R. Johnson, and T. C. Singleton: *Am. Chem. Soc. Dallas Meeting*, 1973, general papers petrochem., p. 193.
128. J. B. Keister and R. Gentile: *J. Organomet. Chem.* **222**, 143 (1981).
129. J. W. Rathke and H. M. Feder: *J. Am. Chem. Soc.* **100**, 3623 (1978).
130. W. Keim, M. Berger, and J. Schlupp: *J. Catal.* **61**, 359 (1980).
131. A. Aguilo and T. Horlenko: *Hydr. Proc.* **59** (11), 120 (1980).
132. C. D. Chang (Mobil Oil Corp.): Ger. Patent 2,615,150 (1976).
133. H. Kolbel and F. Engelhardt (Rheinpreussen A.G.): U.K. Patent 733,792 (1955).
134. C. M. Bartish (Air Prod. & Chem. Inc.): U.S. Patent 4,171,461 (1979).
135. R. T. Clark (Celanese Co.): U.S. Patent 3,972,952 (1976).
136. M. Novotny and L. R. Anderson (Allied Chem. Co.): U.S. Patent 4,126,752 (1978).
137. W. J. Ball and D. G. Stewart (British Petroleum Co.): U.K. Patent 2,053,915 (1980).
138. H. W. Sternberg and I. Wender: *Coordination Chem.* **35** (1959).
139. I. Wender: *Catal. Rev. Sci. Eng.* **14**, 97 (1976).
140. F. Calderazzo, R. Ercoli, and G. Natta: *Metal Carbonyls: Preparation, Structure and Properties* (Organic Syntheses via Metal Carbonyls, vol. 1, eds. I. Wender and P. Pino), pp. 1–272. Wiley (1968).
141. I. Wender, H. W. Sternberg, and M. Orchin: *J. Am. Chem. Soc.* **74**, 1216 (1952).
142. A. Bortinger, P. J. Busse, and M. Orchin: *J. Catal.* **52**, 385 (1978).
143. P. S. Braterman, B. S. Walker, and T. H. Robertson: *J. Chem. Soc., Chem. Commun.* **651** (1977).
144. O. Hecht and H. Kroeper: *Naturforsch. u. Medizin in Deutschland* 36 (I), 143 (1939–1947).
145. Y. Takahashi, N. Tomita, N. Yoneda, and A. Suzuki: *Chem. Lett.* **997** (1975).
146. D. Forster: *J. Mol. Catal.* **17**, 299 (1982).
147. J. Falbe and C. D. Frohning: *J. Mol. Catal.* **17**, 117 (1982).
148. R. F. Heck and D. S. Breslow: *J. Am. Chem. Soc.* **85**, 2779 (1963).
149. F. Ungvary and L. Markò: *J. Organomet. Chem.* **193**, 379 (1980).
150. B. D. Dombek: *J. Am. Chem. Soc.* **101**, 6466 (1979).
151. D. L. Thorn: *J. Mol. Catal.* **17**, 279 (1982).
152. D. F. Fahey: *J. Am. Chem. Soc.* **103**, 136 (1981).
153. J. W. Rathke and H. M. Feder: *Mechanism of the Homogeneous Hydrogenation of Carbon Monoxide* (Catalysis of Organic Reactions, vol. 5, ed. W. R. Moser), pp. 219–234. Dekker (1980).
154. I. Wender, S. Friedman, W. A. Steiner, and R. B. Anderson: *Chem. Ind.*, 1694 (1958).
155. W. Rupilius and M. Orchin: *J. Org. Chem.* **37**, 936 (1972).
156. M. Orchin and W. Rupilius: *Catal. Rev.* **6**, 85 (1972).
157. Z. Nagy-Magos, G. Bor, and L. Markò: *J. Organomet. Chem.* **14**, 205 (1977).
158. H. Berke and R. Hoffman: *J. Am. Chem. Soc.* **100**, 7224 (1978).
159. J. Demuynck, A. Strich, and A. Veillard: *Nouveau J. Chim.* **1**, 217 (1977).
160. J. P. Collman, J. N. Cawse, and J. I. Brauman: *J. Am. Chem. Soc.* **94**, 5905 (1972).
161. J. P. Collman, R. G. Finke, J. Cawse, and J. Brauman: *J. Am. Chem. Soc.* **100**, 4766 (1978).

162. S. Beda-Butts, E. M. Holt, S. H. Strauss, N. W. Alcock, R. S. Stimson, and D. F. Shriver: *J. Am. Chem. Soc.* **101**, 5864 (1979).
163. S. Beda-Butts, T. G. Richmond, and D. F. Shriver: *Inorg. Chem.* **20**, 278 (1981).
164. M. Bianchi, U. Matteoli, P. Frediani, and F. Piacenti: *J. Organomet. Chem.* **120**, 97 (1976).
165. R. F. Heck: *Synthesis and Reactions of Alkylcobalt and Acylcobalt Tetracarbonyls* (Adv. Organomet. Chem., vol. 4, eds. F. G. A. Stone and R. West), pp. 243–266. Academic Press (1966).
166. P. Pino, F. Piacenti, and M. Bianchi: *The Hydroformylation (Oxo) Reaction* (Organic Syntheses via Metal Carbonyls, vol. 2, eds. I. Wender and P. Pino), pp. 43–229. Wiley (1977).
167. R. W. Goetz and M. Orchin: *J. Org. Chem.* **27**, 3698 (1962).
168. L. Markò: *Proc. Chem. Soc.* 67 (1962).
169. C. L. Aldridge and H. B. Jonassen: *J. Am. Chem. Soc.* **85**, 886 (1963).
170. D. S. Breslow and R. F. Heck: *Chem. Ind.*, 467 (1960).
171. F. Piacenti, M. Bianchi, and E. Benedetti: *Chim. Ind. (Milan)* 49, 245 (1967).
172. N. S. Imianitov, N. M. Bogoradovskaia, and T. A. Semionova: *Kinet. Catal.* **19**, 573 (1978).
173. D. Forster, A. Hershman, and D. E. Morris: *Catal. Rev.* **23**, 89 (1981).
174. M. J. Chen, H. M. Feder, and J. W. Rathke: *J. Mol. Catal.* **17**, 331 (1982).
175. F. Wada and T. Matsuda: *J. Organomet. Chem.* **61**, 365 (1973).
176. L. P. Hammett and H. L. Pfluger: *J. Am. Chem. Soc.* **55**, 4079 (1930).
177. F. Zanni, A. M. Raspolli, G. Braca, and G. Sbrana, unpublished results.
178. D. Forster: *Mechanistic Pathways in the Catalytic Carbonylation of Methanol by Rhodium and Iridium Complexes* (Adv. Organomet. Chem., vol. 17, eds. F. G. A. Stone and R. West), pp. 256–267. Academic Press (1979).
179. J. Jeffery and R. J. Mawby: *J. Organomet. Chem.* **40**, C 42 (1972).
180. C. F. J. Barnard, J. A. Daniels, and R. J. Mawby: *J. Chem. Soc., Dalton* 1331 (1979).
181. W. R. Roper, G. E. Taylor, J. M. Waters, and L. J. Wright: *J. Organomet. Chem.* **182**, C 46 (1979).
182. J. M. Manriquez, D. R. McAlister, R. D. Sanner, and J. E. Bercaw: *J. Am. Chem. Soc.* **100**, 2716 (1978).
183. P. John: *Chem. Ber.* **103**, 2178 (1970).
184. J. Chatt, B. L. Shaw, and A. Field: *J. Chem. Soc.* 3466 (1964).
185. A. Dobson and S. D. Robinson: *Inorg. Chem.* **16**, 137 (1977).
186. H. Imai, T. Nishiguchi, and K. Fukuzumi: *J. Org. Chem.* **41**, 665 (1976).
187. G. Henrici-Olivé and S. Olivé: *Angew. Chem. Int. Ed.* **15**, 136 (1976).
188. R. A. Sanchez-Delgado, G. S. Bradley, and G. Wilkinson: *J. Chem. Soc. (A)* **399** (1976).
189. C. P. Casey, M. A. Andrews, and D. R. McAlister: *J. Am. Chem. Soc.* **101**, 3371 (1979).
190. J. Hjortkjaer and V. W. Jensen: *Ind. Eng. Chem., Prod. Res. Dev.* **15**, 46 (1976).
191. J. Hjortkjaer and O. R. Jensen: *Ind. Eng. Chem., Prod. Res. Dev.* **16**, 281 (1977).
192. G. R. Clark, C. E. L. Headford, K. Marsden, and W. R. Roper: *J. Organomet. Chem.* **231**, 335 (1982).
193. C. P. Casey, M. A. Andrews, D. R. McAlister, W. D. Jones, and S. G. Harsy: *J. Mol. Catal.* **13**, 43 (1981).
194. G. Jenner, H. Kheradmand, A. Kiennemann, and A. Deluzarche: *J. Mol. Catal.* **18**, 61 (1983).
195. A. Aquilo, J. S. Alder, D. N. Freeman, and R. J. H. Voorhoeve: *Hydr. Proc.* **62** (3), 57 (1983).
196. W. R. Pretzer and M. M. Habib: *Symposium on Catalytic Conversion of Synthesis Gas and Alcohols*, 17th Middle Atlantic Regional Meeting Am. Chem. Soc., Apr. 6, 1983; Plenum Press, in press.
197. M. M. Habib and W. R. Pretzer (Gulf): U.S. Patent 4,361,706 (1982).

198. M. M. Habib and W. R. Pretzer (Gulf): U.S. Patent 4,361,707 (1982).
199. W. R. Pretzer, T. P. Kobylinski, and J. E. Bozik (Gulf): U.S. Patent 4,239,704 (1980).
200. W. R. Pretzer, Y. P. Kobylinski, and J. E. Bozik (Gulf): U.S. Patent 4,239,705 (1980).
201. D. Forster (Monsanto): U.S. Patent 4,190,729 (1980).
202 W. E. Walker (Union Carbide): U.S. Patent 4,337,365 (1982).
203. M. M. Habib, W. R. Pretzer (Gulf): U.S. Patent 4,352,947 (1982).
204. J. F. Knifton and L. Jiang-jen (Texaco): Eur. Patent Appl. 56679 (1982).
205. B. Cornils, C. D. Frohning, E. Wiebus, and H. Bahrmann (Ruhrchemie): Ger. Patent 3,042,434 (1982).
206. G. Doyle: *J. Mol. Catal.* 18, 251 (1983).
207. M. Hidai, M. Orisaku, M. Ue, Y. Koyasu, T. Kodama, and Y. Uchida: *Organometal.* 2, 292 (1983).
208. W. R. Pretzer, T. P. Kobylinski, and J. E. Bozik (Gulf): U.S. Patent 4,339,610 (1982).
209. W. R. Pretzer, T. P. Kobylinski, and J. E. Bozik (Gulf): U.S. Patent 4,339,611 (1982).
210. See references [3], [8], [9] and [79] of the main text.
211. R. V. Porcelli (Halcon): Fr. Patent 2,497,794 (1981).
212. N. Isogai, T. Okawa, M. Hosokawa, N. Wakui, and T. Watanabe (Mitsubishi Gas Chem.): U.K. Patent 2,078,219 (1981).
213. G. Jenner, H. Keradmand, A. Kiennemann, and A. Deluzarche: *J. Mol. Catal.* 18, 61 (1983).
214. G. Braca, G. Sbrana, A. M. Raspolli, and F. Zanni (C.N.R.): Ital. Pat. Appl. 21645 A/83 (1983).
215. J. F. Knifton: *J. Catal.* 79, 147 (1983).
216. J. F. Knifton (Texaco): U.S. Patent 4,270,015 (1981).
217. R. D. Dombek: *J. Organomet. Chem.* 250, 467 (1983).
218. J. F. Knifton: *J. Catal.* 76, 101 (1982).
219. J. F. Knifton (Texaco): U.S. Patent 4,268,689 (1981).
220. R. Kummer, V. Taglieber, and H. W. Schneider (B.A.S.F.): Eur. Pat. Appl. 56488 (1982).
221. N. Isogai, T. Okawa and N. Wakui (Mitsubishi): Ger. Pat. 3,016,715 (1980).
222. N. Isogai, T. Okawa, M. Hosokawa, N. Wakui and T. Watanabe: Ger. Pat. 3,134,747 (1982).
223. M. T. Barlow, and D. G. Stewart (B.P.): Eur. Pat. Appl. 36724 (1981).
224. K. H. Keim and J. Korff (Union Rheinische Braunkohlen Kraftstoff): Ger. Pat. 3,045,891 (1982).
225. R. V. Porcelli (Halcon): Ger. Pat. 2,952,517 (1980).

INDEX